Comprehensive Manuals in Pediatrics

Series Editors

Michael Katz

E. Richard Stiehm

Pediatric Sports Medicine
for the Practitioner
From Physiologic Principles to Clinical Applications

Oded Bar-Or

Professor of Pediatrics
Department of Pediatrics
Director, Children's Therapeutic
Exercise and Health Centre
McMaster University and
Chedoke-McMaster Hospitals
Hamilton, Ontario, Canada

With 124 Illustrations

Springer-Verlag
New York Berlin Heidelberg Tokyo

Oded Bar-Or, M.D., Chedoke-McMaster Hospitals, Chedoke Division, Evel 4, Box 2000, Station "A", Hamilton Ontario L8N 3Z5, Canada

SERIES EDITORS

Michael Katz, M.D., Reuben S. Carpentier Professor and Chairman, Department of Pediatrics, College of Physicians and Surgeons, Columbia University; and Director, Pediatric Service, Presbyterian Hospital (Babies Hospital), New York, New York, U.S.A.

E. Richard Stiehm, M.D., Professor of Pediatrics, UCLA School of Medicine, UCLA Hospitals and Clinics, Los Angeles, California, U.S.A.

Library of Congress Cataloging in Publication Data
Bar-Or, Oded.
Pediatric sports medicine for the practitioner.
(Comprehensive manuals in pediatrics)
Bibliography: p.
Includes index.
1. Pediatric sports medicine—Handbooks, manuals, etc. I. Title. II. Series. [DNLM: 1. Sports medicine—In infancy and childhood. 2. Exertion—In infancy and childhood. QT 260 B223p]
RC1218.C45B36 1983 617'.1027'08054 83-10437

Softcover reprint of the hardcover 1st edition 1983

Typeset by Bi-comp, Incorporated, York, Pennsylvania

9 8 7 6 5 4 3 2 1

ISBN-13: 978-1-4612-5595-6 e-ISBN-13: 978-1-4612-5593-2
DOI: 10.1007/978-1-4612-5593-2

To Marilyn, who made it all possible and to
Amit, Yuval and Tali, who made it worthwhile.

Contents

6 Nutritional Diseases 192

7 Neuromuscular Diseases 227

Series Foreword

Comprehensive Manuals in Pediatrics are designed to broaden the practitioner's clinical scope by providing a wide range of diagnostic and management skills ordinarily considered to be the exclusive domain of the specialists. Although the series as a whole constitutes a comprehensive text in pediatrics, each volume stands on its own as a self-contained reference for the busy practitioner.

In order to maintain a uniform style and coverage of each subject, each manual is usually written by no more than one or two authors. Each author is an acknowledged expert in his or her field and provides a comprehensive, up-to-date account of the topic under discussion. Practically oriented, each volume offers concise guidelines and courses of treatment.

Michael Katz
E. Richard Stiehm

Preface

Much knowledge has been generated in recent years by scientists investigating the triad: child—exercise—health. Yet little of this information is available in pediatric textbooks, for *application* by the clinician. This book is intended to bridge the resulting gap.

Until the mid-1960's, exercise-related research focused on the young adult or the middle-aged individual. Recent years have witnessed a surge of interest among exercise scientists in the child and young adolescent. A major impetus for such interest has been the increasing involvement of children in advance-level athletics. Today's pre-adolescent athlete is often exposed to training regimens which only a decade ago were considered too demanding even for adult athletes. As a result, in such sports as gymnastics and swimming, for example, champions and record-holders are younger than ever before. In other sports, such as tennis, excellence is usually reached in the third decade but systematic training from childhood is a prerequisite. The emerging popularity of jogging and other recreational activities attracts participants of all ages. A few years ago, any long-distance running for children was discouraged. Today a 10-year-old child who completes a marathon race is no longer a novelty.

But pediatric aspects of exercise-related research are by no means limited to the young athlete. The relevance of exercise to clinical pediatrics has been gaining increasing attention, be it in diagnosis, prevention, management or etiology. Pediatric clinics and exercise laboratories in various countries are now using exercise for diagnosis, or in clinical management, of several pediatric diseases.

The primary boost to pediatric exercise research was given by the International Group of Pediatric Work Physiology. This group, which comprises clinicians and physiologists, was established in 1967. It has since convened ten international symposia in which some 350 papers have been presented and published.[1–7] However, as seen from the references used in this book, the number of publications on children and exercise far exceeds that number. In 1981–1982, journals of pediatrics, physiology, or sports medicine published three to five times as many articles on the exercising child as were published ten years earlier.

To what extent has this body of knowledge become accessible to the practitioner? A scan of recent textbooks of pediatrics or pediatric rehabilitation reveals little useful information on exercise. Entries such as effort, energy expenditure, fatigue, fitness, physical activity or sports do not appear in the indexes of most of these books. In chapters on asthma, for example, some mention is made of exercise-induced bronchoconstriction, but the information is too abstract for the reader who wishes to administer an exercise-provocation test. Nor is there practicable information on recommended or contraindicated activities for the asthmatic child. Chapters on diabetes mellitus include statements on the importance of daily physical exercise, but no information is available on the calorie equivalence of activities, nor on the concept of exercise "dosage."

Among paramedical disciplines, pediatric physiotherapy could obviously benefit from the principles of exercise physiology. Nevertheless, leading textbooks in this area do not deal with such basic concepts as warm-up, fatigue, or stamina. Nor is there any mention of ergometric tests, which have been specifically designed for the physically disabled child. Obsolete and non-standardized tests (such as jumping on a trampoline or running around the block) are still recommended for assessing the physical working capacity of children.

As with other clinical tests, evaluation by exercise requires safety precautions, standardized protocols, and criteria for interpretation. As in other modes of treatment, therapy by exercise must be based on clear indications and contraindications. An exercise prescription can, and must, be quantified in terms of intensity, frequency, and duration of each activity.

The purpose of this monograph is to acquaint pediatricians, general practitioners, physiatrists, and physiotherapists with such considerations and to equip them with tools for applying exercise in health care.

In keeping with its title, this book discusses both the physiologic principles of the child's response to exercise and their clinical implications. Chapter 1 introduces the reader to pediatric exercise physiology, with emphasis on the effects of growth and maturation on responses to acute exercise and to conditioning. An overview of the relevance of such principles to the health care of children is presented in Chapter 2. This is followed by a detailed analysis of interactions between activity and pediatric diseases in Chapters 3 to 8.

Children are often exposed to the combined stresses of exercise and environmental heat. These may induce a marked strain on the thermoregulatory and other body systems and affect the well-being of the child. This is the topic of Chapter 9.

To provide the reader with concrete tools for clinical use, appendices have been included. Appendix I presents "norms" of some measures of physical working capacity. Appendix II outlines the methodology of exercise testing, including suggested protocols and means of precaution. The other appendices comprise some useful tables, an activity questionnaire, and a glossary of terms used in this book. This glossary is intended for clinicians who are not conversant with the terminology and jargon of the exercise physiologist or physical educator.

As far as possible, the data quoted have been directly obtained from children and adolescents. Occasionally, however, it has been necessary to extrapolate from information on adults. Such cases have been specified in the text.

While not a systematic textbook of exercise physiology, this monograph contains ample information on the physiologic responses of *healthy* children to exercise. These are meant as a background for the clinically-oriented reader, but may also be of use to students of exercise physiology or physical education who are interested in the exercising child.

Without a doubt, orthopedic aspects are part and parcel of any discussion on child health and exercise. Their omission from this book reflects a shortcoming on the part of the author, who has limited the scope of this volume to a physiology-based orientation.

The inclusion of exercise-related issues in the curriculum of medical schools is only now emerging. To many clinicians exercise, as a clinical entity, is still a "black box," yet to be opened. If this book stimulates health professionals to integrate exercise into their overall strategy of child-health care, it will have fulfilled its purpose.

References

1. Bar-Or O (ed): Pediatric Work Physiology. Proceedings of the IVth International Symposium. Natanya, Wingate Institute, 1973.
2. Berg K, Eriksson BO (eds): Children and Exercise IX. Baltimore, University Park Press, 1980.
3. Borms J, Hebbelinck M (eds): Children and Exercise. Proceedings of the Vth International Symposium on Pediatric Work Physiology. Acta Paed Belg (Suppl 28), 1974.
4. Borms J, Hebbelinck M (eds): Pediatric Work Physiology. Proceedings of the VIth and VIIIth International Symposia. Basel, Karger, 1978.
5. Ilmarinen J, Välimäki I (eds): Pediatric Work Physiology X. Berlin, Springer-Verlag, 1983.
6. Lavallée H, Shephard RJ (eds): Frontiers of Activity and Child Health. Proceedings of the VIIth International Symposium of Pediatric Work Physiology. Quebec, Pélican, 1977.
7. Thorén C (ed): Pediatric Work Physiology. Proceedings of the Third International Symposium. Acta Paed Scand (Suppl 213), 1971.

Acknowledgments

Although written by one author, this book reflects data, ideas, and support generated by a large group of individuals. Most of the original research on children's response to exercise was done with my colleagues Dr. Omri Inbar and Raffy Dotan at the Department of Research and Sports Medicine, Wingate Institute, Israel. Drs. Norman L. Jones, Alvin Zipursky and Angus B. MacMillan generated the support needed to launch my writing. Special thanks go to Angus B. MacMillan who, while playing the role of devil's advocate, gave me much needed encouragement in moments of despair. Raffy Dotan, Drs. Mohan K.R. Pai, Hugh M. O'Brodovich, John R. Sutton, R. Clifford Way, and William M. Wilson went over sections of the manuscript and provided most constructive criticism and advice. Last, but not least, my deep appreciation to Audrey Hitchcock and Rosemary Phillis, who, with much devotion, managed to transform my scribble into a neatly typed manuscript.

This book was written in part while the author was a visiting professor at McMaster University, Hamilton, Canada. Support was given by the Hospital for Sick Children Research Foundation in Toronto and by the Faculty of Health Sciences at McMaster University.

1

Physiologic Responses to Exercise of the Healthy Child

Introduction

In order to understand the sick child's response to exercise, one must first be familiar with the "normal" physiologic response to exercise. Whether adapting to a single bout of exercise, or to repeated exercise stimuli, the child—like the adult—undergoes physiologic changes. The basic premise of this chapter is that, although such changes take place at all ages, there are growth- or development-related differences in response to exertion. We do not expect a 6-year-old child to run as fast or as far as a teenager who, in turn, is slower and weaker than a young adult. Nor can a child share the same muscle strength as the more mature individual. On the other hand, children recover faster from the strain of exercise and are ready for another bout sooner than do adolescents and adults.

Physiologic capacities have long been recognized as dependent on body and system *dimensions*. Without morphologic growth of the myocardium, for example, its contractile force cannot be high enough to pump sufficient blood to the growing periphery. Similarly, when the bone scaffolding of a teenager is growing, body strength will not follow suit unless muscle mass also develops, and this in itself may depend on hormonal and other pubertal changes.

It is beyond the scope of this chapter to methodically analyze the physiologic changes that occur with exercise in each and every system. Nor shall we recite the fundamentals of growth and development, for which the interested reader may resort to standard texts of exercise physiology[17,181] or of growth and development.[160] In the following sections, we shall present general concepts of exercise physiology and amplify the *differences* in physiologic responses to exercise between children and older age-groups. We shall also indicate how such differences may determine the physical capability of the child and set a limit on his or her performance.

Our understanding of children's physiologic reactions to exercise is still deficient. More so than with adults, we are limited by ethical considerations and by methodologic constraints. There are very few investigators who would, for example, puncture the artery of a child, take a needle biopsy of his muscle, or push a thermocouple against his tympanum merely to satisfy curiosity. To study environmental effects, we cannot readily expose any child to "hostile environments" where extreme cold, high heat and humidity, or hypoxic conditions prevail. Nor can adequate animal models be set up to study age-related differences in trainability, motor learning, or thermoregulation. The pediatric exercise physiologist is still seeking instruments and protocols appropriate for body size, body proportions, level of motivation, and attention span of the young child, especially the preschooler. Many studies have borrowed concepts, methods, and instruments that are suitable for adults but not for children. There is a definite need for methodology-oriented research in pediatric exercise physiology.

Due to the above constraints, the state of the art is still limited to some knowledge of the cardiovascular and pulmonary systems, to general concepts of energy transfer and metabolism, and to some phenomena of thermoregulation and body fluid shifts. Very few data are available on hormonal, muscle cell, or subcellular phenomena. Nor have we begun to understand processes within the central or peripheral nervous system of the exercising child.

This chapter therefore focuses on the metabolic, cardiovascular, and pulmonary systems. Some mention is also made of exercise perception. A detailed discussion of thermoregulation is presented in Chapter 9.

Response to Acute Exercise

Metabolic Responses to Exercise in Children

For mechanical energy to be released at the myofibrillar level and effect muscle contraction, splitting of adenosine triphosphate (ATP) must take place. This high-energy compound is available in small quantities (about 4–5 mmol/kg wet weight) in the resting muscle. However, once contractions start, there is an immediate need for reinforcement of ATP. This can be supplied from: 1) limited stores of creatine phosphate, 2) glycolysis, or 3) the tricarboxylic acid (Krebs) cycle. The former two sources do not require addition of O_2 and are therefore called *anaerobic.* The latter requires O_2 and is termed *aerobic.* Muscle contractions that result from anaerobic reactions cannot be sustained longer than 40–50 sec. In contrast, muscle contractions utilizing aerobic energy turnover can last many minutes or even hours. Even though most activities utilize both aerobic and anaerobic pathways, in the jargon of sports scientists physical tasks are subdivided into "aerobic"-type and "anaerobic"-type activities. The former including long-distance running, swimming, cycling, cross-country skiing, and other endurance-requiring tasks. The latter include sprinting, jumping, throwing, and other sports where the required power intensity is high and the duration short. A question commonly asked is whether, compared with adults, children are characteristically aerobic or anaerobic performers. We shall try to offer an answer in the following sections.

Maximal Aerobic Power. The most commonly used index of maximal aerobic power has been maximal O_2 uptake, which is the highest volume of oxygen that can be consumed by the body per time unit. This value reflects the highest metabolic rate made available by aerobic energy turnover. One liter of O_2 is equivalent to about 5 kcal or 21 kJ.

Maximal O_2 Uptake. Fig. 1.1 presents the relationship between maximal O_2 uptake and chronologic age in 3910 girls and boys, 6 to 18 years old. It is evident that, with the growth of the child, there is a concomitant increase in his or her maximal O_2 uptake. Until age 12, values grow at the same rate in both sexes, even though boys have higher values as early as age 5.[209] While maximal O_2 uptake of boys keeps increasing until the age of about 18, it hardly develops beyond age 14 in girls.

Maximal aerobic power is strongly related to lean body mass,[45,76,156] which explains, to a great extent, the above sex-related difference. Such a difference practically disappears once maximal O_2 uptake is related to the muscle mass that performs the activity, as shown in Fig. 1.2. When maximal O_2 uptake is plotted against lean leg volume, the regression line is practically identical for both sex groups.[57] Whether this finding means muscle mass is the limiting factor in maximal aerobic power of children has yet to be shown.

When the maximal O_2 uptake of adolescents of different ages but of the same body weight or body height is compared, it is positively related to age.[190] Thus, maximal aerobic power depends also on the maturity of the individual and not only on his or her body dimensions.

Maximal O_2 Uptake per Kilogram Body Weight. Based on the above discussion, one might conclude that maximal aerobic power is less developed in children than at older ages. For tasks that involve moving the whole body from one place to another, however, the child, whose body mass is smaller, may not *need* as high an absolute maximal O_2 uptake as that of the heavier adolescent or adult. To compare the maximal aerobic power of individuals who differ in body mass, one should therefore express

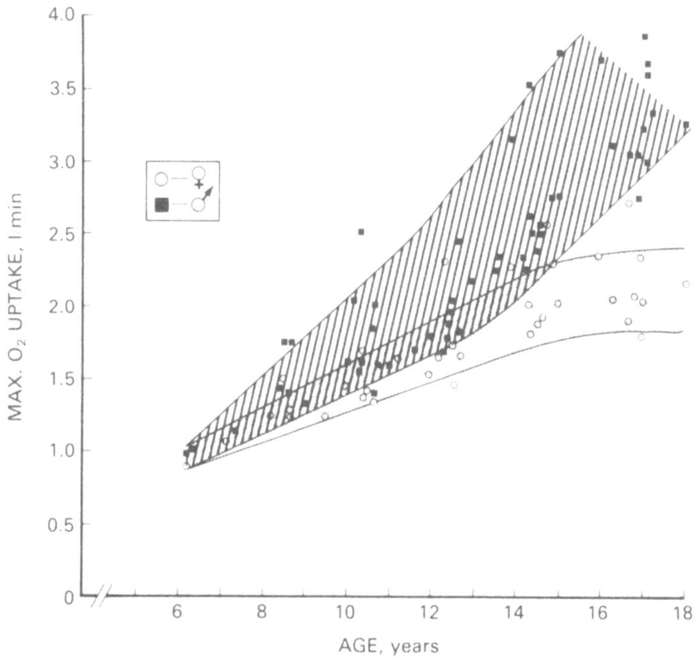

Figure 1.1. Maximal aerobic power and age. Absolute values of maximal O_2 uptake in girls (n = 1730) and boys (n = 2180) 6- to 18-years old. Each dot represents a mean of a group. The shaded areas were constructed by eye-ball technique to indicate a general trend. Data by Andersen and Magel,[7] Andersen et al,[10] Åstrand,[15] Bar-Or and Zwiren,[32] Bar-Or et al,[31] Chatterjee et al,[46] Ekblom,[61] Gaisl and Buchberger,[83] Hermansen and Oseid,[102] Ikai et al,[106] Kobayashi et al,[117] MacDougall et al,[133] Máček et al,[134] Mocellin,[145] Nagle et al,[151] Robinson,[162] Seliger,[179] Shephard et al,[182] and Thorén.[194]

maximal O_2 uptake in *relative* rather than absolute terms, e.g., per body height, surface area, lean body mass, or body weight.

Although theoretically not the method of choice, the most common way of expressing maximal O_2 uptake for comparative purposes has been per kilogram body weight. Such a comparison is shown in Fig. 1.3. While there is hardly any age-related change in maximal O_2 uptake in the boys, it continuously *declines* among the girls. Such a decline may reflect an increase in body adiposity (and hence a relative decrease in lean mass) of girls during adolescence.

The Dimensionality Theory and Maximal O_2 Uptake. Opposition has grown in recent years to the use of body weight as the basis for scaling down maximal O_2 uptake and other physiologic functions.[14,33,76,166,200] An alternative approach is based on the dimensionality theory, assuming that proportions among body segments remain fairly constant during late

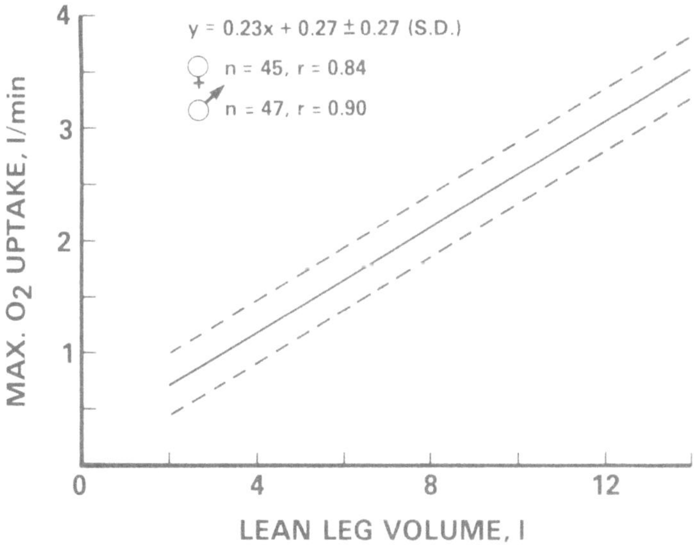

Figure 1.2. Maximal aerobic power and lean leg volume. Maximal O_2 uptake, determined by a progressive upright cycle-ergometer test, in relationship to lean leg volume as assessed by length and circumference measurements and corrected for skinfold thickness. Subjects were 92 girls and boys, 6- to 16-years old. Adapted from Davies et al.[57]

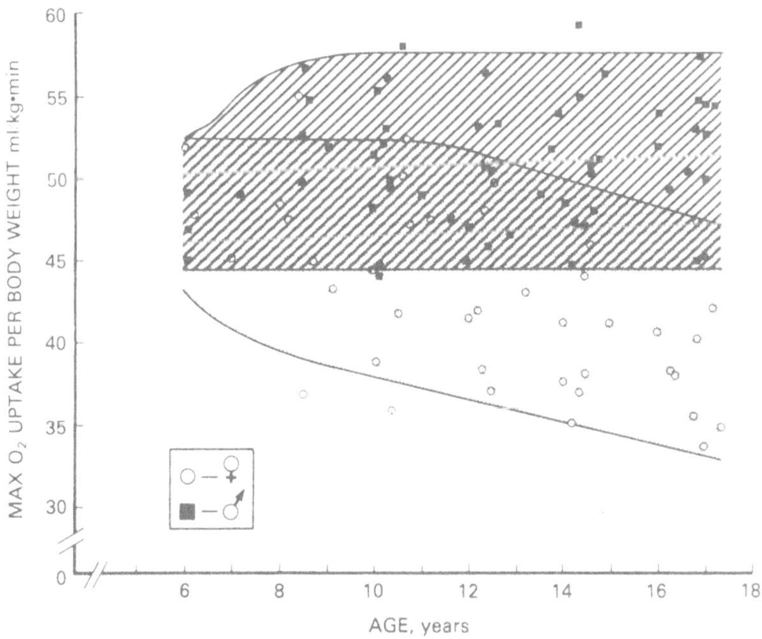

Figure 1.3. Maximal aerobic power and age. Maximal O_2 uptake per kilogram body weight in girls (n = 1730) and boys (n = 2180). Subjects, symbols, and sources as in Fig. 1.1.

childhood, adolescence, and adulthood.[13] Using this theory one can predict growth-related changes in body dimensions and functions by monitoring the respective changes in body length (L). For example, the length of body segments and organs will be proportional to L, their surface area or cross-sectional area to L^2, and their volume or mass to L^3. The same rationale applies to physiologic functions. Muscle force or strength, for example, should be scaled to L^2 (strength being proportional to cross-sectional area of muscle), lung volume, stroke volume, or work to L^3. Time is proportional to L.[200] Therefore, power, which is work/time, should be scaled to $L^3/L = L^2$, and so should maximal O_2 uptake, which is volume/time.

Will scaling of maximal O_2 uptake to L^2 modify our interpretation of the growth-related changes in maximal aerobic power? An answer can be illustrated by the following example. The maximal O_2 uptake of an 117 cm tall 6-year-old boy is 1.0 liter/min. Scaled to L^2, his expected maximal O_2 uptake at age 18, when 176 cm tall, will be 2.3 liters/min. This is contrasted with a value of 3.0 liters/min commonly found in a healthy nonathletic young adult. Thus, if we were to infer the maximal aerobic power of children merely from the principles of dimensionality, it would be *lower* than that of adults. Is the theoretical exponent of 2 actually found experimentally? Various longitudinal studies have shown that maximal O_2 uptake changes with height-exponents, which range from 1.51 to 3.21.[12,18,184,190] Thus the application of dimensionality principles to the prediction of maximal aerobic power (and probably other fitness items) has not been confirmed experimentally. It has not shown any *practical* advantage over the use of body weight or lean body mass for growth-related comparisons.

Mechanical Efficiency and Economy of Movement. When muscle contraction results in movement, external mechanical work is produced. The calorie equivalent of this work is only some 20–25% of the chemical energy utilized during the contraction. The other 75–80% is converted into heat.

Mechanical efficiency (ME) has been defined as the ratio between external mechanical work (W) produced by the muscle and the chemical energy (E) utilized during the contraction. Because some energy (e) is required by the muscle during its resting state, the *net* energy utilized during contraction is $E - e$. ME is usually expressed in percentage of net energy as follows:

$$ME = \frac{W \times 100}{E - e}$$

When work is done during a known period of time, power units rather than work units are used in the numerator and denominator.

To assess the mechanical efficiency of the body as a whole, one must calculate the mechanical power produced by the body and the chemical power needed for that activity. The latter is done conveniently by measuring O_2 uptake and assigning 5 kcal (21 kJ) for each liter of O_2. (This value varies somewhat, depending on the fuel source. It is 4.70 kcal/liter for fats and 5.05 kcal/liter for carbohydrates.) O_2 uptake at rest must be subtracted from that measured during exercise to obtain *net* O_2 uptake. Thus:

$$ME = \frac{\text{Mechanical power output}}{\text{Exercise metabolic rate} - \text{Resting metabolic rate}}$$

The measurement of mechanical power poses methodologic problems. While one can determine it with good accuracy during cycle ergometry (assuming a negligible power loss between the pedals and the flywheel), its assessment in other activities is harder. Although equations are available to calculate power output during walking or running, based on speed, slope, and body weight, these equations disregard differences in gait. Such differences exist, for example, in the extent of vertical displacement of the body (or its segments) and in the degree of lateral pelvic tilt, which are not included in most calculations of power output.

Thus, interindividual differences in O_2 uptake during walking or running do not necessarily denote differences in mechanical efficiency at the cellular level but, rather, a difference in the *economy* of locomotion.[112] The same concept applies in other activities, such as swimming, where O_2 uptake varies markedly among individuals due to different levels of economy of motion.

In this book we shall refer to mechanical efficiency only when both the numerator and the denominator in the equation are measurable. "Economy of motion" will be used instead for data in which the numerator has not been measured. The concept of economy of motion is important to understanding the differences in exercise performance between children and other groups, as well as among children with various disabilities.

Mechanical efficiency of cycling is similar in children, adolescents, and adults, ranging between 18 and 30% (average 25%).[19,35,89,179,182,193,204] In contrast, O_2 cost during walking and running is higher among children, when expressed per kilogram body weight.[15,89,123,133,162,187] This is shown in Fig. 1.4, which presents submaximal O_2 uptake of 5- to 18-year-old girls and boys who ran on a treadmill at various speeds.[15] At 10 km/hr, for example, there was an 8 ml/kg × min (20%) greater metabolic cost for the 5-year-old child than for the 17-year-old adolescent. A similar difference is shown in Fig. 1.5 for walking at different inclines. The higher O_2 cost in young children cannot be explained

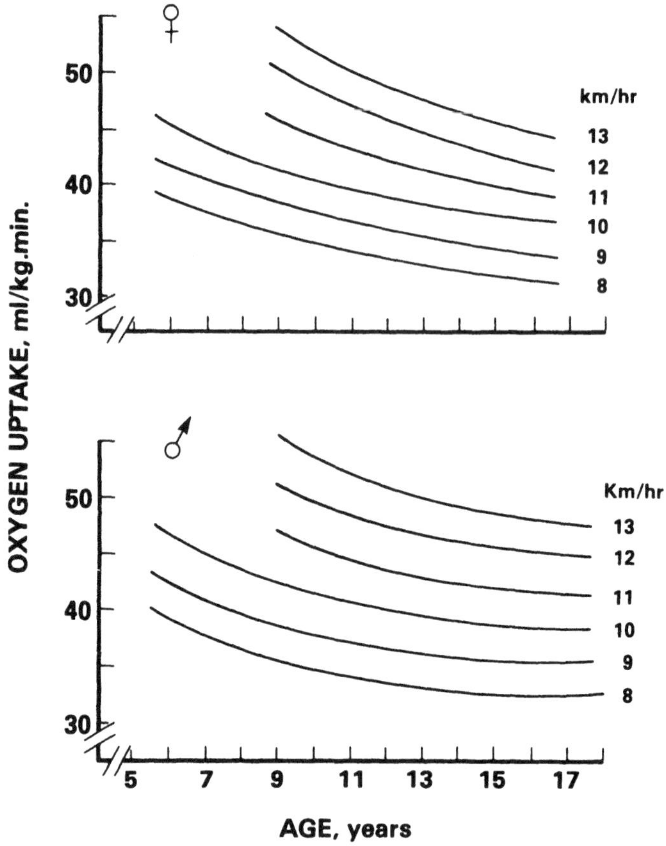

Figure 1.4. Submaximal O_2 uptake and age. Sixty-seven girls and 72 boys, 4- to 18-years old walked on a treadmill at various speeds. Based on Åstrand.[15]

merely by a difference in resting metabolism, which is only some 1–2 ml/kg × min, but rather by their relatively "wasteful" gait while walking or running. With training, running becomes more economical in children[55] and adolescents,[112] as reflected by a lower O_2 cost.

Whatever its underlying mechanism, the high metabolic cost of walking or running makes the small child less of an "aerobic machine" than might be expected from his high maxmal O_2 uptake. If one takes the difference between maximal O_2 uptake and the O_2 uptake needed for a given task to represent metabolic *reserve,* one can see that children are at a disadvantage. This is displayed schematically in Fig. 1.6. The 8-year-old child who runs at 180 m/min is operating at 90% of maximal aerobic power, while the 16-year-old requires at the same running speed only 75% of maximum. Thus, the "reserve" range of running speeds becomes higher with age, which explains why children are less capable

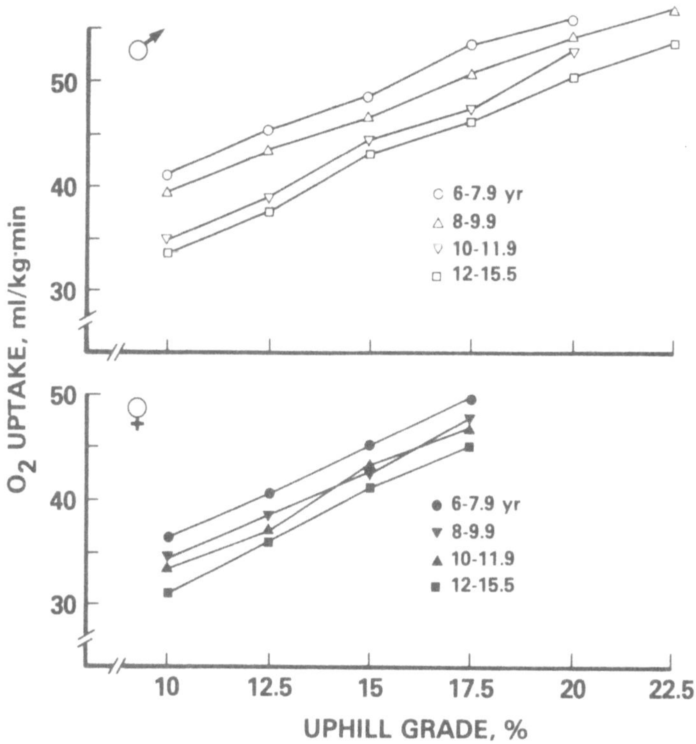

Figure 1.5. Oxygen cost of walking at various slopes. Mean values for 6- to 15-year-old girls (n = 64) and boys (n = 83) subdivided into four age-groups. Subjects walked at 5–6 km/hr on a motor-driven treadmill. Based on data by Skinner et al.[187]

than adolescents and adults of competing over long distances, even though they can maintain a slow speed for long periods.

Anaerobic Characteristics. *Mechanical Power Output.* The ability of children to perform *anaerobic*-type activities (i.e., supramaximal tasks that last 1 min or less) is distinctly lower than that of adolescents and adults.[57,109,126] Figures I.9 and I.10 (Appendix I) summarize the anaerobic capacity of girls and boys of different ages, as measured by the Wingate anaerobic test. For a description, see Appendix II, Examples of Exercise Protocols.[23,25] "Anaerobic capacity" in this test is defined as the mean mechanical power (or total mechanical work) produced in 30 sec. As expected, performance expressed in absolute power units is positively related to age (Fig. I.9). Even when normalized for body weight, however, the power produced by an 8-year-old boy is still only 70% of that generated by an 11-year-old boy.

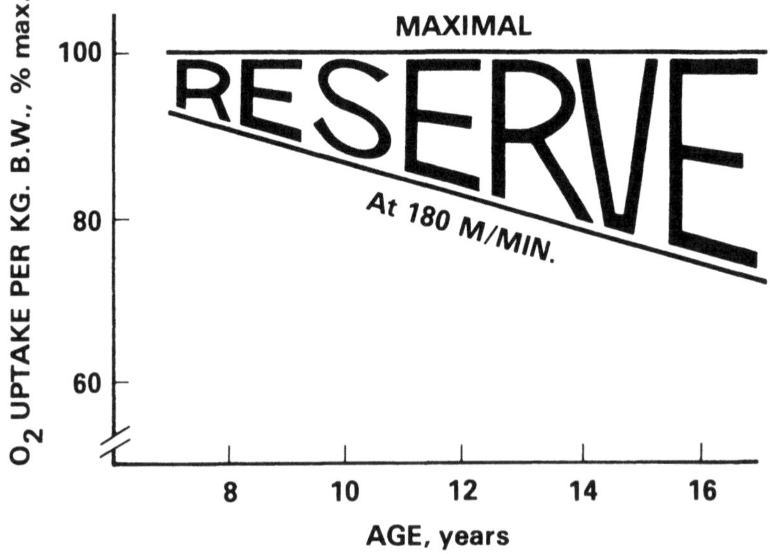

Figure 1.6. Aerobic "reserve" and age. Maximal O_2 uptake and O_2 uptake during a treadmill run at 180 m/min in 134 girls and boys 7–16 years old. Based on data by MacDougall et al.[133] Reproduced with permission from Bar-Or.[26]

Another index of the Wingate anaerobic test is the *peak* power produced at any 5 sec interval. This index characterizes the explosive nature of the subject's muscle contraction. As shown in Figs. I.11 and I.12, the peak anaerobic power of young children is definitely lower than in adolescents, whether in absolute power units (Fig. I.11) or when normalized for body weight (Fig. I.12).

A similar age-related progression in peak anaerobic power is found by the Margaria step-running test.[57,58,126] As shown in Fig. 1.7 the peak energy turnover, whether expressed in absolute units or per kilogram body weight, is distinctly lower, the younger the subject.

To sum up: whether or not children have inferior maximal aerobic power is still debatable; our conclusion will depend on the basis we choose for comparison. This is not the case with anaerobic performance, which is lower in the child in absolute and relative terms alike, whether scaled to body weight, height squared, or lean body mass. A graphic comparison of the growth-related difference between aerobic and anaerobic performance is shown in Fig. 1.8. To use a common scale, values are shown as a percentage, taking the value at 18 years as 100%. While maximal aerobic power does not change (in boys) or even decreases (in girls) with age, there is a progressive growth-related increase in anaerobic performance. The same *relative pattern* in the growth of the two fitness components will be obtained, whichever mode one uses to normalize the data.

Underlying Biochemical Characteristics. As discussed above, differences in maximal *aerobic* power can be accounted for by the mass of active muscle tissue.[56,57] This is not so with *anaerobic* capacity. The markedly lower anaerobic capacity of the young child reflects, to a great extent, a *qualitative* deficiency in his muscle (or in the recruitment of motor units within the muscle), which reduces its ability to perform anaerobically. There are a number of findings that support this notion: Table 1.1 summarizes characteristics of biochemical substrates within the muscle that are utilized for muscle contraction. The main age-related difference is in glycolytic capacity: resting concentration of glycogen, and especially the rate of its anaerobic utilization, are lower in the child, who is therefore at a functional disadvantage when performing strenuous activities that last 10–60 sec.

One way of assessing glycogen utilization is by measuring lactate concentration in the blood or, preferably, in the muscle. Lactate is the end product of anaerobic glycolysis, and its maximal concentration in muscle reflects, in part, the maximal rate of glycolysis. Maximal lactate levels in the blood[15,39,142,149,162,206] and in the quadriceps muscle[69,70,72,114] are lower

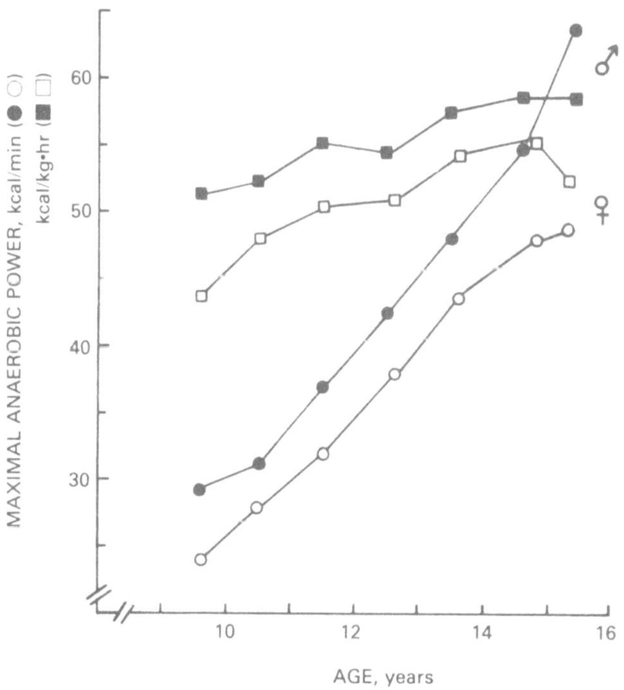

Figure 1.7. Maximal anaerobic power and age. Chemical power expended in the Margaria step-running test of 294 9- to 16-year-old girls and boys. Mean value of age-groups represented in absolute terms (○ = girls, ● = boys) and per kilogram body weight (□ = girls, ■ = boys). Based on data by Kurowski.[126]

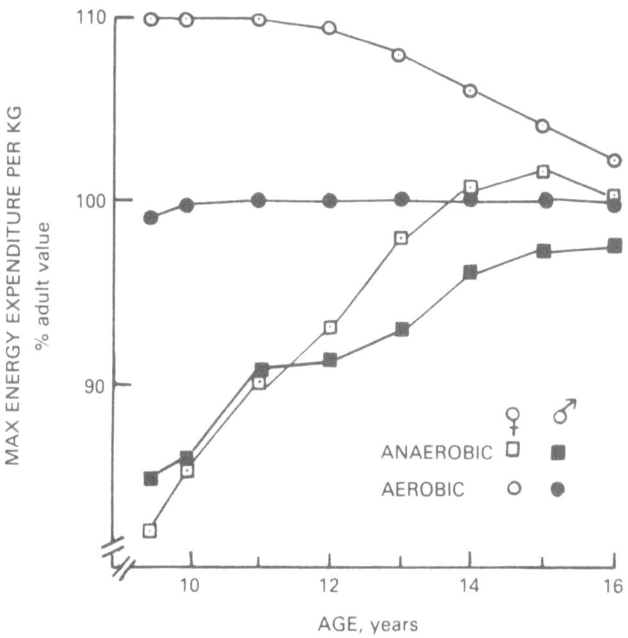

Figure 1.8. Development of aerobic and anaerobic characteristics. Maximal O_2 uptake and maximal performance in the Margaria step-running test in 9- to 16-year-old girls and boys. Mean values are percentages, taking the value at 18 years as 100%. Based on Kurowski[126] and Bar-Or.[26]

in children than in older subjects. An example is shown in Fig. 1.9, where adolescent boys were compared with young men at various intensities of cycling exercise. At each exercise level the adolescents had a lower muscle lactate concentration, maximal level being about 35% lower than in the adults.[72] Studies in rats have shown that lactate production is related to the level of circulating testosterone.[124] It has been suggested, but not confirmed, that the ability of boys to produce lactate during maximal exercise depends on their sexual maturity.[72] The lower anaerobic performance of the mature female when compared with the male and the lesser age-related difference among females are in line with this notion. It is, however, premature to state that the difference between rate of glycosis of boys and men is explained by differences in male hormone activity.

The rate of glycolysis is limited by the activity of such enzymes as phosphorylase, pyruvate dehydrogenase, and phosphofructokinase. The latter enzyme has been found less active in the muscle cells of 11- to 13-year-old boys,[69,70] or 16- to 17-year-old boys,[78] than in young adults. This finding suggests one reason for the slower rate of glycolysis in the younger individual.

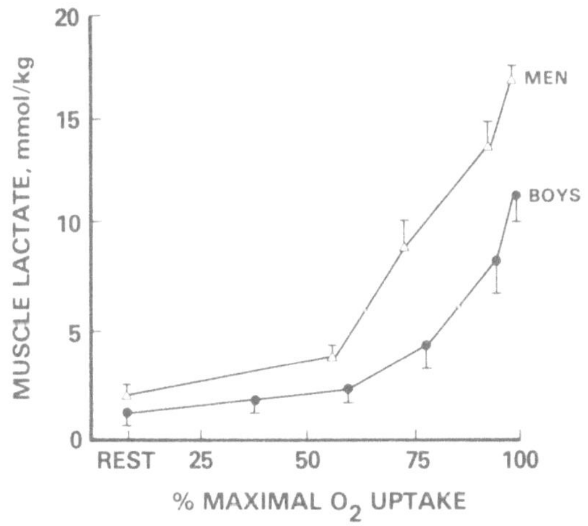

Figure 1.9. Muscle lactate of children and adults. Comparison between 13.5- to 14.8-year-old boys and young men of muscle lactate concentration (per wet tissue) at different cycling intensities. Vertical lines denote 1 S.E.M. Redrawn by permission from Eriksson.[72]

An additional indicator of anaerobic capacity is the degree of acidosis at which the muscle can still contract. Some trained adult athletes can push themselves to exercising at arterial blood pH as low as 6.80,[115] which is equivalent to pH = 6.60 or less in the active muscle cell. Untrained individuals, on the other hand, can seldom sustain exercise when their arterial blood pH reaches 7.20. Children do not reach as high levels of acidosis as can adolescents or young adults,[83,115,142,199] as shown in Figs. 1.10A and 1.10B. In Fig. 1.10A are plotted mean blood pH values of different groups of male subjects, all of whom performed all-out tasks. One of these[142] was followed up longitudinally for four consecutive years. Irrespective of the absolute pH values, there is an age-related increase in maximal acidosis at the rate of 0.01–0.02 pH units per year. A similar age-related pattern is seen with base-excess during all-out cy-

Table 1.1 Substrate Availability and Utilization in Muscles of Preadolescent Boys*

Substrate	*Resting Values*		*Utilization Rate During Exercise*
	Concentration in Muscle nmol/kg Wet Weight	*Compared with Older Individuals*	
ATP	3.5–5	No change with age	Same as adults
CP	12–22	Lower in children	Same or less than adults
Glycogen	45–75	Lower in children	Much less than adults

* Based on Eriksson and Saltin,[75] Eriksson,[65] and Karlsson.[114]

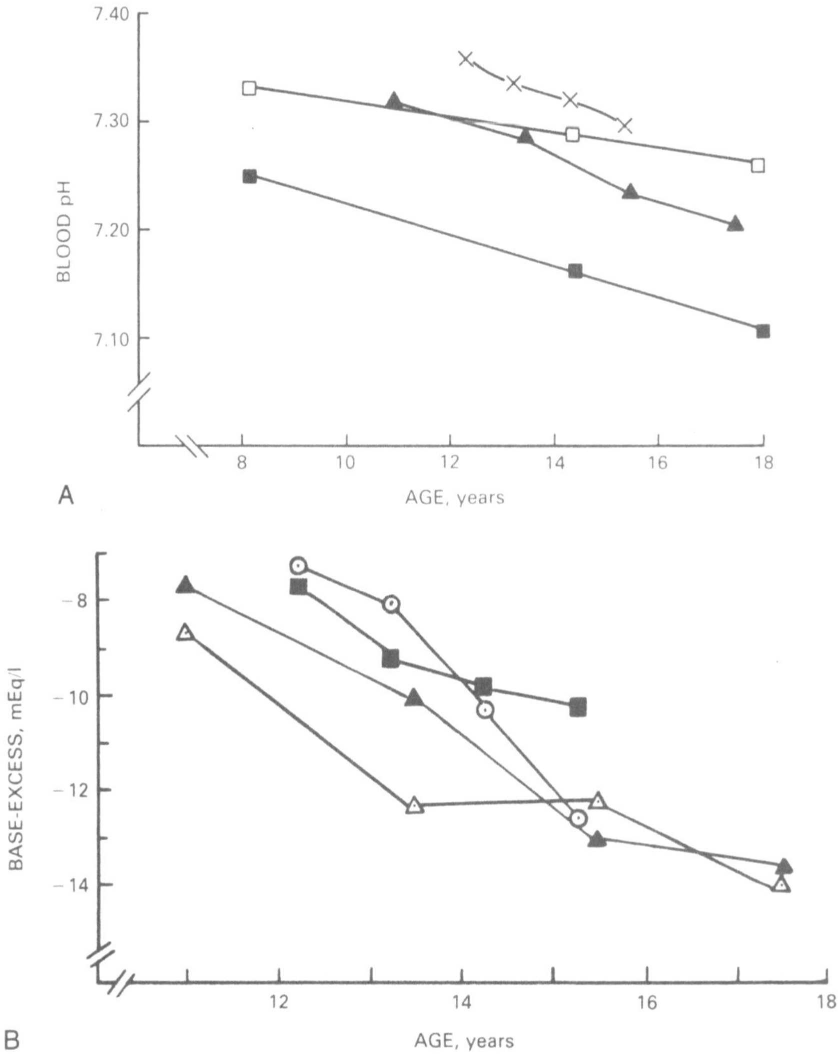

Figure 1.10. **(A)** Acid-base balance and age. Mean values of blood pH following all-out cycle ergometer exercise (□, ▲, X) or 300-m run (■). Based on data by Kindermann et al,[115] Matějková et al,[142] and Von Ditter et al.[199] **(B)** Acid-base balance and age. Mean values of base-excess following all-out cycle ergometer exercise in athletic girls (◬) and in athletic (▲, ⊙) and nonathletic (■) boys. Based on data by Von Ditter et al,[199] and Matejková et al.[142]

cling (Fig. 1.10B). The rate of decrease in base-excess is about 1–1.5 mEq/liter/yr.

Anaerobic Threshold. During a progressively increasing exercise a point is reached at which lactate production exceeds its elimination from the blood. At this point, which has been termed "anaerobic threshold," blood lactate starts accumulating appreciably. Related to such accumulation, there is an increase in pulmonary ventilation which is disproportionate to the increase in metabolic demands (see section entitled Pulmonary Ventilation, below). Some physiologists use the exercise level at which pulmonary ventilation accelerates, rather than the point of increase in blood lactate, as an index of anaerobic threshold.

It has been suggested that the anaerobic threshold of adults better reflects their cardiopulmonary endurance (e.g., performance in long-distance running) than does maximal O_2 uptake. This has been confirmed for children and adolescents in some studies,[186] but not in others (Dotan et al, unpublished). It has further been suggested that, among prepubescents, anaerobic threshold may better reflect conditioning-related improvement in maximal aerobic power than does maximal O_2 uptake. This has not been confirmed in a recent study from the author's laboratory in Israel (Dotan et al, unpublished).

The anaerobic threshold of children and adolescents, when expressed as a percentage of maximal O_2 uptake, is somewhat higher than in adults. There does not seem, however, to be an age-related difference among children or adolescents.[128]

O_2 Uptake Transients. In line with the notion that children are less capable than adults of exerting anaerobically is the pattern of their O_2 uptake transients when they increase metabolic rate. Any individual, on transition from rest to exercise or from a certain exercise level to a higher one, increases his or her metabolic rate. At first, the aerobic supply of additional energy lags behind the actual demands for energy and an "O_2 deficit" is contracted. The balance of chemical energy at these initial stages is facilitated by anaerobic pathways. During submaximal exercise, the aerobic energy supply will gradually catch up with the demand and a new metabolic "steady-state" will be established within 2–5 min.

Children have shorter O_2 uptake transients than adults.[80,137,138,162] As seen in Fig. 1.11, the O_2 deficit of 10- to 11-year-old boys is smaller than among young adults. Even more marked is the high rate at which boys increase their O_2 uptake compared with adults: the children in Fig. 1.11 reached 55% of their final O_2 uptake within 30 sec and a steady state within 2 min. The adults reached only 33% of their final O_2 uptake within the first 30 sec and required some 3–4 min to reach a steady state.[137]

A positive linear relationship exists between age and the time needed to reach 50% of steady-state O_2 uptake (Fig. 1.12).[80] An interesting question is whether, due to their shorter O_2 transients, children *do not need* to resort much to anaerobic pathways (hence the smaller O_2 deficit and lactate production), or whether these shorter transients are *compensatory* for their low glycolytic capacity. This question is still unresolved. Another possibility, not yet tested, is that the shorter transients in chil-

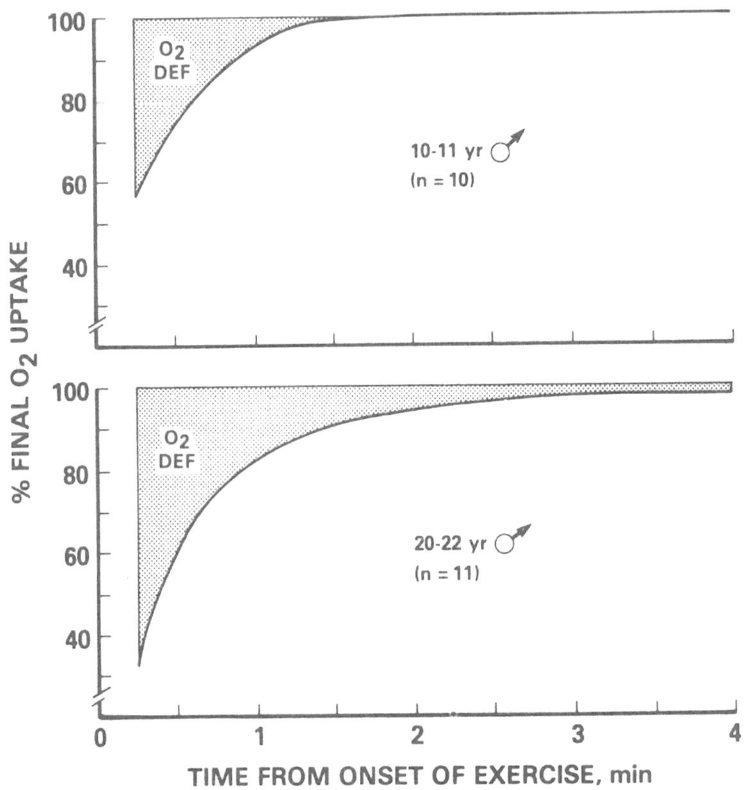

Figure 1.11. Oxygen deficit of children and adults. Oxygen uptake transients of 10- to 11-year-old boys and of 20- to 22-year-old men who cycled at 90–100% of their predetermined maximal power loads. Adapted from Máček and Vávra.[137] Reproduced with permission from Bar-Or.[26]

dren are a reflection of a smaller body and the resulting shorter circulation time.[51]

Metabolic "Specialization." Morphologic and functional characteristics of physically active adults, especially high-level athletes, are often highly specialized. Some of these characteristics are acquired, such as increased muscle mass and strength following weight training. Others are predominantly inherent and constitute one's "talent," such as tall stature, high maximal aerobic power, or short reaction time. Specificity of characteristics in adult athletes is obvious *between* sports (e.g., weight lifters vs. runners) but is also apparent *within* sports. Among runners, for example, sprinters are mesomorphic, with a well-developed musculature, but they have only average maximal aerobic power. In contrast, long-distance runners are exceptionally thin, with a high maximal aero-

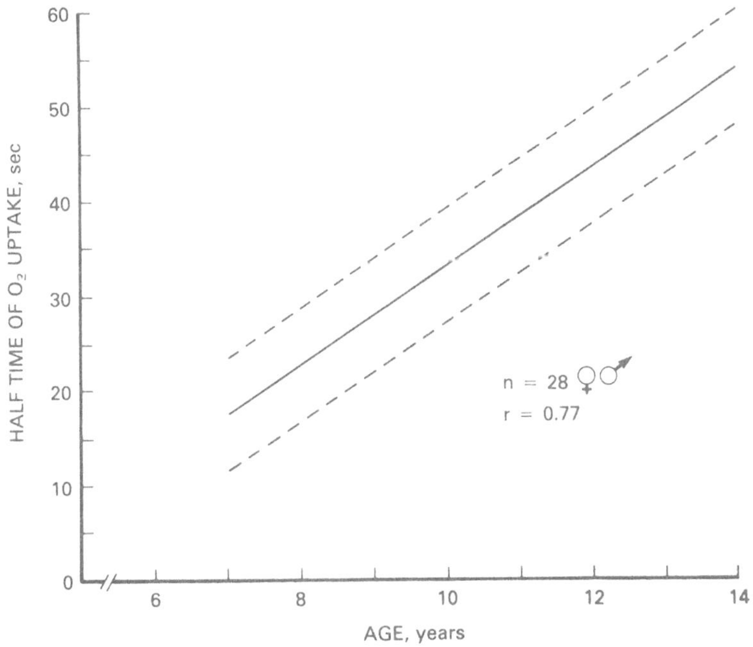

Figure 1.12. Oxygen uptake on-transients and age. Half-time of O_2 uptake during a submaximal cycle exercise (59 Watt) in 28 7- to 14-year-old boys and girls. Broken lines denote 1 S.E.M. Adapted from Freedson et al.[80]

bic power and low explosive strength. Sprinters have about equal distribution of fast- and slow-twitch muscle fibers. Marathoners, on the other hand, have as much as 80–90% slow-twitch fibers with a highly oxidative biochemical profile.

Does such specialization exist as early as during childhood? Only scant information is available to answer this question. Based on morphologic characteristics and performance, children do *not* seem to be specialists. The somatotype, for example, of children who are successful athletes seldom reaches the extremes found among adult athletes. Functionally, a child who is the sprinting "star" of his class is often also above average in long-distance running and successful in a variety of team sports. In the laboratory, children who possess a high maximal O_2 uptake also perform above average anaerobically.[29] An example is a study in which aerobic and anaerobic characteristics were assessed in 16 8- to 11-year-old boys, six of whom were elite U.S. long-distance age-group runners.[143] As seen in Fig. 1.13, those boys who scored well in the Wingate anaerobic test also had a high maximal O_2 uptake, which suggests a lack of specialization. We have found similar results in 8- to 12-year-old swimmers.[108]

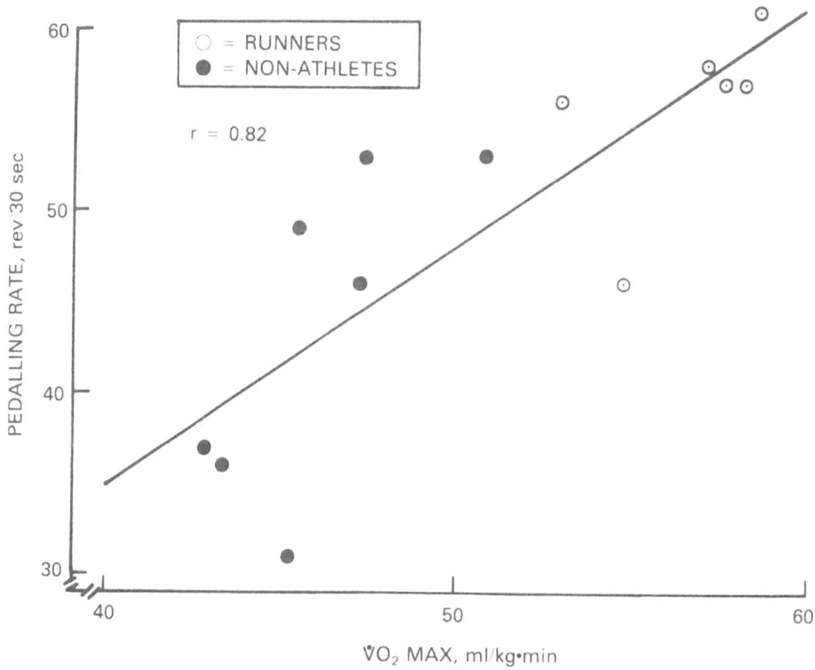

Figure 1.13. The child as a physiologic "nonspecialist." Individual performance (pedaling rate) in the Wingate Anaerobic Test plotted against maximal aerobic power ($\dot{V}O_2$ max). Data are for 13 boys, 8- to 11-years old, six of whom were elite cross-country runners and the others nonathletes. Based on Mayers and Gutin.[143]

Muscle fiber type differentiation takes place by the first year of life. It will be of the utmost interest to establish whether the specialization apparent in fiber types of adult athletes also exists in children. So far, however, data are available only for 6-year-old nonathletes. Their fiber type distribution is much the same as in sedentary adults.[34,132] Fiber type distribution of 15-year-old boys who had been participating in an endurance running program was not different from the distribution found in the same boys three years earlier.[109]

In conclusion: based on somewhat fragmentary information, it appears that prepubertal children are "metabolic nonspecialists," even when engaging in specialized sports.

Cardiovascular Response to Exercise

The main role of the cardiovascular system during exercise is in the transport of additional O_2 to the exercising muscles and of CO_2 from the muscles. Other roles include transport of nutrients, metabolites, and hormones, retention of osmotic and acid-base balance, and convection of heat from body core to periphery.

Changes that facilitate a greater O_2 supply to the muscle can best be described through Fick's principle:

$$\dot{V}O_2 = \dot{Q} \times (Ca_{O_2} - C\bar{v}_{O_2})$$

in which $\dot{V}O_2$ is O_2 uptake, $\dot{Q}$ is cardiac output, and $Ca_{O_2} - C\bar{v}_{O_2}$, the difference in O_2 content of arterial and mixed-venous blood. $\dot{Q}$ is the product of heart rate (*HR*) and stroke volume (*SV*). Therefore

$$\dot{V}O_2 = HR \times SV \times (Ca_{O_2} - C\bar{v}_{O_2})$$

For O_2 uptake to increase, either cardiac output or arteriovenous O_2 difference must rise. In fact, both rise during exercise—cardiac output through an increase in heart rate and in stroke volume and arteriovenous O_2 difference through an increase in muscle blood flow, which causes a decrease in mixed-venous O_2 content.

Table 1.2 summarizes the differences in hemodynamic response to exercise between children and adults.

Cardiac Output and Stroke Volume. As in adults, the cardiac output of children rises at the beginning of exercise, or upon transition to a higher level of exercise. A new steady-state cardiac output is established within a few minutes. Typically, children have a markedly lower stroke volume than do adults, at all levels of exercise. This is compensated for, but only in part, by a higher heart rate. The end result is a somewhat lower cardiac output at each metabolic level, as shown in Fig. 1.14.[30,60,63,71,73,146,164] It is not clear whether the somewhat lower cardiac output of chil-

Table 1.2. Central and Peripheral Hemodynamic Response to Exercise. A Comparison Between Children and Adults

Function	*Children's Response (compared with adults')*
Heart rate—submax	Higher, especially at first decade
Heart rate—max	Higher
Stroke volume—submax and max	Lower
Cardiac output—submax	Somewhat lower
Arterio-mixed venous O_2 difference—submax	Somewhat higher
Blood flow to active muscle	Higher
Systolic and diastolic blood pressure—submax and max	Lower

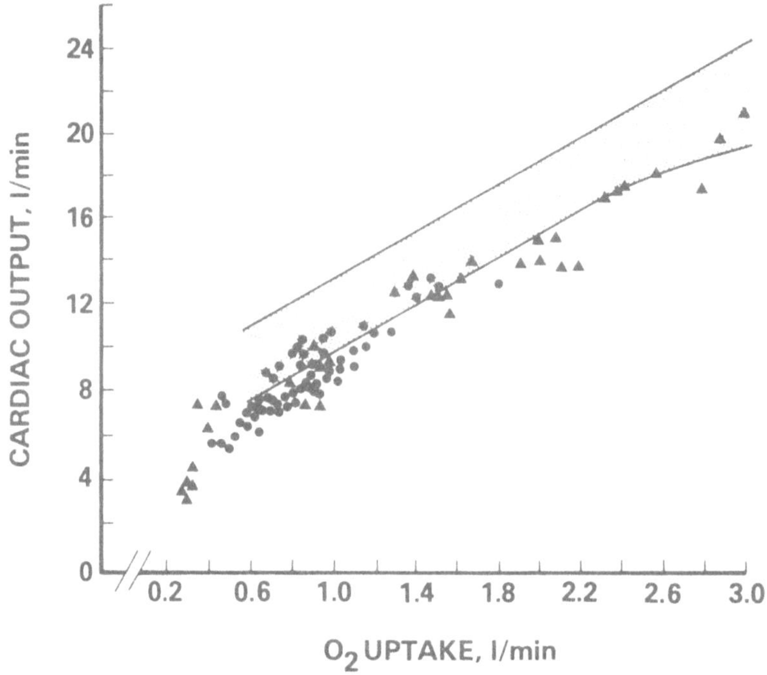

Figure 1.14. Cardiac output of the exercising child. Individual values for boys performing upright submaximal and maximal cycle-ergometer exercise. ● = data by Godfrey et al[93] on boys with body height of 110 to 154 cm, tested by CO_2 rebreathing. ▲ = data by Eriksson on boys 13–14 years old, tested by dye dilution.[63] The shaded area represents young adults performing upright exercise. Compiled from the literature by Bar-Or et al[30].

dren is of any biologic significance. Quite possibly the concomitant higher arterio-mixed venous O_2 difference is sufficient to compensate the O_2 transport system during submaximal exercise. A potential handicap due to low cardiac output may exist, however, during *maximal* exercise when peripheral O_2 extraction can no longer rise[73] or when the child is exposed to combined stresses of exercise and extreme heat. In the latter case, the circulation is called upon to simultaneously support the increased metabolic needs and the need for heat convection to the periphery. Such demands may not be met in full, and exercise-in-the-heat cannot be sustained.[24,60] (See also Chapter 9.) At all levels of exercise, stroke volume in boys is somewhat higher than in girls.[30,49,93,164]

The degree of hypokinetic response is dependent upon the age, or the developmental stage, of the child. Fig. 1.15 depicts a situation in which younger children had a lower stroke volume at a given exercise level than did older ones. This was compensated for by a higher heart rate, such that cardiac output of the young children was only a little lower than that of their older counterparts[146] (see also Refs. 88, 164).

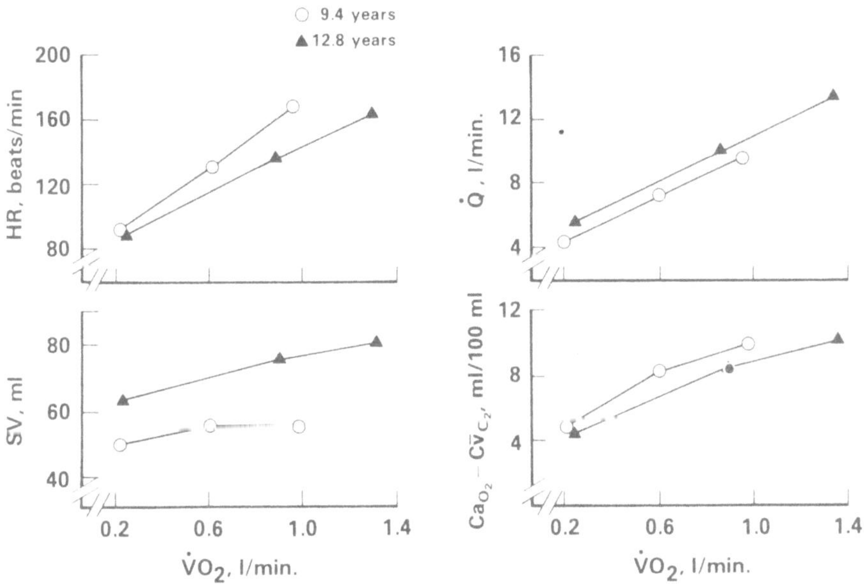

Figure 1.15. Hemodynamic changes during upright cycling in two age-groups of boys. Cardiac output was determined by the dye-dilution method. Mean values for ten 8- to 11.5-year-old, and for twelve 11.5- to 14-year-old boys. Data by Mocellin et al.[146]

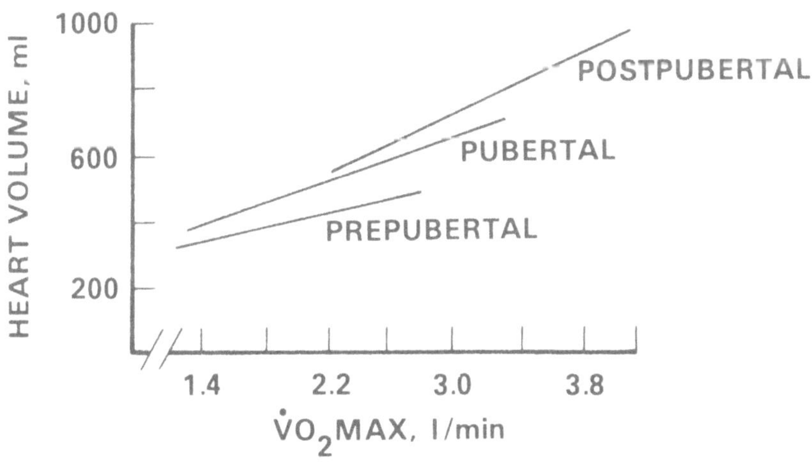

Figure 1.16. Heart volume, maximal aerobic power ($\dot{V}O_2$ max), and biologic development. Fifty-one male and female swimmers were subdivided into prepubertal (10.6 ± 0.4 yr), pubertal (12.5 ± 0.3), and postpubertal (16.4 ± 0.7) groups. Subjects performed an all-out cycle-ergometer test. Heart volume was assessed at rest. Regression lines reproduced with permission from Wirth et al.[206]

A related finding is presented in Fig. 1.16. Girls and boys who performed maximal exercise were subdivided into prepubertal, pubertal, and postpubertal groups. At an equal maximal O_2 uptake, heart volume (an important factor in the ability to raise stroke volume during exercise[104]) of the prepubertal children was the smallest.[206] It is possible that the prepubescents also had a lower maximal cardiac output and, true to Fick's principle, had to resort to higher peripheral O_2 extraction to achieve a certain level of maximal O_2 uptake.

Heart Rate and Exercise. Thanks to its relative ease of monitoring, heart rate has been the most commonly analyzed variable in exercise physiology. Its measurement has proven valuable for monitoring the cardiovascular response to exercise. In addition, its close relationship to metabolic level has made heart rate a useful indirect indicator of energy expenditure, as well as a means for prediction of maximal O_2 uptake. Due to its marked sensitivity to any increase or decrease in conditioning, heart rate has become a valuable gauge for determination of fitness and of compliance to exercise programs.

In this section, we shall analyze the relationship between heart rate and other physiologic and psychologic variables before, during, and following exercise. Special emphasis will be given to factors that modify the heart rate response to exercise in children, as summarized in Table 1.3. The practitioner who wishes to include the measurement of heart rate in his diagnostic repertoire should be thoroughly familiar with such factors.

Age. Submaximal heart rate in children declines with age.[8,15,43,93,162,168,196,206,207] As shown in Fig. 1.17, heart rate can be as much as 30–40 beats/min higher in an 8-year-old child than it is in an 18-year-old performing the same absolute task. Such a difference is partially due to the greater *relative* exercise intensity performed by the younger children in Fig. 1.17, but it is also found at equal relative metabolic loads.[206] The higher heart rate among young children is biologically sound, as it compensates for a lower stroke volume.

The *maximal* heart rate of children and adolescents ranges between 195 and 215 beats/min[8,15,30,162] and starts declining with age only after maturity has been reached. Such a decline is independent of sex, level of conditioning, climate, or other environmental conditions. It is equivalent to 0.7–0.8 beats/min per year.[28] If one takes the difference between submaximal and maximal heart rate as reflecting a certain "cardiac reserve," a 16-year-old adolescent has a distinctly greater reserve than a 6-year-old child. This is shown graphically in Fig. 1.18 and is in line with the smaller metabolic reserve of children, as discussed above, under Metabolic Responses to Exercise in Children.

Table 1.3. Factors Known to Affect Heart Rate Response to Exercise Among Children and Adolescents

Factor	*Submaximal HR (at a given exercise)*	*Maximal HR*
Age	Young > old	No effect
Sex	Females > males	No effect
Adiposity	Obese > lean	No effect
Climatic stress	↑	No effect
Emotional stress	↑	No effect
Active muscle mass	Small > large	Large > small
Body position	Upright > supine	Upright > supine
Conditioning	↓	No effect or slight ↓
Deconditioning	↑	No effect
Heat acclimatization	↓	No effect
Habituation	↓	No effect
Diseases		
Anemia	↑	No effect
Anorexia nervosa	↑	↓
A-V block	↓	↓
Cyanotic heart defects	↑	↓
Dysrhythmias	↑↓	Various
Fever	↑	No effect
Muscle dystrophy and paralysis	↑	↓
Neurocirculatory asthenia	↑	No effect
Medication		
β blockers	↓	↓
Methylphenidate	↑	No effect
β_2 sympathomimetics	↑	No effect
Thyroid hormone	↑	No effect

Sex. As a rule, women have a higher heart rate than men at a given exercise level. Traditionally, this difference has been attributed to the low blood hemoglobin concentration of the mature female.[15] However, a higher heart rate is also found among preadolescent girls whose hemoglobin level is not different from that of boys[30,135,168,202] and even among children as young as six years old.[11,93,158] The degree of relative tachycardia in females ranges between 10 and 20 beats/min. Fig. 1.19 is a demonstration of such a comparison in which 6- to 7-year-old girls and boys carried various loads on their backs while walking at a constant speed. Irrespective of the load, heart rate of the girls was by some 20 beats/min higher than among the boys. The cause of high exercise heart rate in the young female is not clear. It could result from a lower stroke

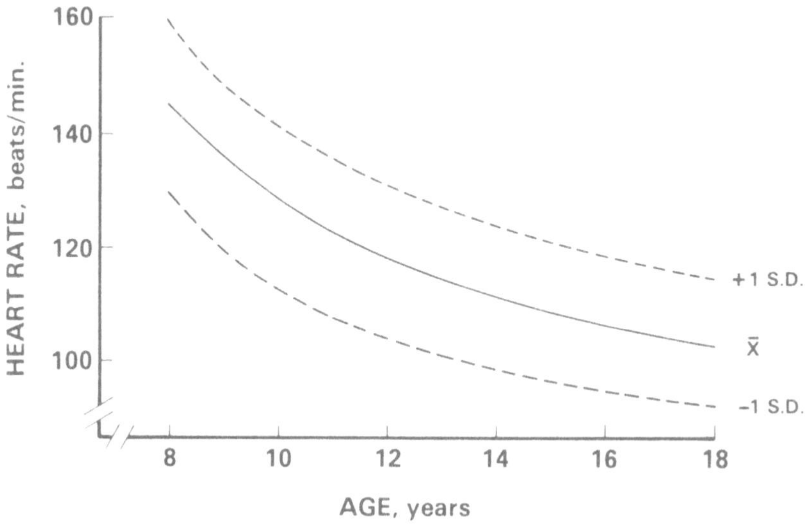

Figure 1.17. Exercise heart rate and age. Subjects were 237 boys, 8–18 years old, who participated in a growth and maturity study in West Germany. They all performed a cycle-ergometer task at a power load of 29.4 Watts. Adapted from Bouchard et al.[43]

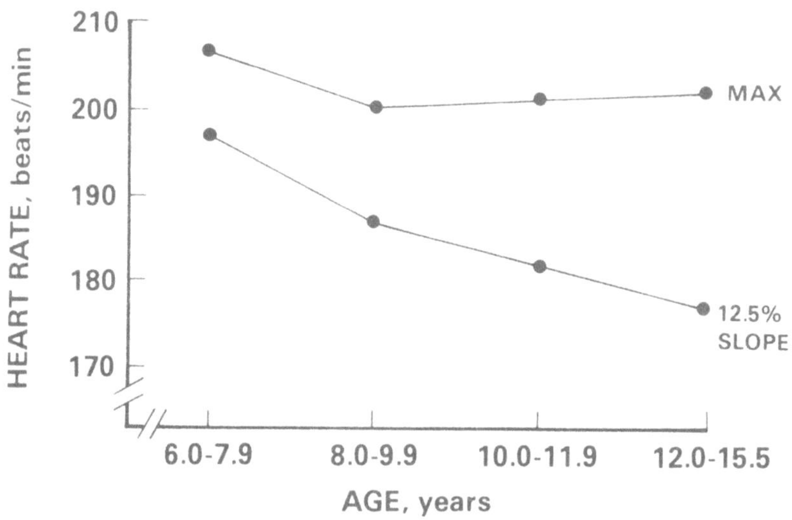

Figure 1.18. Heart rate "reserve" and age. Sixty-one girls and 83 boys, 6–15.5 years old walked on a treadmill at 5.6 km/hr, performing a progressive all-out protocol. Their mean heart rate is shown at a treadmill slope of 12.5% and at the highest attainable slope (max). Based on data by Bar-Or et al[31] and Skinner et al.[187]

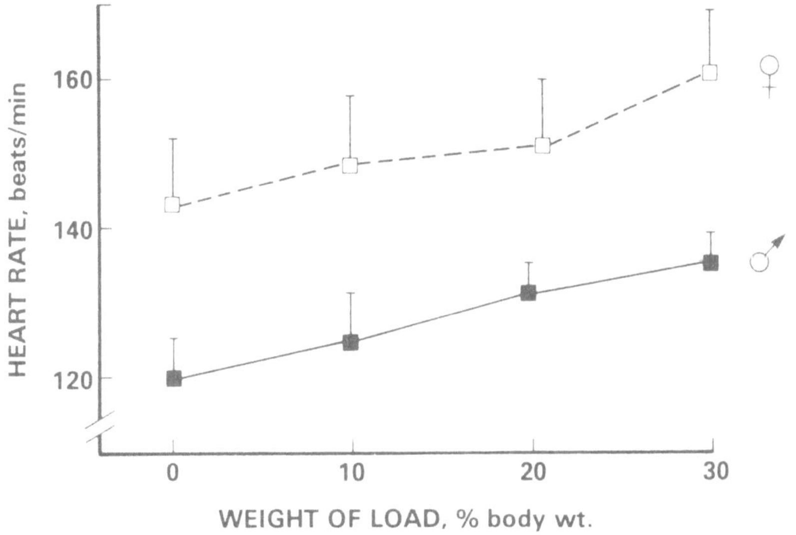

Figure 1.19. Exercise heart rate and the sex of the child. Nine girls and 6 boys, 6–7 years old, walked on a treadmill at 4 km/hr, without carrying any load and while carrying on their back a school-bag weighing 10%, 20%, or 30% of their body weight. Heart rate was monitored at the end of a 4-min walk in each condition. Vertical lines denote 1 S.E.M. Data from the author's laboratory in Israel.

volume[30] or from lower levels of habitual activity.[87] One cannot rule out sex-related differences in autonomic cardiac regulation, but there are no data to substantiate such a notion. Young boys have a faster *post-exercise* decline of heart rate than do girls,[158] which may reflect quantitative differences in regulatory mechanisms.

Adiposity. Obese children have a higher submaximal heart rate than do lean ones (Chapter 6, section entitled Response to Acute Exercise).

Climatic Stress. During exercise in hot or humid climates, heart rate is higher than in a neutral environment. An elevation above comfortable ambient conditions (23–24°C, 60–70% relative humidity) of 2–3°C or of 15–20% relative humidity may cause an increase of 10 or more beats per minute and render false information. To a clinician who wishes to test children by exercise in his office, this is an important factor to remember (see Chapter 9).

Emotional Stress and Habituation. As in adults, the heart rate of children is increased with excitement or fear. It can rise by as much as 30–50 beats/min above the "real" value. Such an increase is apparent mostly at rest and during mild levels of exercise. To a healthy or a sick child the first encounter with a laboratory setup, its instruments, and the unfamiliar personnel and tasks can all induce apprehension. In research, a session of habituation is mandatory, during which the child should undergo all procedures and be famliarized with the equipment. Data from that session should be discarded. If in a clinical setup, reassure the child, take measurements, but discard heart rate data from low exercise levels if they look out of line with those obtained at higher power loads. One way to habituate a young child is to let him play prior to testing with such equipments items as the mouthpiece, nose-clip, or stethoscope.

Active Muscle Mass. When a given mechanical power is produced by small muscles, heart rate will rise more than when the same power is produced by a larger muscle mass. A case in point is arm ergometry, where heart rate can be 20–30 beats/min higher than that obtained at identical loads during leg ergometry.[21,90] Thus, when children with such diseases as cerebral palsy, poliomyelitis, muscular dystrophy, or spina bifida are tested by arm cranking, allowance must be made for the small muscle mass involved. Norms and nomograms designed for leg ergometry should not be used. Unfortunately, normative data for arm ergometry in children are not available.

Conditioning and Deconditioning. The heart rate response to exercise is a highly sensitive gauge of the level of aerobic conditioning (see Conditioning and the Cardiovascular System, below).

Acclimatization to Heat. The difference in heart rate between the acclimatized and nonacclimatized state can be as much as 15–20 beats/min.[24] Such a difference can easily mask any other effects on heart rate. Special care should be taken during changes of weather or when newcomers to a hot region are to be tested.

Diseases. Some diseases can induce specific changes in the submaximal or maximal heart rate (Table 1.3). For details see the respective chapters.

Medication. Pharmacologic agents that affect the autonomic nervous system or the metabolic rate induce changes in heart rate. At present, specific information is scant: propranolol, a commonly used β-blocker, was found to reduce both the submaximal and maximal heart rate of healthy children, by some 15–30 beats/min (Fig. 1.20). Methylphenidate, which is recommended for "hyperactive" children, induced an increase of some 10 beats/min in submaximal heart rate, but very little in maximal heart rate (Fig. 1.21). The same changes have been obtained with theophylline and β_2-sympathomimetics, often prescribed in asthma. Although not documented, it is highly conceivable that other drugs would also affect children's hemodynamic response to exercise.

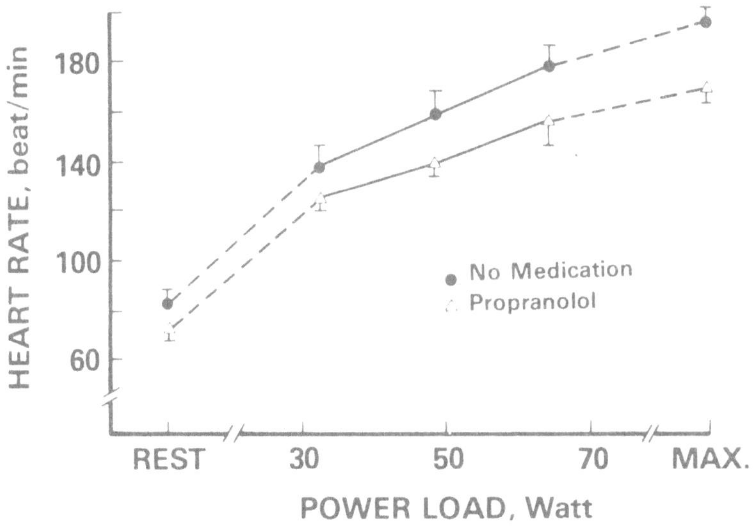

Figure 1.20. Beta-adrenergic blockers and exercise heart rate. Changes in heart rate with and without orally administered 10 mg propranolol. The subjects were six 11-year-old healthy boys who performed a progressive supine cycle test. Vertical lines denote 1 S.E.M. Data by Thorén.[194]

Body Position. Supine exercise testing often has to be resorted to, especially in a clinical setting. In this body position, which is more favorable for venous return, resting submaximal and maximal heart rates are lower than during upright exercise.[50] For example, among 8- to 15-year-old healthy boys, maximal heart rate at upright cycling was higher by 9 beats/min than during supine exercise.[49]

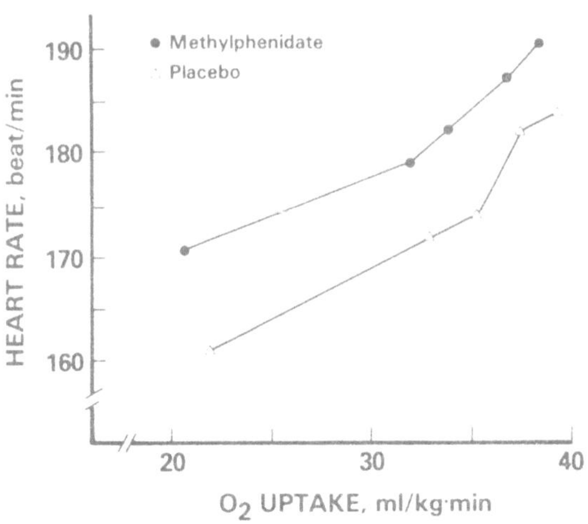

Figure 1.21. Methylphenidate and exercise heart rate. Comparison between the effect of methylphenidate and placebo pills on twenty 6- to 12-year-old "hyperactive" girls and boys. Adapted from Boileau et al.[40]

Muscle Blood Flow. In addition to an increase in cardiac output, there is a dramatic redistribution of regional blood flow during exercise. In adults, an increase takes place in blood flow to the exercising muscles (skeletal and respiratory), to the myocardium, and, at some power intensities, to the skin. Concurrently, there is a marked compensatory blood flow decline to the kidneys, splanchnic area, and nonexercising muscle. Data on peripheral blood flow of children are rather limited. Compared with young adults, 12-year-old boys had a higher muscle blood flow after exercise.[118] This difference diminished when the same boys were retested one year later,[119] and again four years[120] later.

The higher muscle blood flow of children may represent a more favorable peripheral distribution of blood during exercise. Such distribution facilitates a greater O_2 transport to the exercising muscle (in spite of a lower cardiac output), and a greater decline in the O_2 content of mixed-venous blood. As in adults, muscle blood flow of boys seems to reach a new plateau 30–40 sec after the onset of a new exercise level. Such adjustment is faster than that of O_2 uptake or cardiac output. Apparently muscle blood flow is not a limiting factor in the child's O_2 transport capacity.

Arterial Blood Pressure. *Rhythmic Exercise.* The higher contractile forces of the myocardium during exercise cause an increase in intraventricular systolic pressure. This is one mechanism by which more blood is made available to the periphery. Such an increase is manifested, among other functions, by a rise in arterial systolic blood pressure. In contrast, diastolic blood pressure, which depends primarily on the peripheral vascular resistance, changes little with exercise and its direction cannot be predicted. When rhythmic exercise is performed, the rise in systolic blood pressure is proportional to exercise intensity and to the overall metabolic level.[73] Such a pattern is operative in all healthy individuals, irrespective of age. There are, however, age- or size-related *quantitative* differences: for a given level of exercise, a small child responds with lower systolic and diastolic pressures than does an adolescent.[111,125,161,192] This pattern is shown in Fig. 1.22.

A lower exercise blood pressure in the young child is in line with the lower cardiac output and stroke volume (*vide supra*). It may also reflect a lower peripheral vascular resistance in children, due to their shorter blood vessels. There is no reason to assume that such an age-related pressure difference is either materially beneficial or detrimental to the working capacity of the young child.

Within a given age-group, boys have a higher peak systolic blood pressure than do girls,[161] probably because of higher maximal stroke volume in the boys. For reasons unknown, black children respond to exercise with higher arterial blood pressure than do white ones.[3,192]

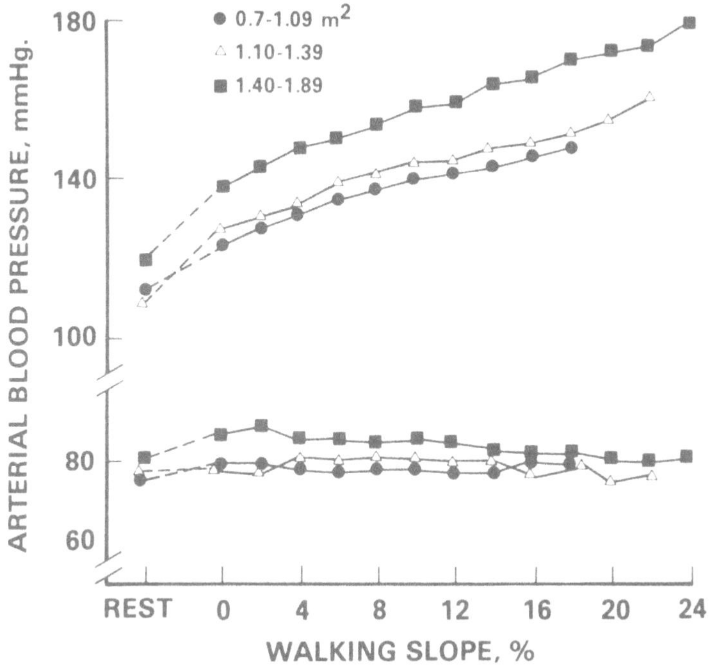

Figure 1.22. Blood pressure during a walking exercise in children and adolescents. Subjects were 274 girls and boys who had been referred to an outpatient pediatric clinic for reasons other than organic cardiovascular disease. Walking was done on a treadmill at 3.5 km/hr. The slope was raised each minute by 2%. The participants are subdivided into three groups according to body surface area. Adapted from Riopel et al.[161]

Static Exercise. The blood pressure responses to static contraction are different from those that occur during rhythmic contractions. Most striking is the dissociation between metabolic demands and blood pressure: both systolic and diastolic pressures increase above and beyond those expected merely from an increase in metabolism. Whenever the static effort exceeds some 20% of the maximal voluntary contraction of the respective muscle group, a steady-state cannot be reached. Fatigue keeps increasing and exhaustion takes place within minutes, even though heart rate may remain as low as 110–120 beats/min.[28,130] Apparently, some pressor response is activated during static contraction such that within 1 min (50% of maximal voluntary contraction) systolic pressure of adults may reach near maximal values, with a concomitant rise in diastolic blood pressure. In adults, the rise during static contraction is not related to the absolute force, but to the intensity of contraction of the respective muscle group, relative to the maximal voluntary contraction of *that muscle group*. Thus, a hand-grip at 50% of maximal intensity will cause a similar blood pressure rise as a contraction of both quadriceps muscles at 50% of their maximal voluntary contraction.

Although it has been assumed that children and adolescents respond like adults to static exercise, specific data on the blood pressure response of children are scant and limited to hand-grip only.[77,174,192] The peak systolic blood pressure during rhythmic cycling is *higher* in children than that achieved by exhausting hand-grip efforts of various combinations of intensity and duration. No studies are available where the metabolic levels of rhythmic and static efforts were equated to obtain a valid comparison of their effect on blood pressure.

There is a definite need for more data on the blood pressure response of children to static activities. Such information is relevant, for example, to the hypertensive child and to patients who, due to locomotor disability, resort to static activities.

Pulmonary Response to Exercise in Children

In addition to its role in the increased O_2 and CO_2 transport during exercise, the respiratory system affects acid-base balance by controlling body stores of CO_2. To facilitate the increase in O_2 and CO_2 exchange, pulmonary ventilation rises, bringing about a rise in alveolar ventilation. In addition, blood flow through the pulmonary capillaries increases proportionately.

Fewer data are available on the respiratory than on the circulatory system of the exercising child. These suggest that children's pulmonary response to exercise is quite similar to that of adults, with some *quantitative* differences. These are summarized in Table 1.4.

Pulmonary Ventilation. In absolute terms, ventilation grows with age. Whereas a 6-year-old child may reach a maximal pulmonary ventilation of 30–40 liters/min, a young adult can reach 100–120 liters/min and more.[15,162] When expressed per kilogram body weight, maximal ventilation is about the same in children, adolescents, and adults. Submaximal ventilation per kilogram, on the other hand, is higher in children, diminishing regularly with age. At 6 years, it can be 50% higher than at 17 years.[162] This suggests a lower "ventilatory reserve" at a young age.

During progressive exercise, pulmonary ventilation increases linearly with the metabolic rate until about 60% of maximal O_2 uptake when a ventilatory "breaking point" is reached, beyond which ventilation rises in a more accelerated manner. Such accelerated ventilation parallels, and is presumably triggered by, an increasing concentration of blood lactate and hydrogen ions. The ventilatory breaking point is related to, and often coincides with, the "anaerobic threshold" (see Anaerobic Threshold, above). As shown in Fig. 1.23, ventilation at any given O_2 uptake is higher in children, and the ventilatory breaking point appears earlier, than in adolescents or adults. One might conclude that children's ventilation is more wasteful. When percent maximal ventilation is plotted

Table 1.4. Respiratory Function During Exercise. A Comparison Between Children and Adults

Function	*Children's Response (compared with adults)*
Ventilation per kilogram body weight—maximal	Same
Ventilation per kilogram body weight—submaximal	Higher
Ventilatory "breaking point"	Earlier or same
Respiratory rate—maximal and submaximal	Higher
Tidal volume/vital capacity—maximal	Lower
Tidal volume/vital capacity—submaximal	Same or lower
Ventilatory equivalent—maximal and submaximal	Higher
Dead space/tidal volume	Same
Partial pressure of arterial CO_2	Somewhat lower

against percent maximal O_2 uptake, however, there is no age-related difference in the ventilatory breaking point.[169] This suggests that the control of pulmonary ventilation in the exercising child is similar to that of older individuals.

Respiratory Frequency and Tidal Volume. Exercising children have a markedly higher respiratory frequency than older individuals who perform the same task, maximal or submaximal. For example, during a walk at 5–6 km/hr, 8.6% incline, a 6-year-old child respires at a rate of some 50 cycles/min, while a 25-year-old man uses only 25 cycles/min.[162] During a *maximal* running test, respiratory frequency was 70 cycles/min in 5-year-old children compared with 50/min at 17 years.[15] Age- or

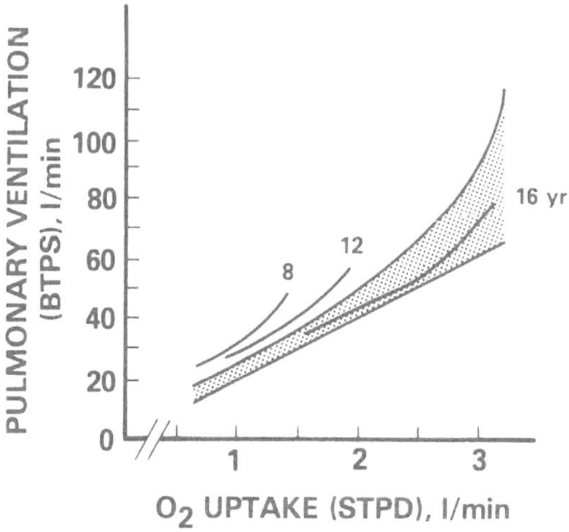

Figure 1.23. Pulmonary ventilation during exercise. Subjects were 83 Norwegian boys, subdivided into three age-groups, who cycled at two submaximal and one maximal loads. BTPS = body temperature, ambient pressure, saturated with water vapor; STPD = standard temperature and pressure, dry air. The gray area represents data for young adults.[17] Data for the children are by Andersen et al.[9]

maturation-related differences in respiratory frequency during cycling may be less distinct than during running or walking, but are still apparent.[43,92,169,177]

Does their relative tachypnea reflect shallow breathing in children? One way to express depth of ventilation is by the ratio tidal volume/vital capacity. During submaximal tasks this ratio is marginally lower in children than in adults. During maximal exercise, however, it is only 0.42–0.48 in children, 0.48–0.54 in adolescents and 0.56–0.59 in adults.[15,92,162,169] Thus we can conclude that, compared with adolescents and adults, children respond to exercise with relative tachypnea and with shallow respirations.

Ventilatory Equivalent. This ratio, which represents the number of liters of air needed to ventilate the lungs in order to supply 1 liter of O_2, is a numerical expression of ventilatory efficiency. A high ratio reflects low efficiency.

In a number of studies,[9,15,117,163] but not in others,[162,169] ventilatory equivalent has been shown to decrease with age, whether during submaximal or maximal exercise. This trend, which suggests a less efficient ventilation among children, is shown in Fig. 1.24. It has been suggested[15] that girls have a higher ventilatory equivalent than age-matched boys. This has not been confirmed by a longitudinal study in which some 120 girls and boys were followed up for 5–7 years.[169]

The major implication to children of less efficient ventilation is the greater O_2 cost of respiration. This, to some extent, could explain their relatively high metabolic demands during submaximal tasks.

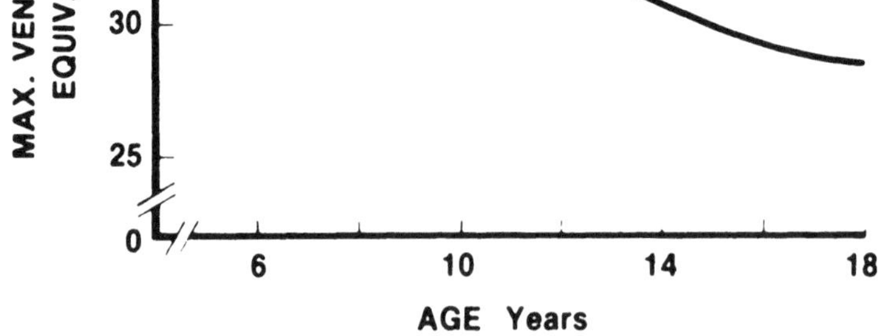

Figure 1.24. Ventilatory equivalent during maximal exercise, as related to age. A schematic presentation based on data by Andersen et al,[9] Åstrand,[15] and Kobayashi et al.[117]

Alveolar Ventilation and Gas Exchange. It is the alveolar rather than the pulmonary ventilation that determines gas exchange in the alveoli. In spite of their shallow breathing pattern, children's physiologic deadspace is smaller than in adults, and alveolar ventilation is adequate for gas exchange.[93,183] In fact, children have a mild *alveolar hyperventilation,* as judged from their arterial P_{CO_2} (Pa_{CO_2}). During various levels of treadmill[11] or cycling[92,93] exercise, children's arterial P_{CO_2} has been lower (33–36 mm Hg) than among adults (40 mm Hg).

There are no systematic data to indicate age-related differences during exercise in pulmonary diffusion. It appears, though, that during near maximal exercise young adolescents have a lower diffusion capacity at a given level of functional residual capacity than do adults.[121]

Vital Capacity and Exercise Performance. Vital capacity is strongly related to body size, particularly height, of children and adolescents.[159] Historically, this function was considered an index of fitness, based on high vital capacity among some groups of athletes. However, when body size is partialled-out, correlation between athletic performance of healthy children and their vital capacity is low.[48]

Vital capacity of girl swimmers[16,62] and boy swimmers[66] was higher than expected from their body size, but no correlation was shown between swimming performance of these children and their vital capacity. Only in advanced lung or chest-wall disease will vital capacity become a limiting factor in exercise performance (see Chapters 3 and 7).

Prolonged Activities

"Prolonged activities" will be defined as those that last 30 min or more and are continuous in nature. Although such activities are contrary to a child's pattern of spontaneous exercise, many children have been pursuing them since long-distance running has become a popular children's sport. Prolonged activities are also practised by other young athletes and dancers, as part of their training regimen.

Some research, mostly in Czechoslovakia, has been done on children's adjustment to prolonged activities. Basically, the responses of children and adolescents, whether exposed to neutral environments or to heat stress,[141] are not different from those of adults. When exercise intensity is some 60–70% of maximal O_2 uptake and it lasts about 1 hour, heart rate will reach at first a new plateau, but then will gradually increase and may be some 10–15% higher after 60 min than at 10 min of exercise.[136,139] Pulmonary ventilation may rise during that time by some 2 liters/min, and O_2 uptake by 1–2 ml/kg × min. Whereas the rise in heart rate and ventilation can be explained by an increase in core temperature and a mild level of dehydration, the rise in O_2 uptake may be due to a shift from carbohydrate to fat utilization by the active muscle. Such a shift is also reflected by a concomitant reduction in the respiratory quotient (i.e., the ratio CO_2 production/O_2 uptake).

The main age-related differences in the response to prolonged exercise are a lower accumulation of blood lactate,[116,139] a milder increase in serum potassium,[36] and a milder decrease in plasma volume[140] of children. Whether lower accumulation of blood lactate signifies less production in the muscle or a faster removal from the blood is not clear. The former is more plausible, based on data regarding short-term exercise.

In conclusion, apart from low economy of locomotion (see Mechanical Efficiency and Economy of Movement, above), there do not seem to be any underlying physiologic factors that would make children less suitable than adults for prolonged continuous exercise. A child's preference for activities of shorter duration must be explained by psychologic causes, such as shorter attention span, the need for recreational stimuli, and lower socially induced motivation for long-term exercise.

Warm-up Effect

Warming-up is a common practice by athletes and dancers. It *should* be practiced by anyone who exercises. The dividends are both improvement of performance and prevention of injuries. There are a number of possible physiologic mechanisms by which warm-up exerts its benefits: a faster impulse propagation along neurons; faster muscle contraction; a "shift to the right" in the oxyhemoglobin dissociation curve (thus, greater O_2 availability to tissues); improved coronary blood flow and myocardial oxygenation (preventing myocardial ischemia at the start of intense exercise); decrease in O_2 deficit at the beginning of exercise; increased mechanical efficiency of the muscle and reduction of viscosity within the muscle, between tendons and their sheath, and inside the joints.

Proponents of psychologic mechanisms claim that much of the improved performance is due to the attitude of the athlete who will not perform "all-out" without warming up.

To circumvent such a psychologic mechanism, a study was done in the author's laboratory with young children who had been found ignorant of the concept of warm-up. These 7- to 9-year-old nonathletic boys performed a 4-min "aerobic" cycling task and a 30-sec anaerobic task (the Wingate anaerobic test) with and without warming up. Warm-up consisted of 15-min intermittent treadmill runs raising rectal temperature by 0.52°C and elevating HR to 150 ± 10 beats/min. The time interval between this prior exercise and the criterion task was 4 min. The results are shown in Figs. 1.25 and 1.26. Warming up helped the children utilize greater aerobic resources and reach steady-state faster than without prior exercise. It also improved the score in the Wingate test, increasing the mean power output by 7%.[107] Intermittent warm-up in

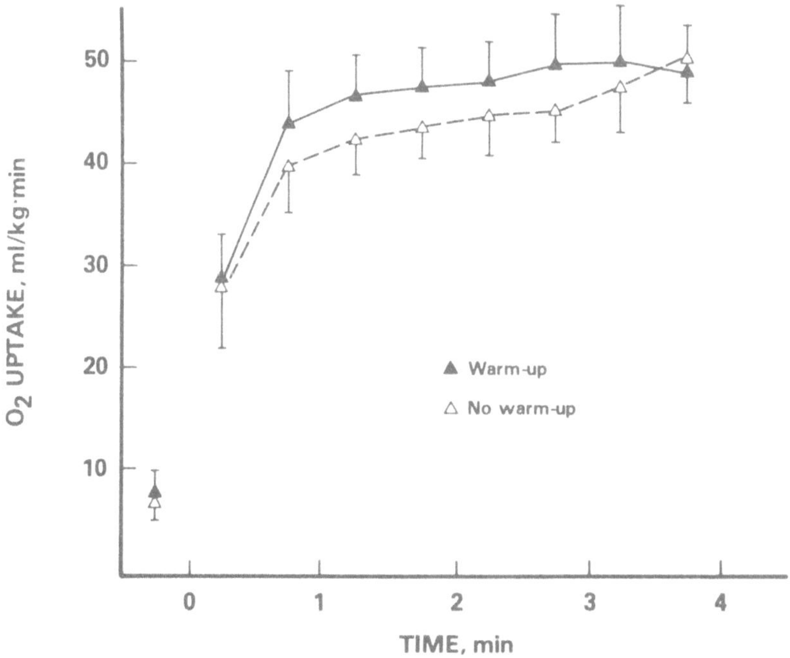

Figure 1.25. Warm-up effect on O_2 uptake in strenuous exercise. Twelve 7- to 9-year-old boys performed a 4-min one-stage cycling task with and without prior warm-up. Vertical bars denote 1 S.D. See text for procedures. Reproduced by permission from Inbar, Bar-Or.[107]

these children was found more effective than continuous warm-up, even when both induced an identical increase in metabolic level.

Various protocols can be used for warming up, depending on the task that follows, on climatic conditions, and on individual habits. One should, however, attempt to include the following three components in any warm-up: nonspecific activities that raise core temperature, stretching exercises, and specific activities (e.g., a pitcher in baseball should throw the ball, using the actual pitching motion). The optimal duration of warming up at the start of a nonspecific activity session is 8–12 min.

Exercise Perception and Age

As discussed in Chapter 2, children are habitually more active than adults. One possible explanation for this phenomenon is that adults prefer to be sedentary because they perceive exercise as more fatiguing than do children.

To analyze this possibility, we conducted a study with some 1300 seven- to 68-year-old males, all of whom performed an identical cycle

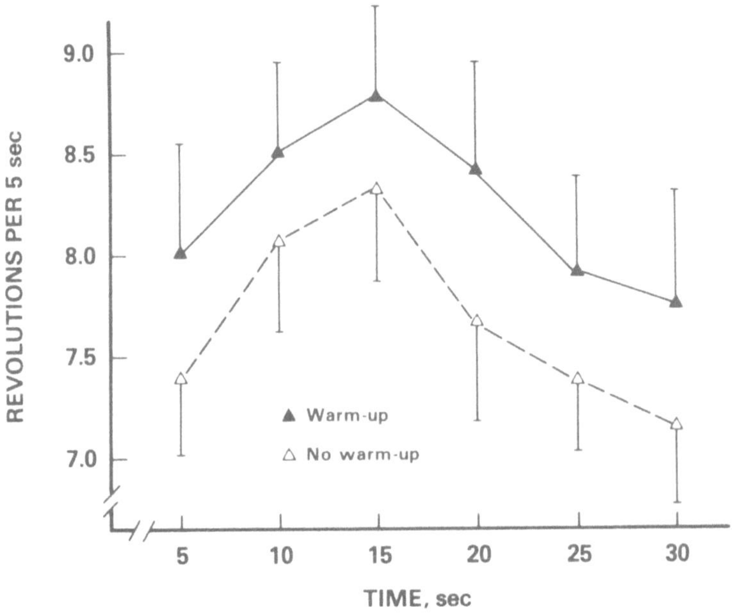

Figure 1.26. Warm-up effect on anaerobic performance. Twelve 7- to 9-year-old boys performed the 30-sec cycling Wingate Anaerobic Test with and without prior warm-up. See text for procedures. Vertical bars denote 1 S.D. Reproduced by permission from Inbar, Bar-Or.[107]

ergometry test.[22] Among other measurements, rating of perceived exertion (RPE) was determined at each power load, using the Borg scale[41] (Appendix II, section entitled Rating of Perceived Exertion, under Measurements Taken During Exercise). If submaximal heart rate is taken as an index of cardiovascular strain, the ratio RPE/heart rate can represent the subjective strain at a given level of "objective" strain. Fig. 1.27 summarizes this ratio for various age-groups at different exercise levels. Excluding the youngest group of 7- to 9-year-old gymnasts, all other subjects had an age-related increase in RPE/heart rate. Thus, at a given cardiovascular strain, children rated exercise to be lighter than did the older subjects. Because submaximal heart rate at a given exercise level is higher in young children (see Fig. 1.17), RPE was also analyzed at a certain *percentage* of maximal heart rate. This is a more prudent way for comparing cardiovascular strain in people who vary in age. The results are shown schematically in Fig. 1.28, in which the ratio RPE/percent maximal heart rate is lower in children than in adolescents and even more so than adults. These data suggest that, indeed, exercising at a certain physiologic strain is perceived to be easier by children than by older individuals. The reason for such regularity is not clear. Although

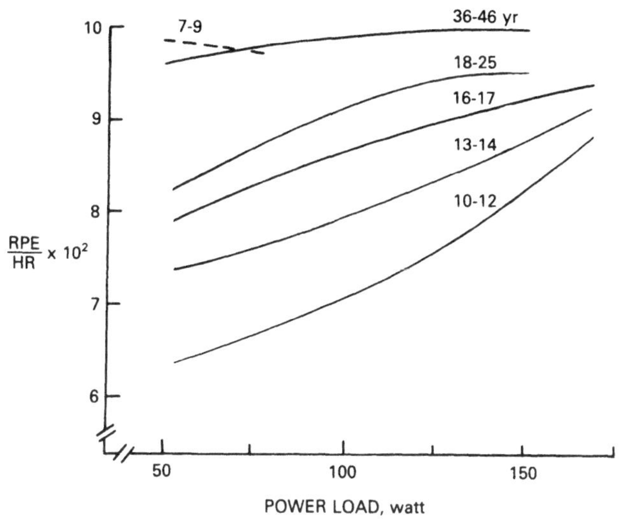

Figure 1.27. Development of exercise perception with age. The ratio of rating of perceived exertion (RPE) to heart rate (HR) at various power loads. Data on 904 seven to 46-year-old males who exercised on the cycle ergometer. Subjects are subdivided into six age-groups. Based on Bar-Or.[22]

this phenomenon is in line with the lower blood lactate and O_2 deficit levels of children, there are no data to confirm a cause-and-effect relationship between lactate concentration and RPE. There is, in fact, a suggestion that RPE in 16-year-old boys is *not* dependent on blood lactate level but rather on power load during cycling.[170]

People with experience in pediatric exercise testing often note the fast rate of *recovery* following exercise in the young subjects. Adults, for example, who complete an all-out maximal aerobic power test are usu-

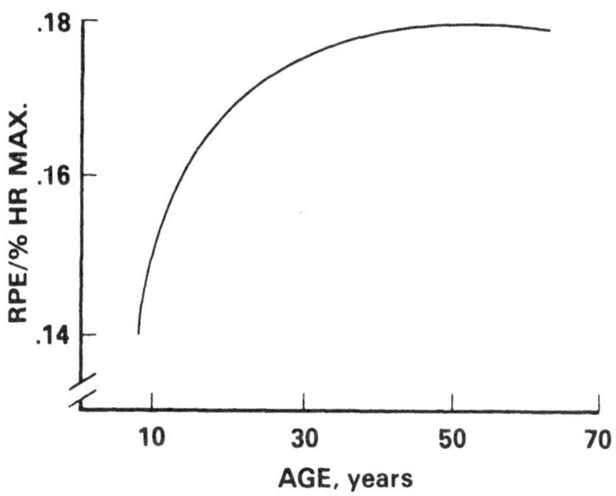

Figure 1.28. Exercise perception and age. Rating of perceived exertion (RPE) at a given percentage of maximal heart rate ($\%HR_{max}$) in some 1300 9- to 68-year-old males who exercised at 100 Watts. Schematic presentation, based on Bar-Or.[22] Reproduced with permission from Bar-Or.[26]

ally exhausted and quite reluctant to pursue further activities for several hours. This is not the case with children, some of whom often amaze us with a request to repeat the all-out task 30–45 min after its completion, because they think they could improve on their previous performance! There is no systematic comparison on the actual rate of physiologic recovery in children and adults. It is possible, though, that children's eagerness to repeat a strenuous task is related to their low rating of perceived exertion.

Another age-related difference in exercise perception has been found following conditioning or heat acclimatization. This point is further discussed in Chapter 9.

Conditioning and Training

Introduction

Physical conditioning is the process by which exercise, repeated during weeks and months, induces morphologic and functional changes in body tissues and systems. Mostly affected are the skeletal muscles, myocardium, blood vessels, adipose tissue, bones, ligaments, tendons, and the central nervous and endocrine systems.

"Conditioning" and "training" are often used interchangeably—and rather loosely—in reference to long-term exercise regimens and their effects. "Conditioning" will be used in this text for intervention programs in which emphasis is given to a general increase of the metabolic level (e.g., a mixture of running and calisthenics). "Training," on the other hand, will refer to programs that emphasize a *specific* fitness component or a specific part of the body (e.g., sprint training, or weight training of the elbow flexors). "Trainability" will be used to denote the responsiveness of an individual to either general conditioning or a specific training program.

Methodologic Constraints. Research on the conditioning and training of children has inherent methodologic obstacles and pitfalls. As in other age-groups, studies should include an intervention program and a longitudinal follow-up. In adults, changes in function between pre- and post-intervention can be attributed with fair certainty to the conditioning program. Not so with children or adolescents. Here, changes due to growth, development, and maturation often outweigh and mask those induced by the intervention. It is intriguing that many of the physiologic changes that result from conditioning and training also take place in the natural process of growth and maturation. Such is the case, for example, with the reduction of submaximal heart rate and ventilatory equivalent, or the increase of stroke volume, maximal blood lactate, muscle strength, or economy of running.

The development of techniques to partial-out the growth and development effect from the conditioning effect has been a major challenge to physiologists and kinanthropometrists.[166] In recent years, increasing emphasis has been given to dimensional principles (see Dimensionality Theory and Maximal O_2 Uptake, above) by which size-related changes can be predicted and accounted for in any conditioning study.[33,61,152,200] This approach, while useful in the absence of a control group, is not an adequate *substitute* for a control group, as it does not allow one to partial-out such effects as habituation or learning.[97] Furthermore, the validity of the theoretical exponents of body height is questionable for some variables.[18,181]

In various intervention studies "controls" have been included, but very seldom have children been assigned randomly to the activity and the control groups. Thus, an *a priori* selection cannot be ruled out. Such preselection is especially obvious when members of a sports club, with presumably above average talent and motivation, are compared with nonathletic children. Even studies on the effects of "sports classes" within a school (using other classes as controls) seldom escape preselection: children who are assigned to such sports classes often volunteer and need parental consent. They, therefore, may have different attitudes toward sports than their unselected counterparts.

Due to the above constraints, the current state-of-the-art on conditioning of children is still fragmentary. We must look upon the data as changes that *occur with* conditioning and training but do not necessarily *result from* them.

Principles of Physical Conditioning

The clinically oriented reader who may wish to incorporate exercise into a therapeutic regimen is not expected to master the *techniques* of conditioning or training. This role should be left to experts in "adapted physical education" (a subspecialty of physical education, dealing with the clinically disabled child) or to physiotherapists with a special interest in fitness and sports. It is important, however, that a clinician who considers prescribing an activity be conversant with the *rationale, principles,* and *terminology* of conditioning. These will be presented below. Discussion of a more technical nature can be found in other textbooks.[17,79,181]

Specificity of Training. The changes that take place in body tissues as a result of chronic exercise are stimulus-specific: a particular activity may induce a change in one tissue and not in another. Furthermore, a tissue may respond in one way to a certain exercise stimulus and in another to a different one.[101] Long-distance running, for example, will induce myocardial hypertrophy, improvement of the oxidative capacity within muscle fibers of the quadriceps, and preferential hypertrophy of the slow-twitch fibers. In contrast, knee extension against high resistance will barely affect the myocardium, and will induce hypertrophy of the quadriceps (mostly fast-twitch fibers) and a lesser improvement of the oxidative capacity in that muscle. An example of specificity of training is a study by Fournier et al,[78] of 16- to 17-year-old boys. A 5-month sprint training increased the activity of muscle phosphofructokinase (an "anaerobic" enzyme) by 20.6%, with no change in succinyl-dehydrogenase (an oxidative enzyme). In contrast, endurance training induced a 42% increase in succinyl-dehydrogenase activity and only 6% in phosphofructokinase activity.

Specificity of the stimulus can be nicely illustrated in strength training.[171] Each of the following factors will modify the response to a weight-training regimen:

1. The muscle(s) involved.
2. Type of contraction (concentric, eccentric, isometric).
3. Intensity of contraction and number of repetitions.
4. Velocity of contractions.
5. Joint angle at which contraction is performed.
6. Movement pattern.

The fine nuances of specific responses to training are especially important to athletes, who strive to attain perfection. They are less crucial to those interested in the general improvement of physical working capacity. Nevertheless, an intelligent approach to prescribing exercise to a disabled or detrained child should tailor the type of activity to his or her specific needs. While some children, especially those with musculoskeletal disabilities, may require prescription of activities "custom made" to their residual ability, others can benefit from sports practiced by the general population.

Different sports develop different components of physical fitness. Table 1.5 summarizes the major benefits to fitness that various sports can induce. The fitness components are presented in this table in popular rather than physiologic terms. "General stamina," for example, stands for maximal aerobic power and "local muscular endurance" for anaerobic capacity. The importance assigned to each sport in improving a certain fitness component is the most *typical* effect, but there are many variations. For example, volleyball is not the sport of choice for weight control. It can, however, be practiced in such a way that calorie expenditure is high, and body fat is reduced. Similarly, soccer is not the most suitable sport for developing strength of trunk muscles. However, by incorporating certain elements (calisthenics, weight training) a soccer player *can* greatly develop these muscles.

Dosage of Conditioning. In analogy to other forms of prescription, conditioning can be characterized by its intensity, frequency, duration of each session, and the overall duration of the program.

Intensity. Intensity of an activity is determined by either the metabolic demands (e.g., O_2 uptake), the strain on the cardiovascular system (e.g., heart rate), or in the case of strength training by the weight to be lifted, pushed, or pulled. Although often described in absolute terms, intensity should also be described in relationship to the current fitness level of the individual. For example, to a disabled (or a small) child with maximal O_2 uptake of 1.0 liter/min, an activity that requires 0.75 liter/min of O_2 is more intense than it is to a healthy or a larger child with maximal O_2 uptake of 1.5 liters/min. Similarly, the lifting of 20 kg is less intense for a muscle with a maximal voluntary contraction of 40 kg than it is for a muscle with a maximum of only 30 kg. It is therefore advisable to describe the intensity of an aerobic (or weight training) task as a *percentage* of the individual's maximal aerobic power (or maximal voluntary contraction).

It has long been established that the *conditioning effect* (i.e., the *extent* of changes that take place as a result of conditioning) is positively related to the intensity of the activities performed. This is especially true for maximal aerobic power and strength but also applies to other fitness components.

An important concept to recognize is that of the *intensity threshold.* This is the intensity of exercise below which little or no conditioning effect can be discerned. For young adults an intensity threshold of maximal aerobic power is 60–70% of maximal O_2 uptake (or 70–80% of maximal heart rate). No specific data are available for children, but experience shows that their aerobic intensity threshold is at least as high as that for adults. The threshold for developing strength is about 60–65% of maximal voluntary contraction.

A conditioning or training program must be *progressive* in its demands. Once an individual has improved a certain fitness component, the intensity previously sufficient to induce changes may now be below the threshold and may no longer be sufficient. An example is a muscle with maximal voluntary contraction of 30 kg trained by lifting 20 kg. As a result, it becomes stronger and its maximal voluntary contraction is now 35. At this level, to achieve a further training effect, the load may have to be increased to 23 or 25 kg.

Frequency. The number of exercise sessions per week may determine the success of a program. Because of interaction with the effects of intensity and duration of each session, there is no single optimal frequency that is suitable for all programs. While some conditioning effect can be achieved by one weekly session, it is generally advisable to practice at least two or three times per week. Some athletes, children included, train as often as 12–15 times per week! Such a regimen is obviously not recommended for a child who is undergoing a therapeutic program or merely wishes to improve his or her fitness. The frequency of sessions

Table 1.5. Typical Effects of Various Sports on Fitness

Type of Sport	*General Stamina*	*Local Muscular Endurance*	*Muscle Strength*	*Speed*	*Agility*	*Flexibility*	*Body Weight Control*
Individual Sports							
Boxing	+++	+++	++	+++	+++	–	+++
Cycling (long and middle distances)	+++	+++	++	++	–	–	+++
Figure skating	++	+++	++	+	+++	+++	++
Golf	–	–	+	+	–	+	++
Gymnastics	+	+++	+++	++	+++	+++	++
Horseback riding	–	++	+	–	–	–	–
Jumping (track and field)	+	+	+++	+++	+	+++	+
Rowing	+++	+++	+++	+	–	+	+++
Running							
sprint	+	+	+++	+++	++	+	+
middle distances	+++	+++	++	++	+	+	+++
long distances	+++	+	+	+	–	–	+++
Sailing	+	+++	++	+	++	–	+
Skiing—downhill and slalom	++	+++	++	++	+++	+	+

Skiing—cross country	+++	++	++	+	+	−	+++
Swimming	+++	+++	++	++	+	++	+++
Tennis (squash)	++	++	++	++	+++	+	++
Throwing (discus, etc.)	+	++	+++	+++	++	++	+
Walking	++	+	+	−	−	−	++
Weight-lifting	−	++	+++	+++	+	+	+
Wrestling, judo	++	+++	+++	++	+++	+++	++
Team Sports							
Baseball	+	++	+	++	+++	+	+
Basketball, Soccer	++	++	+	++	+++	+	++
Football (American)	++	+++	+++	+++	++	+	++
Ice hockey	++	+++	++	+++	+++	+	++
Volleyball	+	++	++	++	+++	++	+
Waterpolo	+++	+++	++	++	++	++	++

* − = Hardly any effect, + = some effect, ++ = much, +++ = very much.

for such children is also dependent on logistic limitations (e.g., ability of parents to drive the child to the gymnasium).

In our therapeutic programs we encourage children to participate three times per week in the hope that they will attend *at least* two weekly sessions. When exercise is prescribed for home, the recommended frequency is four to five times per week.

Duration of Sessions. As with frequency, there is no single optimal duration. One should allow some 10 min for warming up at the beginning and 5–7 min for cooling down at the end of each session. The main portion of the session, during which the intensity threshold is exceeded, should last 15–30 min. Thus, the overall duration will be about 35–45 min. When the main part of the session is too long, the child may become overly fatigued. The duration should also be tailored to the attention span of the participants. A preschooler, or a mentally retarded child, may lose interest within 10–15 min. A more mature or better motivated child can maintain interest for a much longer period.

Duration of the Program. As with any therapeutic program, there is a minimum period of time for which intervention should be sustained for it to yield any effect. While some conditioning effects can be attained within 1–2 weeks, most programs last at least 6–8 weeks. As a rule, the longer the program the more effective it is (providing the principle of progression is adhered to). Overall duration should also be determined by the specific goals. A weight-reducing regimen, for example, may require 6 months to achieve noticeable effects. In contrast, the strengthening of a specific muscle group may need only 1 month.

An important aim of exercise programs for healthy and disabled children alike is to form habits, to acquire skills, and to enjoy sports. Such educational goals, and not only the physiologic ones, should also dictate the duration of the program.

Conditioning Effect Related to Preconditioning Fitness. Preconditioning fitness of children and adults often determines their response to conditioning or to specific training: those having a lower initial fitness level respond with greater changes than less fit individuals.[113,168,173,180] The implications of this phenomenon are obvious for the morale and motivation of the unfit participant.

Trainability of Muscle Power and Force in Childhood and Adolescence

Are there specific developmental stages at which the growing individual is most—or least—responsive to conditioning stimuli? This question is of paramount importance to the coach and the young athlete, who wish to maximize performance. It should also interest the physical therapist, physiatrist, and pediatrician who wish to include physical conditioning in their management repertoire. In a broader theoretical sense such a topic relates to the basic question: are there phases in the formative years when certain characteristics are "fixed," or are most (or least) sensitive to environmental stimuli?

Trainability of motor skills in the growing child is discussed in textbooks of motor development and will not be dealt with here. Our purpose is to focus on recent evidence regarding trainability of maximal aerobic power at different developmental phases. Some mention will also be made of other fitness components.

Maximal Aerobic Power. Among adults who undergo an aerobic conditioning program, there is an age-related trend: the younger the individual, the more trainable he or she is.[172] By extrapolating from such a trend, one might expect children to be even more trainable than young adults, but this is not the case.

In some studies adolescents and children responded in a predictable manner to either general conditioning or specific training regimens.[20,59,61,97,98,185,189,203] There is, however, growing evidence to suggest that aerobic trainability in prepubescents, particularly in those less than 10 years old, is lower than expected, even though their athletic performance may improve.[6,27,32,52,54,55,86,100,117,147,157,167,176,185,188,191,208] An example is a study performed in our laboratory in which 91 nine- to 10-year-old girls and boys performed interval running two, three, or four times a week, for 9 weeks. Although markedly improving their running performance, they had no improvement in maximal aerobic power.[32] Similar results were obtained for 5-year-old children who ran 750 to 1500 meters five times per week during 14 months[208] or for other prepubescents who practiced middle- and long-distance running.[147,191]

How can one reconcile an improvement in long-distance running with no concomitant increase in maximal O_2 uptake? One possibility[55] is that the child improves his running *style* such that his movements are mechanically more economical. Another explanation is an improvement in *anaerobic* capacity, even though the conditioning program is considered "aerobic."[97] It is also possible that maximal aerobic power of the prepubescent does improve but maximal O_2 uptake is not a valid or sensitive indicator to show it.[52,143,167,175] Alternative indicators should be sought, one of which is the anaerobic threshold. A final reason for the apparent lack of conditioning effects in prepubescents could be the high habitual activity of the "controls." It is quite possible that even though they do not participate in the regimented program, these children are so active in their free time that intergroup differences are quite negligible.[53,100]

Children's maximal aerobic power may respond to some, but not all, training protocols known to be effective in adults. "Logging" miles by noninterrupted running, for example, may be more effective than interval training,[131] and swimming, more effective than other sports.[197]

At what developmental stage does aerobic trainability reach the extent found among young adults? Reports in this regard are conflicting. One might attempt to reconcile such findings by aligning the subjects according to the age at which they reach their peak height velocity. Based on this approach, Kobayashi et al.[117] suggested that the effectiveness of aerobic conditioning was greatest at, or around, peak height velocity. Another study, using a similar technique, could not identify such a critical stage in aerobic trainability.[5]

Anaerobic Capacity. As seen in Table 1.6, conditioning of 11- to 13-year-old boys increases the concentration of muscle ATP, creatinine phosphate and glycogen, the activity of phosphofructokinase, and the rate of glycogen utilization.[64] It is not known whether such changes reach maximum at a certain developmental stage.

Dynamic endurance of the arm flexors was studied among 9- to 17-year old children before and after training. Among the boys this component was *least* responsive to training half a year before peak height velocity. Among girls, a dip in trainability occurred at peak height velocity or half a year later. At other ages, improvement in dynamic muscular endurance was quite uniform.[127] These results contradict an earlier study in which age 11–13 years was the *most* sensitive stage for training of dynamic endurance of the elbow flexors.[105]

Muscle Strength. Muscle strength of 8-year-old girls and boys was found somewhat more trainable than among adults who were given the

Table 1.6. Training of the Anaerobic System in Children

Variable (in muscle)	*Pretraining Level (vs. adults)*	*Response to Training*
% FT fibers	Similar	No change (?)
CP (rest)	Similar	Increase
ATP (rest)	Similar	Increase
Glycogen (rest)	Low	Increase
PFK	Low	Increase
Glycogen depletion with exercise	Less	Increase
Lactate maximum	Low	Increase

Data from Eriksson.[64]
n = 12 boys 11–15 years old with aerobic and anaerobic training for 6–16 weeks.

same relative training stimulus.[165] Similarly, girls 7–13.4 years old had a greater increase in isometric strength than did 13.5- to 19-year-old girls, all training for 5 weeks.[152] In contrast, postpubescent boys had a greater response to a strength training program than did prepubescent boys,[201] and adolescents of both sexes had a lesser improvement in isometric strength than did young adults.[103]

To sum up, the above data are inconclusive. The definitive study on the trainability of children has yet to be performed. Such a study will have to include children, adolescents, and young adults, who, at the start of the program, will all be at the same fitness level. In addition, the conditioning dosage will have to be carefully equated among all groups. Until such a project is launched, our knowledge on the optimal ages of trainability will be tentative at best.

Physiologic Effect of Detraining and Deconditioning

Performance and physiologic functions deteriorate fast whenever one's activity level is appreciably reduced. This is the reverse process of conditioning and training. It often occurs as a result of bed rest, immobilization by cast, or during an off-season period when an athlete reduces his or her training load.

The understanding of the processes that underlie detraining and deconditioning is not only important to athletes and coaches. It is as important to the clinician, who often must decide whether to prescribe bed rest, cast, cessation of training, or other means of immobilization.

Studies with healthy adults have shown that bed rest can induce, within 4 to 7 days, a dramatic deterioration in function and performance.[82] For example, when young adult volunteers underwent bed rest for 7 days, complying with the routine of a clinical ward, there was a 6–7% reduction in maximal O_2 uptake, peak power performed on the cycle ergometer, total hemoglobin, and plasma, blood, and cell volumes. There was also a drop in the resting urinary excretion of norepinephrine,[82] but not in muscle strength.[81] When bed rest is superimposed on a surgical trauma, deterioration is even faster.

The degree of deconditioning, according to some but not all[96] studies, is in direct relationship to the initial level of fitness, those who are more fit having a greater drop in performance.

While it can be assumed that children's response to detraining is the same as that of adults, direct information pertaining to children is rather scanty. In one study, high school girls were followed during 23 weeks after cessation of track training.[144] By the seventh week their submaximal heart rate rose by about 10 beats/min, and there was a slowing down of heart rate recovery after exercise. Although the mechanical efficiency of running did not change, there was a gradual decrease in ventilatory efficiency, as shown by a rise in the ratio pulmonary ventilation : O_2

uptake. Aerobic and anaerobic muscle enzymes in adolescent boys decreased 6 months after the termination of an endurance and a sprint program, respectively.[78]

Deconditioning of children is most important in the clinical context. For example, 12-year-old boys who were immobilized by cast or by traction in an orthopedic ward were compared to boys who were hospitalized, but not immobilized. Arterial blood pressure of the former was higher, and some children of that group had enhanced urinary excretion of calcium.[195] Detrimental effects of deconditioning have also been shown for children with muscle dystrophy. At one stage of their progressive disease, these children can still walk, but the slightest further diminution in activity may render them immobile for the rest of their life. For such children unjustified bed rest may be catastrophic, as discussed in Chapter 7. Similar findings have been reported for children with rheumatoid arthritis.[110]

Adolescent athletes who had stopped training and were retested a few years later had a level of maximal aerobic power similar to that of those who were never athletes.[76,150] In contrast, the heart volume of previously elite girl swimmers remained above normal.[67] When such girls underwent a swimming program at ages 29–31 years, their increase in maximal O_2 uptake was similar to that of women who had never trained.[68]

Conditioning and the Cardiovascular System

The morphologic and functional changes in the cardiovascular system that take place with conditioning are summarized in Table 1.7.

Morphologic Changes. Myocardial mass, depicted by echocardiography, and heart volume (by X-rays) are higher in athletic than in nonathletic children and increase as a result of endurance training.[2,61,74,85,129,198] Cardiac hypertrophy results from excessive pressure load and, to a lesser extent, from volume load. Conditioning-induced collateral coronary circulation has been shown in animal studies, but results for adults are equivocal and no data are available for children.

Although stroke volume is related to heart volume, the myocardial mass of children does not seem to affect their athletic capability.[2] Total blood volume and total hemoglobin are higher in trained boys than in untrained ones[122] and they increase as a result of conditioning.[74] A higher blood volume facilitates a higher venous return as well as more effective heat convection from body core to periphery. The increase in total hemoglobin raises the O_2-carrying capacity of the blood. Hemoglobin *concentration,* on the other hand, does not increase with conditioning. In fact, some endurance athletes have a sports-induced anemia (Chapter 8).

Table 1.7. Cardiovascular Changes That Occur in Children, With Conditioning

Variable	*Change*
Morphologic	
Heart volume	Increase
Myocardial wall	Concentric hypertrophy
Blood volume	Increase
Total hemoglobin	Slight increase
Functional	
Stroke volume—submax, max	Increase
Heart rate—submax	Decrease
Heart rate—max	No change or decrease
Cardiac output—submax	No change or decrease
Cardiac output—max	Increase
Myocardial O_2 requirements	Decrease
Arterio-mixed-venous O_2 difference—submax, max	No change
Muscle blood flow—submax, max	No change
Systolic blood pressure—submax	No change
Systolic blood pressure—max	Increase
Diastolic blood pressure—submax, max	No change or increase
Total peripheral resistance—submax, max	No change

Physiologic Changes. Stroke volume at rest and during all levels of exercise increases with conditioning in pubertal children[74,84,100] and in adults. In contrast, in prepubertal children conditioning does not seem to be accompanied by changes in stroke volume.[100,130] An increased stroke volume may reflect higher blood volume and improved venous return, but it may also reflect improved myocardial contractility.[205]

The heart rate of trained athletes is lower at rest and during all levels of exercise and shows a faster postexertional recovery than in nonathletes. A reduction of resting and submaximal heart rate is a most sensitive response to conditioning[32,44,52,74,100,178] and often precedes any changes in maximal aerobic power.[52] The reduction in maximal heart rate is milder, seldom exceeding 5–7 beats/min.

The mechanism of conditioning-induced bradycardia is not clear. It may be secondary to an increase in stroke volume, but it also reflects a stronger parasympathetic and a weaker sympathetic drive.[73] Whatever its mechanism, a slower heart rate is accompanied by reduced myocardial work and O_2 uptake.

The sensitivity of the heart rate to conditioning and deconditioning has been the basis of fitness tests in which maximal aerobic power is indirectly assessed from submaximal heart rate[1,17,35,42,43,45,84,168] (see Appendix II, section entitled Determination of Maximal Aerobic Power).

With the opposing changes in stroke volume and heart rate, cardiac output changes little at a given metabolic level.[84] *Maximal* cardiac output increases with training in children[74] and in adults, in proportion to the increase in maximal O_2 uptake.

In adults, the maximal arteriovenous O_2 difference increases with conditioning, reflecting an increased peripheral O_2 extraction. In children, however, neither the submaximal[84] nor the maximal[74] arteriovenous O_2 difference changed with conditioning. This discrepancy could result from the arteriovenous O_2 difference—already wider in the untrained child than in the untrained adult.

Neither submaximal nor maximal muscle blood flow changed following a 40-week conditioning program in 12-year-old boys.[119] In adults, blood supply to the active muscles diminishes during submaximal exercise and increases during maximal exercise as a result of conditioning.

Systolic arterial blood pressure at rest decreases in adults following conditioning, by about 7 mm Hg.[181] A similar response has been shown in mildly hypertensive adolescents[99] (see also Chapter 4, section entitled Beneficial Effects of Chronic Exercise). Blood pressure during submaximal exercise does not change with conditioning in prepubertal boys,[130] nor are there conditioning-related changes in total peripheral resistance.[74]

Conditioning and the Pulmonary System

Pulmonary functions, both static (volumes, capacities) and dynamic (flow, diffusion), are strongly related to body size. It is therefore hard to distinguish between growth- and conditioning-related changes in the pulmonary response to exercise. Unfortunately, very few of the intervention studies in this field include an adequate control group to allow for such differentiation. The reported changes as summarized in Table 1.8 therefore need confirmation.

Vital capacity usually does not increase in adults as a result of a running program, and differences in vital capacity between athletes and nonathletes are, to a great extent, due to preselection.[16] In children and adolescents, vital capacity increased following running or swimming programs in some studies,[4,61,62] but not all,[120] above and beyond growth-related changes. Such an increase may reflect stronger respiratory muscles, which help compress the chest at the end of expiration.

With conditioning, ventilation and respiratory rate during a standard task are reduced, and O_2 extraction from the inspired air is greater. This important effect of conditioning may reflect a lower reliance on anaerobic metabolic pathways, a lower O_2 cost of exercise, or a reduction in chemoreceptor sensitivity.[181] In contrast, the maximal ventilation and

Table 1.8. Pulmonary Changes That Occur in Children, with Conditioning

Function	*Change*
Vital capacity	No change (increase in swimming)
Ventilation—submax	Decrease
Ventilation—max	Increase
Respiratory rate—submax	Decrease
Tidal volume—max	Increase
Ventilatory equivalent—submax, max	Decrease or no change
Respiratory muscle endurance	Increase
Pulmonary diffusing capacity	No change

respiratory rates become higher with conditioning[4,189] in proportion to the increase in maximal O_2 uptake.

Specific training of the respiratory muscles, such as isometric expiratory effort, effectively increases endurance of these muscles (i.e., their ability to sustain a high level of ventilation for prolonged periods). Such an effect has been shown among healthy adults and among children with cystic fibrosis (Chapter 3, section entitled Beneficial Effects of Conditioning in CF and Fig. 3.12).

While athletes have a high pulmonary diffusing capacity, this does not increase in adolescents as a result of conditioning, above and beyond the increase in their pulmonary blood flow.[121]

Skeletal Muscle Adaptation to Conditioning

Being the effector of body movement, the skeletal muscle is expected to undergo a major adaptation to conditioning. Since the advent of needle biopsy technique in the early 1960s, much research has been done on adults, demonstrating that such changes, morphologic and biochemical, do take place. Due to ethical limitations, data for adolescents are scant, but they do show a similar adaptation pattern to that of adults. No data are available for children younger than 12 years.

Morphologic Changes. Skeletal muscle hypertrophy, and possibly hyperplasia, occur in the chronically active muscle of adults. Hypertrophy, but not hyperplasia, has been found also among adolescents undergoing endurance conditioning[78,109] but not following sprint training.[78] The cross-sectional area of slow twitch muscle fibers, as well as some of the fast twitch fibers, of 16- to 17-year-old boys increased 10–30% following a 3-month endurance program. There is no evidence for a change in fiber type distribution in children as a result of training.[70,109]

In adults, conditioning induces an increase in muscle capillarization and the creation of collateral circulation to the skeletal muscle. This aspect has not been investigated in children.

Biochemical Changes. Changes related to aerobic and anaerobic energy metabolism are summarized in Table 1.9.

The main conditioning-related change in the organelles of muscle fibers of adults is an increase in the number and volume of mitochondria. While no such information is available for children, their capability for aerobic energy turnover is enhanced following endurance conditioning, as shown by increased glycogen storage and oxidative enzyme activity.[70,72,78] Sprint training does not induce changes in the oxidative enzymes of adolescents,[78] which demonstrates the specificity of the training effect on muscle metabolism.

In adolescents, the activity of phosphofructokinase, which is a rate-limiting enzyme of anaerobic glycolysis, increases as a result of sprinting[78] and endurance cycling[72] but not endurance running.[78] This find-

Table 1.9. Conditioning and Muscle Metabolism in Children and Adults

Variable	*Change*	*Information Available for Children*	*Information on Adults Only*
Aerobic Energy Turnover			
Mitochondria—number	Increase		x
—volume	Increase		x
Glycogen stores	Increase	x	
Triglyceride—stores	Increase		x
—utilization	Increase		x
Myoglobin content	Increase		x
Oxidative enzyme activity (e.g., SDH, cytochrome oxidase, palmityl CoA synthetase)	Increase	x	
Anaerobic Energy Turnover			
ATP—content	Increase	x	
—utilization	Increase	x	
CP—content	Increase	x	
—utilization	Increase	x	
Glycogen utilization	Increase	x	
Lactate—submax	No change	x	
—max	Increase	x	
Anaerobic enzyme activity (e.g., PFK, ATPase, CK, myokinase)	Increase	x	

ing, coupled with evidence of enhanced glycogen utilization,[64,65] demonstrates the sensitivity of the anaerobic system to training. This has also been shown by monitoring the performance of children who underwent intervention programs.[97,98,109]

Conditioning and Body Composition

When conditioning regimens are such that the cumulative energy expenditure is high (as in long-distance running, walking, or swimming) or the effort is intense (as in weight training), changes in body composition may occur. On the one hand, the anabolic effect of exercise induces an increase in the lean body mass (particularly following high-resistance training[94]). On the other, the mass of adipose tissue may decrease, fat being the major source of energy. The overall result is a relative increase in lean body mass and a decrease in adiposity.

The above changes do not occur invariably. Some authors[95,148,155,156] describe a decrease in fatness and an increase in lean body mass following conditioning programs, while others[38,47,91,189] fail to find these changes. The main reason for such conflicting results is that changes in body mass and composition depend also on factors other than energy expenditure. Of special importance are calorie intake and the composition of consumed food.[37] An additional factor to consider is the confounding effect of growth and maturation. Thus, while a certain conditioning regimen effectively modifies body composition, the effects may be masked and even counteracted by growth and by maturation-related changes, as well as by changing dietary habits. In adults, fat loss was enhanced when exposure to a cold climate was superimposed upon conditioning.[153]

Of special importance is the "baseline" adiposity level of the child at the start of the program. Obese youths seem to respond with greater changes in body composition than do nonobese ones[91,148] (see also Chapter 6, section entitled Effectiveness of Conditioning as a Reducing Regimen).

References

1. Adams FH, Bengtsson E, Berven H, et al: Determination by means of a bicycle ergometer, of the physical working capacity of children. Acta Paediatr Scand Suppl 118:120–122, 1959.
2. Allen HD, Goldberg SJ, Sahn DJ, et al: A quantitative echocardiographic study of champion childhood swimmers. Circulation 55:142–145, 1977.
3. Alpert BS, Flood NL, Strong WB, et al: Responses to ergometer exercise in healthy biracial population of children. J Pediatr 101:538–545, 1982.
4. Andersen B, Froberg K: Respiratory functions in highly-trained and normal boys during puberty (abstract). Acta Physiol Scand 105:D35, 1980.

5. Andersen B, Froberg K: Circulatory parameters and muscular strength in trained and normal boys during puberty (abstract). Acta Physiol Scand 105:D36, 1980.
6. Andersen B, Froberg K: Maximal oxygen uptake and lactate concentration in highly trained and normal boys during puberty (abstract). Acta Physiol Scand 105:D37, 1980.
7. Andersen KL, Magel JR: Physiological adaptation to a high level of habitual physical activity during adolescence. Int Z Angew Physiol 28:209–227, 1970.
8. Andersen KL, Seliger V, Rutenfranz J, Berndt I. Physical performance capacity of children in Norway. Part II. Heart rate and oxygen pulse in submaximal and maximal exercises—population parameters in a rural community. Eur J Appl Physiol 33:197–206, 1974.
9. Andersen KL, Seliger, V, Rutenfranz J, Messel S: Physical performance capacity of children in Norway. Part III. Respiratory responses to graded exercise loadings—population parameters in a rural community. Eur J Appl Physiol 33:265–274, 1974.
10. Andersen KL, Seliger V, Rutenfranz J, Mocellin R: Physical performance capacity of children in Norway. Part I. Population parameters in a rural inland community with regard to maximal aerobic power. Eur J Appl Physiol 33:177–195, 1974.
11. Anderson SD, Godfrey S: Cardio-respiratory response to treadmill exercise in normal children. Clin Sci 40:433–442, 1971.
12. Asmussen E: Growth in muscular strength and power. In: Rarick L (ed.) Physical Activity, Human Growth and Development. Academic Press, New York, 1973, pp. 60–79.
13. Asmussen E: Development problems in physical performance capacity. In: Larson L (ed.) Fitness Health and Work Capacity. International Standards for Assessment. MacMillan, New York, 1974.
14. Asmussen E, Heebøll-Nielsen KR: A dimensional analysis of physical performance and growth in boys. J Appl Physiol 7:593–603, 1955.
15. Åstrand PO: Experimental Studies of Physical Working Capacity in Relation to Sex and Age. Munksgaard, Copenhagen, 1952.
16. Åstrand PO, Engström L, Eriksson P, et al: Girl swimmers. With special reference to respiratory and circulatory adaptation and gynecological and psychiatric aspects. Acta Paediatr (Scand) Suppl. 147, 1963.
17. Åstrand PO, Rodahl K: Textbook of Work Physiology, 2nd ed. McGraw-Hill, New York, 1977.
18. Bailey DA, Ross WD, Mirwald RL, Wesse C: Size dissociation of maximal aerobic power during growth in boys. In: Borms J, Hebbelinck M (eds.) Pediatric Work Physiology. Karger, Basel, 1978, pp. 140–151.
19. Bal MER, Thompson EM, McIntosh EH, Taylor CM, MacLeod G: Mechanical efficiency in cycling of girls six to fourteen years of age. J Appl Physiol 6:185–188, 1953.
20. Banister EW: A comparison of fitness training methods in a school program. Res Q Am Assoc Health Phys Educ 36:387–392, 1965.
21. Bar-Or O: Arm ergometry vs. treadmill running and bicycle riding in men with different conditioning levels. In: Hansen G, Mellerowicz H (eds.) Internationale Seminar fur ergometrie. Sports Med Inst, Berlin, 1972.

22. Bar-Or O: Age-related changes in exercise perception. In: Borg G (ed.) Physical Work and Effort. Pergamon Press, Oxford and New York, 1977, pp. 255–266.
23. Bar-Or O: Um nova teste de capacidade anaerobica. Med Esporte Porto Alegre 5:73–82, 1980.
24. Bar-Or O: Climate and the exercising child—a review: Int J Sports Med 1:53–65, 1980.
25. Bar-Or O: Le test anaérobie de Wingate. Charactéristiques et applications. Symbioses 13:157–172, 1981.
26. Bar-Or O: Physiologische Gesetzmässigkeiten sportlicher Aktivität beim Kind. In: Howald H, Han E (eds.) Kinder im Leistungssport. Birkhauser, Basel, 1982, pp 18–30.
27. Bar-Or O: The growth and development of children's physiological and perceptional responses to exercise. In: Ilmarinen Y, Välimäki I (eds.) Pediatric Work Physiology. Springer Verlag, Berlin, in press, 1983.
28. Bar-Or O, Buskirk ER: The cardiovascular system and exercise. In: Johnson WR, Buskirk ER (eds.) Science and Medicine of Exercise and Sport, 2nd ed. Harper and Row, New York, 1974, pp. 121–136.
29. Bar-Or O, Inbar O: Relationships among anaerobic capacity, sprint and middle distance running of schoolchildren. In: Shephard RJ, Lavallée H (eds.) Physical Fitness Assessment. Charles C. Thomas, Springfield, 1978, pp. 142–147.
30. Bar-Or O, Shephard RJ, Allen CL: Cardiac output of 10- to 13-year-old boys and girls during submaximal exercise. J Appl Physiol 30:219–223, 1971.
31. Bar-Or O, Skinner JS, Bergsteinova V, et al: Maximal aerobic capacity of 6- to 15-year-old girls and boys with subnormal intelligence quotients. Acta Paediatr Scand Suppl 217:108–113, 1971.
32. Bar-Or O, Zwiren LD: Physiological effects of increased frequency of physical education classes and of endurance conditioning on 4- to 10-year-old girls and boys. In: Bar-Or O (ed.) Pediatric Work Physiology. Wingate Institute, Natanya, 1973, pp. 183–198.
33. Bar-Or O, Zwiren LD, Ruskin H: Anthropometric and developmental measurements of 11- to 12-year-old boys, as predictors of performance 2 years later. Acta Paediatr Belg 28[Suppl.]:214–220, 1974.
34. Bell RD, MacDougall JD, Billeter R, et al: Muscle fiber types and morphometric analysis of skeletal muscle in six-year-old children. Med Sci Sports Exercise 12:28–31, 1980.
35. Bengtsson E: The working capacity in normal children, evaluated by submaximal exercise on the bicycle ergometer and compared with adults. Acta Med Scand 154:91–109, 1956.
36. Berg A, Keul J, Huber G: Biochemische Akutveränderungen bei Ausdauerbelastungen im Kindes- und Jugendalter. Mschr Kinderheilkd 128:490–495, 1980.
37. Berg K: Body composition and nutrition of adolescent boys training for bicycle racing. Nutr Metabol 14:172–180, 1972.
38. Berg K, Bjure J: Preliminary results of long-term physical training of adolescent boys with respect to body composition, maximal oxygen uptake, and lung volume. Acta Paediatr Belg 28[Suppl.]:183–190, 1974.

39. Blimkie CJR, Cunningham DA, Leung FY: Urinary catecholamine excretion during competition in 11- to 23-year-old hockey players. Med Sci Sports 10:188–193, 1978.
40. Boileau RA, Ballard JE, Sprague RL, et al: Effect of methylphenidate on cardiorespiratory responses in hyperactive children. Res Q Am Assoc Health Phys Educ 47:590–596, 1976.
41. Borg G: Physical Performance and Perceived Exertion. Gleerup, Lund, 1962.
42. Börjeson M: Overweight children. Acta Paediatr Scand Suppl 132, 1962.
43. Bouchard C, Malina RM, Hollmann W, Leblanc C: Submaximal working capacity, heart size and body size in boys 8 to 18 years. Eur J Appl Physiol 36:115–126, 1977.
44. Brown CH, Harrower JR, Deeter MF: The effects of cross-country running on pre-adolescent girls. Med Sci Sports 4:1–5, 1972.
45. Burmeister W, Rutenfranz J, Stresny W, Radny HG: Body cell mass and physical performance capacity ($\dot{W}_{170}$) of school children. Int Z Angew Physiol 31:61–70, 1972.
46. Chatterjee S, Banerjee PK, Chatterjee P, Maitra SR: Aerobic capacity of young girls. Indian J Med Res 69:327–333, 1979.
47. Clarke DH, Vaccaro P: The effect of swimming training on muscular performance and body composition in children. Res Q Am Assoc Health Phys Educ 50:9–17, 1979.
48. Cumming GR: Correlation of physical performance with laboratory measures of fitness. In: Shephard RJ (ed.) Frontiers of Fitness. Charles C. Thomas, Springfield, 1971, pp. 265–279.
49. Cumming GR: Hemodynamics of supine bicycle exercise in "normal" children. Am Heart J 93:617–622, 1977.
50. Cumming GR: Supine bicycle exercise in pediatric cardiology. In: Borms J, Hebbelinck M (eds.) Pediatric Work Physiology. Karger, Basel, 1978, pp. 82–88.
51. Cumming GR: Recirculation times in exercising children. J Appl Physiol: Respir Environ Exercise Physiol 45:1005–1008, 1978.
52. Cumming GR, Goodwin A, Baggley G, Antel J: Repeated measurements of aerobic capacity during a week of intensive training at a youths' track camp. Can J Physiol Pharmacol 45:805–811, 1967.
53. Cumming GR, Goulding D, Baggley G: Failure of school physical education to improve cardiorespiratory fitness. Can Med Assoc J 101:69–73, 1969.
54. Daniels J, Oldridge N: Changes in oxygen consumption of young boys during growth and running training. Med Sci Sports 3:161–165, 1971.
55. Daniels J, Oldridge N, Nagle F, White B: Differences and changes in $\dot{V}O_2$ among young runners 10 to 18 years of age. Med Sci Sports 10:200–203, 1978.
56. Davies CTM: Body composition in children: A reference standard for maximum aerobic power output on a stationary bicycle ergometer. Acta Paediatr Scand Suppl 217:136–137, 1971.
57. Davies CTM, Barnes C, Godfrey S: Body composition and maximal exercise performance in children. Hum Biol 44:195–214, 1972.

58. di Prampero PE, Cerretelli P: Maximal muscular power (aerobic and anaerobic) in African natives. Ergonomics 12:51–59, 1969.
59. Dotan R, Rotstein A, Tenenbaum G, Bar-Or O: Is the anaerobic threshold preferable to maximal O_2 uptake or other indicators for evaluating the effect of aerobic conditioning in prepubescent boys? (in Hebrew). Research Report, Wingate Institute, Natanya, 1982.
60. Drinkwater BL, Kupprat IC, Denton JE, et al: Response of prepubertal girls and college women to work in the heat. J Appl Physiol: Respir Environ Exercise Physiol 43:1046–1053, 1977.
61. Ekblom B: Effect of physical training in adolescent boys. J Appl Physiol 27:350–355, 1969.
62. Engström I, Eriksson BO, Karlberg P, et al: Preliminary report on the development of lung volumes in young girl swimmers. Acta Paediatr Scand Suppl 217:73–76, 1971.
63. Eriksson BO: Cardiac output during exercise in pubertal boys. Acta Paediatr Scand Suppl 217:53–55, 1971.
64. Eriksson BO: Physical training, oxygen supply and muscle metabolism in 11- to 15-year old boys. Acta Physiol Scand Suppl 384:1–48, 1972.
65. Eriksson BO: Muscle metabolism in children—a review. Acta Paediatr Scand Suppl 283:20–27, 1980.
66. Eriksson BO, Berg K, Taranger J: Physiological analysis of young boys starting intensive training in swimming. In: Eriksson B, Furberg B (eds.) Swimming Medicine. University Park Press, Baltimore, 1978, pp. 147–160.
67. Eriksson BO, Engström I, Karlberg P, et al: Physiological analysis of former girl swimmers. Acta Paediatr Scand Suppl 217:68–72, 1971.
68. Eriksson BO, Freychuss U, Lundin A, Thorén CAR: Effect of physical training in former female top athletes in swimming. In: Berg K, Eriksson BO (eds.) Children and Exercise IX. University Park Press, Baltimore, 1980, pp. 116–127.
69. Eriksson BO, Gollnick PD, Saltin B: Muscle metabolism and enzyme activities after training in boys 11 to 13 years old. Acta Physiol Scand 87:485–487, 1973.
70. Eriksson BO, Gollnick PD, Saltin B: The effect of physical training on muscle enzyme activities and fiber composition in 11-year-old boys. Acta Paediatr Belg 28[Suppl.]:245–252, 1974.
71. Eriksson BO, Grimby G, Saltin B: Cardiac output and arterial blood gases during exercise in pubertal boys. J Appl Physiol 31:348–352, 1971.
72. Eriksson BO, Karlsson J, Saltin B: Muscle metabolites during exercise in pubertal boys. Acta Paediatr Scand [Suppl.] 217:154–157, 1971.
73. Eriksson BO, Koch G: Cardiac output and intra-arterial blood pressure at rest and during submaximal and maximal exercise in 11- to 13-year-old boys before and after physical training. In: Bar-Or (ed.) Pediatric Work Physiology. Wingate Institute, Natanya, 1973, pp. 139–150.
74. Eriksson BO, Koch G: Effect of physical training on hemodynamic response during submaximal and maximal exercise in 11–13 year old boys. Acta Physiol Scand 87:27–39, 1973.
75. Eriksson BO, Saltin B: Muscle metabolism during exercise in boys aged 11

to 16 years compared to adults. Acta Paediatr Belg 28[Suppl.]:257–265, 1974.

76. Eriksson BO, Thorén, C: Training girls for swimming from medical and physiological points of view, with special reference to growth. In Eriksson B, Furberg B (eds.) Swimming Medicine IV. University Park Press, Baltimore, 1978, pp. 3–15.
77. Fixler DE, Laird WP, Browne R, et al: Response of hypertensive adolescents to dynamic and isometric exercise stress. Pediatrics 64:579–583, 1979.
78. Fournier M, Ricci J, Taylor AW, et al: Skeletal muscle adaptation in adolescent boys: sprint and endurance training and detraining. Med Sci Sports Exercise 14:453–456, 1982.
79. Fox EL: Sports Physiology. W.B. Saunders, Philadelphia, 1979.
80. Freedson P, Gilliam TB, Sady S, Katch VL: Transient $\dot{V}O_2$ characteristics in children at the onset of exercise. Eur J Appl Physiol, in press, 1983.
81. Friman G: Effect of acute infectious disease on isometric muscle strength. Scand J Clin Lab Invest 37:303–308, 1977.
82. Friman G: Effect of clinical bed rest for seven days on physical performance. Acta Med Scand 205:389–393, 1979.
83. Gaisl G, Buchberger J: The significance of stress acidosis in judging the physical working capacity of boys aged 11 to 15. In: Lavallée H, Shephard RJ (eds.) Frontiers of Activity and Child Health. Pélican, Quebec, 1977, pp. 161–168.
84. Gatch W, Byrd R: Endurance training and cardiovascular function in 9- and 10-year-old boys. Arch Phys Med Rehab 60:574–577, 1979.
85. Geenen DL, Gilliam TB, Steffens C, et al: The effects of exercise on cardiac structure and function in prepubescent children (abstract). Med Sci Sports Exercise 13:93, 1981.
86. Gilliam TB, Freedson PS: Effects of a 12-week school physical fitness program on peak VO_2 body composition and blood lipids in 7 to 9 year old children. Int J Sports Med 1:73–78, 1980.
87. Gilliam TB, Freedson PS, Geenen DL, Shahraray B: Physical activity patterns determined by heart rate monitoring in 6- to 7-year-old children. Med Sci Sports Exercise 13:65–67, 1981.
88. Gilliam TB, Sady S, Thorland WG, Weltman AL: Comparison of peak performance measures in children ages 6 to 8, 9 to 10, and 11 to 13 years. Res Q Am Assoc Health Phys Ed 48:695–702, 1977.
89. Girandola RN, Wiswell RA, Frisch F, Wood K: Metabolic differences during exercise in pre- and post-pubescent girls (abstract). Med Sci Sports Exercise 13:110, 1981.
90. Glaser RM, Laubach LL, Sawka MN, Suryaprasad AG: Exercise stress, fitness evaluation and training of wheelchair users. In: Leon AS, Amundson GJ (eds.) Proceedings First International Conference on Lifestyle Health. University of Minnesota, Minneapolis, 1978.
91. Glick Z, Kaufmann NA: Weight and skinfold thickness changes during a physical training course. Med Sci Sports 8:109–112, 1976.
92. Godfrey S: The growth and development of the cardio-pulmonary responses to exercise. In: Davis JA, Dobbing J (eds.) Scientific Foundations of Paediatrics. W.B. Saunders, Philadelphia, 1974, pp. 271–280.

93. Godfrey S, Davies CTM, Wozniak E, Barnes CA: Cardio-respiratory response to exercise in normal children. Clin Sci 40:419–431, 1971.
94. Goldberg AL, Etlinger JD, Goldspink DF, Jablecki C: Mechanism of work-induced hypertrophy of skeletal muscle. Med Sci Sports 7:248–261, 1975.
95. Goode RC, Virgin A, Romet TT, et al: Effects of a short period of physical activity in adolescent boys and girls. Can J Appl Sport Sci 1:241–250, 1976.
96. Greenleaf JR, Kozlowski S: Reduction in peak oxygen uptake after prolonged bed rest. Med Sci Sports Exercise 14:477–480, 1982.
97. Grodjinovsky A, Bar-Or O: Influence of added physical education hours upon anaerobic capacity, adiposity, and grip strength in 12- to 13-year-old children enrolled in a sports class. In: Ilmarinen Y, Välimäki I (eds.) Pediatric Work Physiology X. Springer Verlag, Berlin, in press, 1983.
98. Grodjinovsky A, Bar-Or O, Dotan R, Inbar O: Training effect on the anaerobic performance of children as measured by the Wingate anaerobic test. In: Borg K, Eriksson BO (eds.) Children and Exercise IX. University Park Press, Baltimore, 1980, pp. 139–145.
99. Hagberg JM, Ehsani AA, Heath GW, et al: Beneficial effects of endurance exercise training in adolescent hypertension (abstract). Presented at 29th Annual Meeting of the American College of Cardiology, 1980.
100. Hamilton P, Andrew GM: Influence of growth and athletic training on heart and lung functions. Eur J Appl Physiol 36:27–38, 1976.
101. Henrikkson KG: Muscle histochemistry and muscle function. Acta Paediatr Scand Suppl 283:15–19, 1980.
102. Hermansen L, Oseid S: Direct and indirect estimation of maximal oxygen uptake in prepubertal boys. Acta Paediatr Scand Suppl 217:18–23, 1971.
103. Hettinger TL: Physiology of strength. Charles C Thomas, Springfield, 1961.
104. Hollmann W: The preventive and rehabilitative role of sport in internal medicine. Das Medizinische Prisma 2:4–28, 1978.
105. Ikai M: Training of muscular endurance related to age. FIEP Bull (Federation Internationale d'Education Physique) 3–4:19–27, 1969.
106. Ikai M, Shindo M, Miyamura M: Aerobic work capacity of Japanese people. Res J Phys Educ 14:137–142, 1970.
107. Inbar O, Bar-Or O: The effects of intermittent warm-up on 7- to 9-year-old boys. Eur J Appl Physiol 34:81–89, 1975.
108. Inbar O, Bar-Or O: Relationships of anaerobic and aerobic arm and leg capacities to swimming performance of 8- to 12-year-old children. In: Lavallée H, Shephard RJ (eds.) Frontiers of Activity and Child Health. Pélican, Quebec, 1977, pp. 283–292.
109. Jacobs I, Sjödin B, Svane B: Muscle fiber type, cross-sectional area and strength in boys after 4 years' endurance training (abstract). Med Sci Sports Exercise 14:123, 1982.
110. Jacobs JC, Dick HM, Downey JA: Weight bearing as a treatment for damaged hips in juvenile rheumatoid arthritis. N Engl J Med 205:409, 1981.
111. James FW, Kaplan S, Glueck CJ, et al: Responses of normal children and young adults to controlled bicycle exercise. Circulation 61:902–912, 1980.
112. Kaneko M, Ito A, Fuchimoto T, Toyooka J: Mechanical work and efficiency of young distance runners during level running. In: Morecki A (ed.) Biomechanics VII. University Park Press, Baltimore, 1981, pp. 234–240.

113. Kaneko M, Otsuka A: Effect of a long-term training on physical fitness in 9–10 year old boys and girls (in Japanese). J Phys Fitness Jpn 7:44–50, 1979.
114. Karlsson J: Muscle ATP, CP and lactate in submaximal and maximal exercise. In: Pernow B, Saltin B (eds.) Muscle Metabolism During Exercise. Plenum Press, New York, 1971, pp. 383–393.
115. Kindermann VW, Huber G, Keul J: Anaerobe Kapazität bei Kindern und Jugendlichen in Beziehung zum Erwachsenen. Sportarzt Sportmed 6:112–115, 1975.
116. Kindermann VW, Keul J, Lehmann M: Ausdauerbelastungen beim Heranwachsenden—metabolische und kardiozirkulatorische Veränderungen. Fortschr Med 97:659–665, 1979.
117. Kobayashi K, Kitamura K, Miura M, et al: Aerobic power as related to body growth and training in Japanese boys: a longitudinal study. J Appl Physiol: Respir Environ Exercise Physiol 44:666–672, 1978.
118. Koch G: Muscle blood flow after ischemic work and during bicycle ergometer work in boys aged 12 years. Acta Paediatr Belg 28[Suppl.]:29–39, 1974.
119. Koch G: Muscle blood flow in prepubertal boys. Effect of growth combined with intensive physical training. In: Borms J, Hebbelinck M (eds.) Pediatric Work Physiology. Karger, Basel, 1978, pp. 39–46.
120. Koch G: Aerobic power, lung dimensions, ventilatory capacity, and muscle blood flow in 12–16-year-old boys with high physical activity. In: Berg K, Eriksson BO (eds.) Children and Exercise X. University Park Press, Baltimore, 1980, pp. 99–108.
121. Koch G, Eriksson BO: Effect of physical training on anatomical R-L shunt at rest and pulmonary diffusing capacity during near-maximal exercise in boys 11 to 13 years old. Scand J Clin Lab Invest 31:95–105, 1973.
122. Koch G, Röcker L: Plasma volume and intravascular protein masses in trained boys and fit young men. J Appl Physiol: Respir Environ Exercise Physiol 43:1085–1088, 1977.
123. Krahenbuhl GS, Pangrazi RP, Chomokos EA: Aerobic responses of young boys to submaximal running. Res Q Am Assoc Health Phys Educ 50:413–421, 1979.
124. Krotkiewski M, Kral JG, Karlsson J: Effects of castration and testosterone substitution on body composition and muscle metabolism in rats. Acta Physiol Scand 109:233–237, 1980.
125. Kulangara RJ, Strong WB: Exercise stress testing in children. Compr Ther 5:51–61, 1979.
126. Kurowski TT: Anaerobic power of children from ages 9 through 15 years. M.Sc. Thesis, Florida State University, 1977.
127. Lammert O, Froberg K, Murer K, Andersen PE: The effect of training in relation to chronological age and developmental stages in children 9 to 17 years of age (abstract). Acta Physiol Scand 105:61A, 1980.
128. Lehmann M, Keul J, Hesse A: Zur aeroben und anaeroben Kapazität sowie Catecholaminexkretion von Kindern und Jugendlichen während langdauernder submaximaler Körperarbeit. Eur J Appl Physiol 48:135–145, 1982.
129. Lengyel M, Gyarfas I: The importance of echocardiography in the assess-

ment of left ventricular hypertrophy in trained and untrained school children. Acta Cardiol 34:63–69, 1979.
130. Lind AR: Cardiovascular responses to static exercise. Circulation 41:173–176, 1970.
131. Lussier L, Buskirk ER: Effects of an endurance training regimen on assessment of work capacity in pre-pubertal children. Ann NY Acad Sci 301:734–747, 1977.
132. MacDougall JD, Bell RD, Howald H: Skeletal muscle ultrastructure and fiber types in prepubescent children. In: Nagle FJ, Montoye HJ (eds.) Exercise in Health and Disease. Charles C. Thomas, Springfield, 1982, pp. 113–117.
133. MacDougall JD, Roche PD, Bar-Or O, Moroz JR: Oxygen cost of running in children of different ages; maximal aerobic power of Canadian schoolchildren (abstract). Can J Appl Sports Sci 4:237, 1979.
134. Máček M, Seliger V, Vávra J, et al: Physical fitness of the Czechoslovak population between the ages of 12 and 55 years. Oxygen consumption and pulse oxygen. Physiol Bohemoslov 28:75–82, 1979.
135. Máček M, Vávra J: Cardiopulmonary and metabolic changes during exercise in children 6 to 14 years old. J Appl Physiol 30:202–204, 1971.
136. Máček M, Vávra J: Prolonged exercise in children. Acta Paediatr Belg 28[Suppl.]:13–18, 1974.
137. Máček M, Vávra J: The adjustment of oxygen uptake at the onset of exercise: a comparison between prepubertal boys and young adults. Int J Sports Med 1:75–77, 1980.
138. Máček M, Vávra J: Oxygen uptake and heart rate with transition from rest to maximal exercise in prepubertal boys. In: Berg K, Eriksson BO (eds.) Children and Exercise IX. University Park Press, Baltimore, 1980, pp. 64–68.
139. Máček M, Vávra J, Novosadová J: Prolonged exercise in prepubertal boys, I. Cardiovascular and metabolic adjustment. Eur J Appl Physiol 35:291–298, 1976.
140. Máček M, Vávra J, Novosadová J: Prolonged exercise in prepubertal boys, II. Changes in plasma volume and in some blood constituents. Eur J Appl Physiol 35:299–303, 1976.
141. Macková J, Sturmová M, Máček M: Prolonged exercise in prepubertal boys in warm and cold environments. In: Ilmarinen Y, Välimäki I (eds.) Pediatric Work Physiology X. Springer Verlag, Berlin, in press, 1983.
142. Matějková J, Kopřivová Z, Placheta Z: Changes in acid-base balance after maximal exercise. In: Placheta Z (ed.) Youth and Physical Activity. J.E. Purkyne University, Brno, 1980, pp. 191–199.
143. Mayers N, Gutin B: Physiological characteristics of elite prepubertal cross-country runners. Med Sci Sports 11:172–176, 1979.
144. Michael E, Evert J, Jeffers K: Physiological changes of teenage girls during five months of detraining. Med Sci Sports 4:214–218, 1972.
145. Mocellin R: Jugend und Sport. Med Klin 70:1443–1457, 1975.
146. Mocellin R, Sebening W, Bühlmeyer K: Herzminutenvolumen und Sauerstoffaufnahme in Ruhe and während submaximaler Belastungen bei 8–14 jährigen Jungen. Z Kinderheilkd 114:323–339, 1973.
147. Mocellin R, Wasmund U: Investigations on the influence of a running-

training programme on the cardiovascular and motor performance capacity in 53 boys and girls of a second and third primary school class. In: Bar-Or O (ed.) Pediatric Work Physiology. Wingate Institute, Natanya, 1973, pp. 279–285.

148. Moody DL, Wilmore JH, Girandola RN, Royce JP: The effects of a jogging program on the body composition of normal and obese high school girls. Med Sci Sports 4:210–213, 1972.
149. Morse M, Schultz FW, Cassels DE: Relation of age to physiological responses of the older boy (10 to 17 years) to exercise. J Appl Physiol 1:683–709, 1949.
150. Murase Y, Kamei S, Kobayashi K, Matsui H: Longitudinal study on aerobic power for super junior athletes. J Phys Fitness Jpn 28:271–279, 1979.
151. Nagle FJ, Hagberg J, Kamei S: Maximal O_2 uptake of boys and girls ages 14 to 17. Eur J Appl Physiol 36:75–80, 1977.
152. Nielsen B, Nielsen K, Behrendt Hansen M, Asmussen E: Training of "functional muscular strength" in girls 7–19 years old. In: Berg K, Eriksson B (eds.) Paediatric Work Physiology, IX. University Park Press, Baltimore, 1980, pp. 69–78.
153. O'Hara W, Allen C, Shepherd RJ, Allen G: Fat loss in the cold: a controlled study. J Appl Physiol 46:872–877, 1979.
154. Oseid S, Hermansen L: Hormonal and metabolic changes during and after prolonged muscular work in prepubertal boys. Acta Paediatr Scand Suppl 217:147–153.
155. Pářizková J: Longitudinal study of the development of body composition and body build in boys of various physical activity. Hum Biol 40:212–225, 1968.
156. Pářizková J: Body Fat and Physical Fitness. M, Nijhoff, The Hague, 1977.
157. Pauer M, Sobolová V, Zelenka V, et al: The effects of intensified school physical education on physical fitness (abstract). Physiol Bohemoslov 29:272, 1980.
158. Pels AE, Gilliam TB, Freedson PS, et al: Heart rate response to bicycle ergometer exercise in children ages 6 to 7 years. Med Sci Sports Exercise 13:299–302, 1981.
159. Polgar G, Promadhat V: Pulmonary function testing in children: techniques and standards. W.B. Saunders, Philadelphia, 1971.
160. Rarick LR (ed.): Physical Activity, Human Growth and Development. Academic Press, New York, 1973.
161. Riopel DA, Taylor AB, Hohn AR: Blood pressure, heart rate, pressure rate product and electrocardiographic changes in healthy children during treadmill exercise. Am J Cardiol 44:697–704, 1979.
162. Robinson S: Experimental studies of physical fitness in relation to age. Int Z Angew Physiol Einschl Arbeitphysiol 10:251–323, 1938.
163. Rode A: Some factors influencing the fitness of a small Eskimo community. Ph.D. Thesis, University of Toronto, 1972.
164. Rode A, Bar-Or O, Shephard RJ: Cardiac output and oxygen conductance. A comparison of Canadian eskimo and city dwellers. In: Bar-Or O (ed.) Pediatric Work Physiology. Wingate Institute, Natanya, 1973, pp. 45–57.
165. Rohmert W: Rechts-links-Vergleich bei isometrischem Armmuskeltrain-

ing mit verschiedenem Trainingsreiz bei achtjährigen Kindern. Int Z Angew Physiol Einschl Arbeitsphysiol 26:363–393, 1968.
166. Ross WD, Drinkwater DT, Whittingham NO, Faulkner RA: Anthropometric prototypes: age six to eighteen years. In: Berg K, Eriksson BO (eds.) Children and Exercise IX. University Park Press, Baltimore, 1980, pp. 3–12.
167. Rost R, Gerhardus H, Hollmann W: Untersuchungen zur Frage eines Trainingseffektes bei Kindern im Alter von 8-10 Jahren im kardiopulmonalen System. Koln, DSHS, 1978.
168. Rutenfranz J: Entwicklung und Beurteilung der körperlichen Leistungsfähigkeit bei Kindern und Jugendlichen. Karger, Basel, 1964.
169. Rutenfranz J, Andersen KL, Seliger V, et al: Exercise ventilation during the growth spurt period: comparison between two European countries. Eur J Pediatr 136:135–142, 1981.
170. Rutenfranz J, Klimt F, Ilmarinen J, Kylian H: Blood lactate concentration during triangular and stepwise loadings on the bicycle ergometer. In: Lavallée H, Shephard RJ (eds.) Frontiers of Activity and Child Health. Pélican, Quebec, 1977, pp. 179–187.
171. Sale D, MacDougall D: Specificity in strength training: a review for the coach and athlete. Can J Appl Sport Sci 5:87–92, 1980.
172. Saltin B: Physiological effects of physical conditioning. Med Sci Sports 1:50–56, 1969.
173. Scheuer J, Tipton CM: Cardiovascular adaptations to physical training. Annu Rev Physiol 39:221–251, 1977.
174. Schieken RM, Geller DF: The cardiovascular effect of isometric exercise in children (abstract). Clin Res 26:741A, 1978.
175. Schmücker B, Dordel J, Hollmann W: The aerobic power of 7–9 years old children participating in a rehabilitative sport programme. In: Lavallée H, Shephard RJ (eds.) Frontiers of Activity and Child Health. Pélican, Quebec, 1977, pp. 307–312.
176. Schmücker B, Hollmann W: The aerobic capacity of trained athletes from 6 to 7 years of age on. Acta Paediatr Belg 28[Suppl.]:92–101, 1974.
177. Schneider EC, Crampton CB: The respiratory responses of pre-adolescent boys to muscular activity. Am J Physiol 117:577–586, 1936.
178. Seliger V: The influence of sports training on the efficiency of juniors. Int Z Angew Physiol Einschl Arbeitsphysiol 26:309–322, 1968.
179. Seliger V: Physical fitness of Czechoslovak children at 12 and 15 years of age. IBP results of investigation 1968–1969. Acta Univ Carol Gymnica 5:6–196, 1970.
180. Shephard RJ: Future research on the quantifying of endurance training. J Hum Ergol 3:163–181, 1975.
181. Shephard RJ: Physiology and biochemistry of exercise. Praeger, New York, 1982.
182. Shephard PJ, Allen C, Bar-Or O, et al: The working capacity of Toronto schoolchildren. Can Med Assoc J 100:560–566, 705–714, 1969.
183. Shephard RJ, Bar-Or O: Alveolar ventilation in near maximum exercise. Data on pre-adolescent children and young adults. Med Sci Sports 2:83–92, 1970.
184. Shephard RJ, Lavallée H, LaBarre R, et al: On the basis of data standard-

ization in prepubescent children. In: Ostyn M, Bennen G, Simons J (eds.) Proceedings 2nd International Seminar on Kinanthropometry. Karger, Basel, 1979.
185. Shephard RJ, Lavallée H, Jéquier JC, et al: Influence of added activity classes upon the working capacity of Quebec schoolchildren. In: Lavallée H, Shephard RJ (eds.) Frontiers of Activity and Child Health. Pélican, Quebec, 1977, pp. 237–245.
186. Sjödin B: The relationships among running economy, aerobic power, muscle power, and onset of blood lactate accumulation in young boys (11–15 yrs). In: Komi PV (ed.) Proceedings International Symposium on Sport Biology, Human Kinetics Ltd, 1981, pp 57–60.
187. Skinner JS, Bar-Or O, Bergsteinová V, et al: Comparison of continuous and intermittent tests for determining maximal oxygen intake in children. Acta Paediatr Scand Suppl 217:24–28, 1971.
188. Sprynarova S: Development of the relationship between aerobic capacity and the circulatory and respiratory reaction to moderate activity in boys 11–13 years old. Physiol Bohemoslov 15:253–264, 1966.
189. Sprynarová S, Pářizková J, Irinová I: Development of the functional capacity and body composition of boy and girl swimmers aged 12–15 years. In: Borms J, Hebbelinck M (eds.) Pediatric Work Physiology. Karger, Basel, 1978, pp. 32–38.
190. Sprynarová S, Reisenauer R: Body dimensions and physiological indicators of physical fitness during adolescence. In: Shephard RJ, Lavallée H (eds.) Physical Fitness Assessment. Charles C. Thomas, Springfield, 1978, pp. 32–37.
191. Stewart KJ, Gutin B: Effects of physical training on cardiorespiratory fitness in children. Res Q Am Assoc Health Phys Ed 47:110–120, 1976.
192. Strong WB, Miller MD, Striplin M, Salehbhai M: Blood pressure response to isometric and dynamic exercise in healthy black children. Am J Dis Child 132:587–591, 1978.
193. Taylor CM, Bal MER, Lamb MW, McLeod G: Mechanical efficiency in cycling of boys seven to fifteen years of age. J Appl Physiol 2:563–570, 1950.
194. Thorén C: Effects of beta-adrenergic blockade on heart rate and blood lactate in children during maximal and submaximal exercise. Acta Paediatr Scand Suppl 177:123–125, 1967.
195. Turner MC, Ruley EJ, Buckley KM, Strife CF: Blood pressure elevation in children with orthopedic immobilization. J Pediatr 95:989–992, 1979.
196. Ulbrich J: Individual variants of physical fitness in boys from the age of 11 up to maturity and their selection for sports activities. Medicina dello Sports 24:118–136, 1971.
197. Vaccaro P, Clarke DH: Cardiorespiratory alterations in 9–11 year old children following a season of competitive swimming. Med Sci Sports 10:204–207, 1978.
198. Van Uytvanck, Vrijens J: Experimentelle Untersuchungen über Anpassungserscheinungen von Adoleszenten mit schwacher Konstitution bei kurzfristiger, genau dosierter Arbeit. Int Z Angew Physiol Einschl Arbeitphysiol 25:310–313, 1968.
199. Von Ditter H, Nowacki P, Simai E, Winkler U: Das Verhalten des Säure-

Basen-Haushalts nach erschöpfender Belastung bei untrainierten und trainierten Jungen und Mädchen im Vergleich zu Leistungssportlern. Sportarzt Sportmed 28:45–48, 1977.

200. Von Döbeln W, Eriksson BO: Physical training, maximal oxygen uptake and dimensions of the oxygen transporting and metabolizing organs in boys 11 to 13 years of age. Acta Paediatr Scand 61:653–660, 1972.

201. Vrijens J: Muscle strength development in the pre- and post-pubescent age. Med Sport (Basel) 11:152–158, 1978.

202. Wasmund U, Nowacki P, Ditter H, Klimt F: Radiotelemetrische Untersuchungen der Herzfrequenz während eines 3000 m-Laufs auf dem Sportplatz und auf dem Laufband bei 10 jährigen Schülern und Schülerinnen. Mschr Kinderkeilkd 126:198–204, 1978.

203. Weber G, Kartodihardjo W, Klissouras V: Growth and physical training with reference to heredity. J Appl Physiol 40:211–215, 1976.

204. Wilmore JH, Sigerseth PO: Physical work capacity of young girls 7–13 years of age. J Appl Physiol 22:923–928, 1967.

205. Winters WG, Leaman DM, Anderson RA: The effect of exercise on intrinsic myocardial performance. Circulation 48:50–55, 1973.

206. Wirth A, Trager E, Scheele K, et al: Cardiopulmonary adjustment and metabolic response to maximal and submaximal physical exercise of boys and girls at different stages of maturity. Eur J Appl Physiol 29:229–240, 1978.

207. Yamaji K, Miyashita M, Shephard RJ: Relationship between heart rate and relative oxygen intake in male subjects aged 10 to 27 years. J Hum Ergol 7:29–39, 1978.

208. Yoshida T, Ishiko I, Muraoka I: Effect of endurance training on cardiorespiratory functions of 5-year-old children. Int J Sports Med 1:91–94, 1980.

209. Yoshizawa S, Ishizaki T, Honda H: Physical Fitness of Children aged 5 and 6 years. J Hum Ergol 6:41–51, 1977.

2

Children and Exercise in a Clinical Context—An Overview

A clinician who wants to incorporate exercise into his diagnostic or management strategy should answer the following questions about each patient:

Is the child sufficiently active?
If not, what is the underlying (physical, psychic, social) cause?
What is the physical working capacity of the child?
Can exercise be of diagnostic value?
Will physical conditioning benefit the health and well-being of the child?
Is exercise detrimental to his health?

This chapter will provide an overview of the relevance of these questions to clinical pediatrics. Details are presented later in the book.

Habitual Activity and Disease

In adults, a sedentary lifestyle may be undesirable, but it is accepted by most societies and does not connote ill-health. In contrast, inactivity in a child almost invariably reflects a deviation from normality, be it physical or psychic malfunction or social maladjustment.

Most adults require extrinsic motivation and frequent reinforcement to become and stay active. Children, on the other hand, need but minimal stimuli to pursue an active life. Even though activity levels vary among children, any healthy child who is given freedom to run, jump, climb, and play will do so without special incentives. Many sick children will behave likewise.

Hypoactivity in this text is defined as an activity level lower than that of healthy peers of similar cultural and socioeconomic background. We by no means imply that today's healthy children are *sufficiently* active.

Reports from several countries[6,22,25,46,51] show that healthy preschoolers and schoolchildren spend only a *few minutes a day* in activities vigorous enough to raise their heart rate (HR) above 160–170 beats/min. The rest of their activities are of an intensity that probably is too low to induce a conditioning effect (see Fig. 2.1).

Theories on psychological mechanisms that underlie children's play and activity are reviewed in a book by Ellis and Schultz.[16] Cultural, social, and educational constraints that shape the activity of today's healthy child are summarized by Shephard.[53] In this section we shall direct our attention to the sick child.

Disease as a Direct and Indirect Cause of Hypoactivity

Table 2.1 is a list of pediatric diseases in which hypoactivity has been documented. The list is divided into two subgroups: one in which hypoactivity is inherent to the disease, and a direct result of it; the other, where hypoactivity is incidental to the disease. While a child with, for example, crippling arthritis or advanced muscular dystrophy is obviously limited in his gait and other movements, the asthmatic or diabetic child *can* be active but often is not. The restrictive effect of ill-health is indirect, often imposed by others, as shown in Fig. 2.2. Factors such as parental overprotection, fear by the child or parents, uneducated attitude of parents and teachers (and, occasionally, health practitioners), and peer-imposed social isolation all lead to hypoactivity. Impaired exercise performance also causes hypoactivity, which becomes part of a vicious circle (discussed in the section entitled Effects of Disease on Physical Working Capacity, below). Specific examples of the indirect effects of disease on activity are listed in Table 2.2.

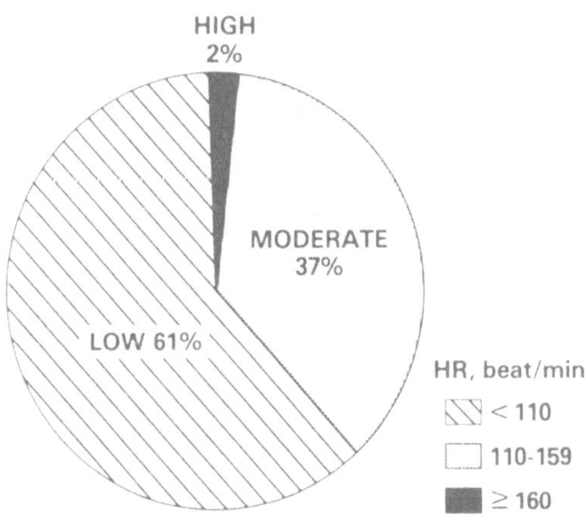

Figure 2.1. Habitual activity of healthy children. An activity profile of 59 nonathletic girls and boys 7–9 years old was derived from continuous heart rate (HR) recording. Values are presented as the percentage of time that each of three HR ranges was recorded during a 12-hour period. Data from Gilliam et al.[23]

Table 2.1. Pediatric Diseases That Are Accompanied by Reduced Habitual Activity

Hypoactivity Inherent to Disease	*Hypoactivity Incidental to Disease*
Arthritis	Bronchial asthma
Cerebral palsy	Cystic Fibrosis—mild and moderate
Cyanotic heart disease	Diabetes mellitus
Cystic fibrosis—severe	Epilepsy
Muscular dystrophies	Gynecomastia
Malnutrition—extreme	Hemophilia
Obesity—severe	Mental retardation
Paralysis	Non-cyanotic heart disease
Scoliosis—severe	Obesity—mild and moderate
	"Non-disease" entities

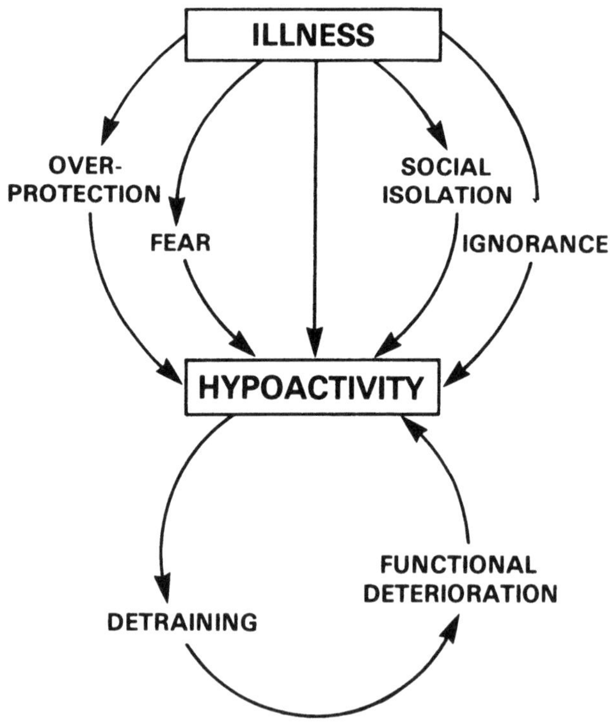

Figure 2.2. Direct and indirect links between illness and hypoactivity.

Table 2.2. Causes of Hypoactivity in Pediatric Diseases, as Stated by Parents or Patients

Disease	*Cause of Hypoactivity*
Bronchial asthma	Fear of post-exercise attack
Diabetes mellitus	Danger of hypoglycemic crisis
Epilepsy	Fear of seizure and injury
Heart diseases	Fear of "heart attack"
Hemophilia	Fear of injury and bleed
Mental retardation	Isolation, social maladjustment
Obesity	Inhibition, social discrimination, low fitness
Innocent murmur	Fear of "heart attack"

"Non-Disease" as a Cause of Hypoactivity

Many children who have a *presumed* illness, with no organic abnormality, are denied sufficient activity. The most common example of such "non-disease" is the innocent murmur.

In a survey of Seattle junior high schools, 93 pupils who, according to the nurse's file, had a "heart disease" were reevaluated by a pediatric cardiologist and their parents interviewed. The study was designed to tell how many of these children really had an organic heart disease and whether the level of activity restriction was related to the presence or absence of disease.[3] As seen in Fig. 2.3, only one in five children had an organic heart disease. The others had either an innocent murmur or no findings at all. The striking result was that in *both* groups, irrespective of evidence of heart disease, activity was restricted among some 40% of the participants! According to most parents, the decision to restrict their child's activity was based on the message that they had received from the physician at the time of the original diagnosis.

If the above findings are typical of other school systems, hypoactive children with cardiac non-disease outnumber those with a confirmed disease. The physician plays a crucial role in the etiology of non-disease and in the resulting hypoactivity. A mere suggestion that the child should refrain from sports, or perceived uncertainty about this issue, may render the child inactive for years.

Assessment of Physical Activity

Physical activity can be defined and characterized in more than one way. To the behaviorist it is overt intentional behavior, expressed primarily

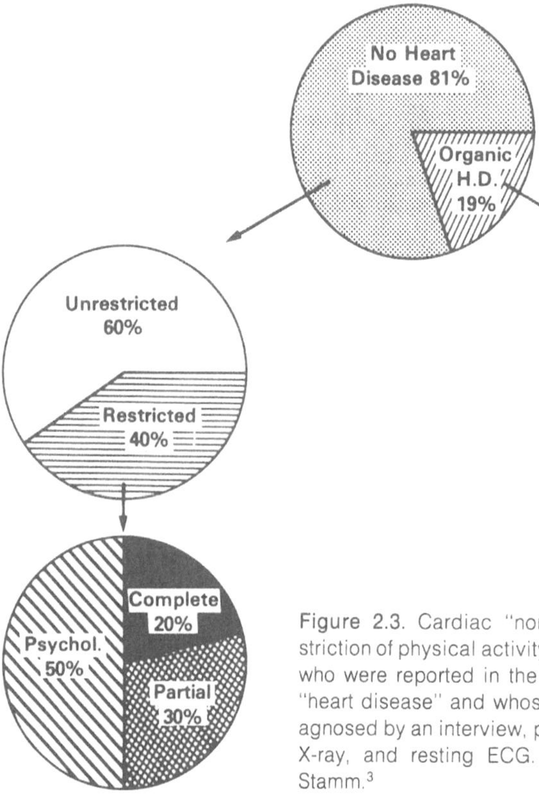

Figure 2.3. Cardiac "non-disease." Degree of restriction of physical activity among 93 school children who were reported in the nurse's file to have had a "heart disease" and whose condition was further diagnosed by an interview, physical examination, chest X-ray, and resting ECG. Data from Bergman and Stamm.[3]

through the musculoskeletal system.[16] Within such a definition the intent and purpose of a child's activity, the stimuli that bring it about, and the amount of social contacts achieved through it are the main foci. To the physiologist, activity is primarily the elevation of metabolic rate through muscular contraction. Here, the effects on various body systems and on body composition are of major interest. Physiologic variables most often monitored are heart rate and energy cost.

For clinical purposes it is important to characterize activity by a combination of the above approaches. One should know the *type* of movements, games, and sports that the child performs habitually (or wishes to perform) and the degree of socialization achieved through them, as well as the calorie equivalent of these activities.

Methodologic details for assessment of activity are available elsewhere.[2,16,43] The following briefly describes methods mentioned elsewhere in this book. Their merits and drawbacks are summarized in Table. 2.3.

Recall questionnaires,[17,47,55] whether administered by an interviewer or filled in by the patient or parents, are inexpensive and can be used in large-scale surveys. They can also serve as a basis for history taking in a pediatric clinic (see Appendix IV for a sample questionnaire). Informa-

Table 2.3. Methods for Assessment of Habitual Activity of Infants and Children

Method	*Merits*	*Drawbacks*
Recall questionnaires	Inexpensive, suitable for large-scale surveys	Low reliability and objectivity
Self-keeping log	Inexpensive	Low objectivity
Observation by investigator	Ability to record complex activities	Time-consuming, activities hard to quantify
Time and motion analysis	Ability to categorize and quantify activities	Time-consuming
Pedometer to count steps	Objective, inexpensive	Nonspecific, low reliability in slow walking or very fast running
Actometer to count limb movements	Objective, inexpensive, good for newborns	Nonspecific, limbs only
Strain gauge to register crib movements	High sensitivity, not connected to subject	Limited to crib, nonspecific
Photography—time lapse or cinematography	Objective, reliable, and valid. Good for complex activities	Limited to confined spaces, expensive, hard to quantify
Heart beat accumulation (electrochemical, electronic)	Objective, inexpensive, long-term observations	Misses activity peaks, low reliability (some instruments), requires individual calibration
Heart rate recording (tape, telemetry)	Objective, reliable, long-term observations, detects activity peaks	Expensive, requires individual calibration
O_2 uptake	Objective, direct relationship to metabolic level	Cumbersome to subject, expensive

tion derived from such questionnaires is based on memory and lacks objectivity. A *log* kept by the patient[14,42,51] does not depend on memory and can be useful if the activities are well defined. This method, however, lacks objectivity and patients may modify their activity during the period of recording. Good objectivity is attained when *observations* are made *by the investigator*.[7,13,59] This method is time-consuming and the end product is often descriptive rather than quantitative. This drawback can be rectified by incorporating a *time-and-motion analysis* (dividing activities into a few basic categories and recording the period spent in each category).[2]

Pedometers are counters triggered by motion. Worn on belt or ankle, they record the number of steps taken.[44,56,61] They are inexpensive and yield objective information. They do not differentiate, though, between walking and running, nor do they detect movements performed by the trunk and upper limbs. Their validity for step counting is poor when the child walks slowly or runs fast.[30,45] *Actometers* are based on a similar principle. They are worn on the wrists and ankles to register limb movements of an infant in the crib.[33,41,48] Another device used with infants is a *strain gauge,* which senses their movements as transferred to the crib.[26] The information is quantifiable, but nonspecific to any given movement.

Photography and *cinematography* are objective and record specific movements and social interaction.[8,10,16,63] Their main disadvantages are the confinement of the child within limited areas and the lack of information on the calorie equivalence of activities.

The continuous *accumulation of heart beats* or *recording of heart rate* is gaining popularity as a method for assessment of the daily metabolic profile. The underlying physiologic principle is a linear relationship between heart rate and O_2 uptake, at submaximal exercise. Once a heart rate vs. O_2 uptake regression line has been established for a child, his heart rate profile during the day can be translated to O_2 uptake and to calorie equivalent.[19,22,26,47,51,59] Disadvantages are the high cost of playback units and the dependence of heart rate on factors other than the metabolic level. This method does not yield information on the *nature* of activity performed by the child.

One can *measure O_2 uptake directly* while the child is active.[50,57] Although this is the most valid method for analysis of energy expenditure (short of direct calorimetry), instruments are expensive and heavy to carry.

There is no single method of assessing the type, intensity, and duration of daily activities. The best approach is therefore to combine a descriptive method (e.g., log-keeping by the child) with one that assesses the calorie equivalence of activities (e.g., heart rate recording). A follow-up should last at least 3 days, including one weekend day.

Effects of Disease on Physical Working Capacity

Disease can cause a decrease in physical working capacity in a number of ways:

Indirectly through hypoactivity and detraining
Lowering of maximal aerobic power
Causing high metabolic cost of submaximal tasks

Hypoactivity—Detraining—Hypoactivity: The Vicious Circle

Disease often causes hypoactivity, which leads to a detraining effect, a reduction in the functional ability of the child, and further hypoactivity. This vicious circle, shown in Fig. 2.2, can occur in any chronic disease, obesity being a typical example. In some patients it follows a short period of bed rest caused by injury, surgery, or acute exacerbation of the chronic disease. An example is a muscle dystrophic patient who can still walk. Following some minor injury he is confined to bed for 2 weeks, and, as a result, can no longer resume a walking status. Detraining can also occur in a healthy individual who, for some reason, reduces his level of activity. For more details on the sequelae of detraining see Chapter 1, section entitled Physiologic Effect of Detraining and Deconditioning.

Reduced Maximal Aerobic Power

Disease can directly affect the O_2 transport system and cause a reduction in the maximal aerobic power. According to Fick's principle, maximal O_2 uptake ($\dot{V}O_2$ max) equals the product of maximal cardiac output ($\dot{Q}$ max) and the maximal arterio-mixed venous difference in O_2 content ($Ca_{O_2} - C\bar{v}_{O_2}$ max. $\dot{Q}$ max is the product of maximal stroke volume (*SV* max) and maximal heart rate (*HR* max). Thus:

$$\dot{V}O_2 \text{ max} = SV \text{ max} \times HR \text{ max} \times (Ca_{O_2} - C\bar{v}_{O_2}) \text{ max}$$

Disease can reduce any of the three functions on the right side of the equation, thereby reducing maximal O_2 uptake. A list of such diseases is presented in Fig. 2.4.

A subnormal maximal stroke volume will result from the following:

Outflow obstruction (aortic stenosis, pulmonary stenosis, tetralogy of Fallot)
Deficient contractility (cardiomyopathy, detraining)
Hypovolemia (hypohydration)
Deficient "forward" stroke volume (ventricular septal defect, tetralogy)

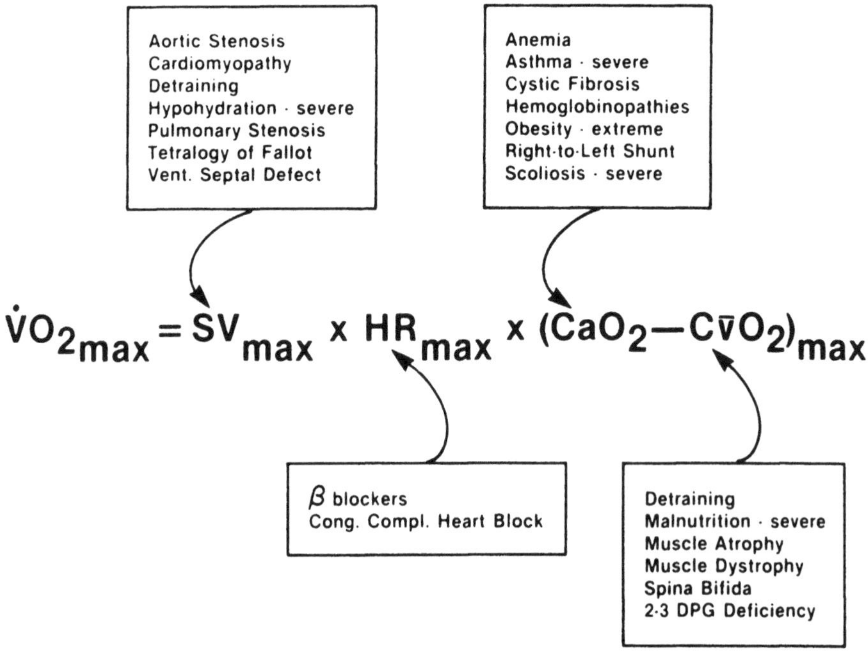

Figure 2.4. Maximal aerobic power ($\dot{V}O_2$ max) and disease. The Fick equation and specific conditions that affect its components, thus reducing $\dot{V}O_2$ max, are shown.

Although peak heart rate in many diseases is lower than the age-predicted maximal heart rate, there is only one disease in which peak heart rate is the *primary* limiting factor: complete congenital heart block. In other diseases a low peak heart rate is a result, rather than a cause, of low maximal aerobic power. A primary low peak heart rate can also result from medication, notably β blockers.

A reduced arterial O_2 content can result from respiratory disorders or from a low O_2-carrying capacity of the blood. The former include lung diseases such as cytsic fibrosis, severe forms of bronchial asthma, or restrictive lung syndromes. They also include chest wall disorders, such as severe scoliosis; or extreme obesity, in which the alveolar ventilation is low. All these respiratory disorders may result in arterial O_2 desaturation. Anemias, hemoglobinopathies, and cyanotic heart diseases are obvious causes of low O_2-carrying capacity of the blood. Mild to moderate activities can be performed well in these disorders because of a compensatory increase in cardiac output. When activity is intense, however, cardiac output can no longer rise and performance is deficient.

A high mixed-venous O_2 content during maximal exercise reflects low O_2 utilization by the periphery. This will happen when muscle blood flow is low relative to blood flow to other organs (e.g., in muscle atrophy, muscle dystrophy, or severe malnutrition). Deficient O_2 utilization by the periphery can also result from reduction in oxidative capacity in the muscle fiber, as occurs in detraining. Erythrocyte 2,3 diphosphoglycerate (2,3 DPG) affects a shift-to-the-right of the O_2 dissociation curve, making more O_2 available to the tissues at a given Po_2. A deficiency of 2,3 DPG might well reduce the availability of O_2 and result in high mixed-venous O_2 content.

High Metabolic Cost of Exercise

Even when maximal O_2 uptake is normal, a high metabolic cost of submaximal activities will leave the individual with a reduced "metabolic reserve" and impair his or her ability to sustain exercise at moderate or intense levels (see Chapter 1, section entitled Metabolic Responses to Exercise in Children). Such is the case, for example, in obesity, where the transport of excessive weight is metabolically expensive. Similarly, excessive demands are apparent during exercise in spastic or athetotic cerebral palsy or in other neuromuscular diseases with incoordination and "wasteful" movements. O_2 uptake can also be excessive because of the high O_2 cost of breathing, as in airway obstructive syndromes, constrictive lung diseases, or chest wall abnormalities.

Based on these facts, it is apparent that some diseases affect exercise performance through more than one mechanism (e.g., obesity, bronchial asthma, tetralogy of Fallot), whereas others cause a specific deficiency. More detailed discussion of the pathophysiology of exercise can be found in the following chapters.

Exercise as a Diagnostic Tool in Pediatrics

Exercise "stress testing" has become a universally accepted tool for assessment of coronary heart disease as well as pulmonary and other disorders in adults.[15,29,60] Exercise in clinical assessment of children, although less publicized, is used in a growing number of pediatric clinics. In fact, there are more pediatric than adulthood disorders in which exercise testing is of clinical relevance. The rationale for the use of exercise in pediatric assessment is summarized in Table 2.4. Methods of testing are outlined in Appendix II.

Children are often referred to the exercise clinic to evaluate their physical working capacity, which may be deficient due to disease or due to inactivity and detraining. For example, children with chronic renal failure, under treatment by hemodialysis, or peritoneal dialysis, were sent to us to evaluate their ability to walk to school rather than take a taxi. We found that the exercise level that they could comfortably sustain exceeded the calorie demands of walking at a regular pace. Our recom-

Table 2.4. Rationale for Exercise Testing in Children

1. Assesses physical working capacity
2. Evaluates *specific* pathophysiologic characteristics
   a. Provides indications for surgery, therapy, or additional tests
   b. Evaluates functional postoperative success
   c. Diagnoses disease
3. Assesses adequacy of medication
4. Assesses "risk" for future disease or for complications in existing disease
5. Instills confidence in child and parents
6. Motivates child for further exercise or weight loss

mendation was—much to the joy of the children—that walking is permitted. Determination of working capacity is also important in order to prescribe activity programs and to periodically assess progress.

Identifying a specific pathophysiologic pattern in a given disease may be of even more clinical relevance than the mere assessment of working capacity. Various patterns can be better revealed during exercise than at rest. The rationale is based on the increase in metabolic demands during exercise that stresses—sometimes to the limit—ventilatory, gas-exchange, cardiac, vascular, neuromuscular, and thermoregulatory functions. A malfunction in a system is more likely to be discovered during stress than during rest, when functional demands are lower. Examples of such specific patterns are the ischemic ECG changes in congenital aortic stenosis (which reflect the degree of narrowing); the appearance of ventricular dysrhythmia in congenital complete heart block or in postoperative tetralogy of Fallot; the high blood pressure in postoperative coarctation of the aorta; or the post-exertional bronchoconstriction in bronchial asthma. Some findings can be used as ancillary indications for surgery (e.g., a rise in right ventricular end-diastolic pressure and a decline in stroke volume in pulmonary stenosis). Others are taken as indications for therapy (cardiac pacing in complete heart block when ventricular rate does not rise sufficiently during exercise; antidysrhythmic drugs in postsurgical tetralogy of Fallot).

A noninvasive exercise test can be used for screening to determine the need for an invasive test (e.g., segmental ST depression of 2 mm or more in aortic stenosis will call for cardiac catheterization). Cardiac or vascular operation may be successful anatomically but less so functionally. This too can be revealed by stress testing (e.g., no rise in stroke volume after pulmonary valvotomy).

While most children arrive at the exercise clinic with a known diagnosis, an exercise test can sometimes establish a diagnosis. This is true for growth hormone deficiency and, occasionally, for asthma (by discovering exercise-induced bronchoconstriction in a child with atypical symptoms).

An area in which exercise testing may prove beneficial is the assessment of the adequacy of drug regimens at different activity levels. While such a consideration may be obvious for insulin, it may also apply to other drugs such as corticosteroids, antihypertensives, or anticonvulsants. Another area still under investigation is the prediction, by exercise, of "risk" for future diseases such as hypertension or coronary heart disease. Also under preliminary investigation is the use of exercise for assessment of subclinical diabetic nephropathy or neuropathy, or of Chagas disease neuropathy.

Many children appearing at the exercise clinic have been inactive for various periods and may have lost confidence in their ability to exercise. The successful completion of a test is, therefore, a revelation to the child, who realizes that he or she *can* exert strenuously. Even more impressive is the realization by some parents that their child can exercise at high intensity with no ill effect. We routinely encourage the presence of parents at the initial exercise test (and at subsequent tests if they wish). They are seated behind the child to avoid distracting him and can watch his reactions. We often hear an amazed parent commenting that he or she has never seen the child work so hard, develop sweat, or become red in the face. The testing situation is a learning experience in itself and can instill confidence in patient and parent alike.

Likewise, the test and its interpretation can be used as vehicles for motivating a child to increase his or her daily activity or to persevere in a weight-reducing program. This is especially evident when a periodic test shows improved performance.

Exercise testing, like other laboratory or clinical tests, should not be used in isolation. It must follow a thorough history taking, including data on the habitual activity of the child and other family members; attitudes at home toward physical activity; and willingness and ability of the parents to spend extra time with the child if additional activity is prescribed (e.g., drive the child to a swimming pool). Physical examination should emphasize the cardiorespiratory and musculoskeletal systems. Other laboratory tests, such as ECG, blood hemoglobin, or pulmonary functions, may often be needed. Conclusions and recommendations must integrate all these with the results of the exercise test.

Exercise as Therapy in Pediatrics

The use of physical activity for treatment is well recognized in physiotherapy, occupational therapy, and adapted physical education. Training can increase muscle strength and range of joint motion, prevent contractures, and improve stamina, ambulation, and various other skills. In the context of pediatric management, conditioning and training are beneficial for additional reasons, as summarized in Table 2.5.

Only seldom will therapy by exercise affect the pathophysiologic process itself. This may happen in obesity, where calorie balance can be directly affected, or in neurocirculatory asthenia, where sympathetic overactivity is ameliorated through physical conditioning. In most other diseases the benefits of exercise are indirect and do not change the basic pathophysiologic process.

In progressive muscular dystrophy, for example, abnormal changes in the affected muscle fibers will continue, but physical conditioning can improve the function of the *residual,* healthy muscle fibers. The end result is an increase in the functional level of the child and prolongation of his walking status. Another example is diabetes mellitus, in which the

Table 2.5. Pediatric Diseases in Which Exercise Is Used as Therapy

Disease	*Benefits*
Pulmonary	
Bronchial asthma	Reduced rate and intensity of EIB*
Cystic fibrosis	Improved airway drainage
Cardiovascular	
Neurocirculatory asthenia	Hemodynamic improvement
Endocrine	
Diabetes mellitus	Better diabetic control
Nutritional	
Anorexia nervosa	Means for behavior modification
Obesity	Weight and fat reduction
Musculoskeletal	
Cerebral palsy	Ambulation, contracture prevention, weight control
Muscular dystrophy	Ambulation, strengthening of residual muscle, weight control
Paralysis	Strengthening of residual muscle
Rheumatoid arthritis	Mobilization, increased range of motion
Other	
Hemophilia	Mobilization, increased range of motion
Mental retardation	Increase of environmental stimuli; socialization

* EIB = exercise-induced bronchoconstriction.

basic endocrine and metabolic deficiencies are not modified by conditioning, but daily diabetic control and physical growth are improved.

Treatment of children by exercise is unique: by prescribing exercise we are signaling to the child that he can, and should, act like his healthy peers. We emphasize thereby his abilities rather than his disabilities. This is in contrast to therapy by medication, diet, or bed rest, where he is made to feel different from others.

Another important characteristic of exercise is that the more it is done the easier it becomes and the greater the sense of accomplishment. The first few sessions are the most difficult in any conditioning program: the child is unfit, lacking in skill, confidence, and motivation. Each new activity may be stressful, causing aches, pains, and frustration. This is the time when professional and parental support is most needed.

The Exercise Prescription

As in other forms of therapy, a conditioning program should be quantified regarding the *intensity, frequency,* and *duration of each session,* the *type* of activities, and the overall *duration of the program* (for details see Chapter 1, section entitled Principles of Physical Conditioning).

The concept of exercise prescription has proven useful for adults in programs for prevention of, or rehabilitation after, coronary heart disease. This concept can be used, with as much success, in pediatrics. Combining the principles outlined in Chapter 1 and specific information from Tables III.1 and 5.3, one can prescribe activities, to be "filled" by a physiotherapist, an exercise therapist, or a physical educator. Simple activities can be supervised by a parent. The following are examples of how an exercise prescription can be prepared.

CASE NO. 1

An 11-year-old mildly obese girl (32% body fat, as determined by skinfold thickness) weighs 43 kg (75th percentile) and is 143 cm tall (50th percentile). She is free of other diseases, taking part in physical education classes, but otherwise is sedentary.

Therapeutic Goal: reduction of fatness to 25%, without interfering with growth.

Analysis of Exercise Requirements: The 7% excess of adipose tissue is equivalent, at the present body weight, to 3.0 kg. To lose this amount the girl must achieve a negative calorie balance of 21,000 kcal (87,800 kJ). Assuming that half of this will be achieved by a mild dietary reduction, the remaining 10,500 kcal must be "burned up" by additional exercise. For 250 kcal per session, 42 sessions will be required. At a rate of three sessions per week, the overall program will last 14 weeks.

The only sport that this girl likes is bowling, but she has a bicycle that she uses for errands. Since the calorie equivalent of bowling is low, the program should be based on cycling. (Jogging is an alternative but is less recreational.) As seen in Table III.1, a 43-kg child consumes about 45 kcal during a 10-min ride at 15 km/hr, on flat terrain. This will be about 50–55 kcal/10 min where the terrain is mildly sloping (as in our patient's neighborhood). Thus, a 45–50 min ride (11.5–12.5 km) at this comfortable pace will be sufficient to consume 250 kcal. Using the principle of progression (Chapter 1, section entitled Principles of Physical Conditioning), this patient should start at a certain distance and gradually increase it.

The Prescription:
Type of exercise—cycling on mildly sloping (2–3%) gradient
Intensity—not important
Frequency of sessions—3 per week
Distance at each session—10 km for the first 3 weeks
11 km for the next 3 weeks
12 km for the next 4 weeks
13 km for the next 4 weeks
Program Duration—14 weeks

Furnished with the above outline, a family member can easily supervise such a program.

CASE NO. 2

A 10-year-old boy with bronchial asthma withdrew one year ago from physical education classes because of breathlessness and wheeziness, triggered by running. He is otherwise asymptomatic and not taking any medication.

Exercise-induced bronchoconstriction (EIB) was documented through an exercise provocation test. In another exercise test, peak power output was 70 watts. This is 1.5 S.D. below mean for age (see Fig. I.3 in Appendix I).

Therapeutic Goals: 1) resumption of normal physical activity; 2) increase of 15% in peak power output.

Analysis of Therapeutic Requirements: The simplest way to enable such a boy to participate in physical education classes is to prescribe medication for prevention of EIB. A $\beta 2$ sympathomimetic (e.g., salbutamol aerosol, one puff), taken just prior to class is a good choice.

Low exercise performance in this child most likely represents detraining. Resumption of activities at school will help improve his fitness to a limited extent only. An additional conditioning program is indicated.

Unlike case No. 1, where the overall calorie expenditure was a primary consideration, here the *intensity* of exercise should be emphasized. To exceed the conditioning threshold (see Chapter 1, section entitled Principles of Physical Conditioning), a heart rate of 160 beats/min or more should be reached. Swimming, the least asthmogenic of sports, is the most appropriate conditioning program for our patient (who swims well).

The Prescription:
Type of exercise—swimming, any style
Intensity—heart rate to exceed 160 beats/min during 15–20 min of each session.
Frequency—2 to 3 times weekly
Duration of Program—2 months
Medication—not needed, unless wheeziness and breathlessness develop. If these occur, salbutamol, 1 puff 5 min before session.

Note that the above prescription does not include guidelines regarding the structure of each session, its duration, or the distance to be covered. A qualified instructor is expected to plan these, according to the prescribed intensity and overall duration.

Deleterious Effects of Exercise

Exertion can be deleterious to a child's health under certain circumstances. The detrimental effects of exercise include sports injuries, overuse syndromes, and abnormal physiologic reactions.

Although injuries result mostly from collision and contact sports such as American football, rugby, ice hockey, and soccer, they can occur in any sport, to any participant, irrespective of his competence or aspiration.[9,20,24,32]

The risk of injury can be reduced by proper conditioning; matching of opponents (by size, maturation, and level of skill); the use of such safety devices as helmets, mouthguards, or knee pads; adaptation of rules; and proper maintenance of sports facilities. An annual preparticipation examination is recommended,[1,52] although its value in preventing sports injuries is still debatable.[21]

Overuse syndromes may result when a movement is repeated, usually at high intensity, over months or years, causing excessive mechanical stress on bone, cartilage, tendon, or muscle. They are seen mostly in ambitious young athletes who practice often and intensively.[38] Overuse syndromes typical for children and adolescents are "little league elbow" in baseball pitchers;[58] low back pain in oarsmen, girl gymnasts, and horseback riders;[27,28,39,40] and shoulder pain in swimmers.[12]

In this book we shall limit our discussion to the nontraumatic deleterious effects of exercise, as manifested by abnormal physiologic responses. These are found mostly in children with specific diseases, but can occur also in healthy youngsters, as outlined in Table 2.6. Responses range from the benign, such as proteinuria[49] or hematuria,[4] to such life-threatening conditions as ventricular tachycardia[62] or heatstroke.[18]

Table 2.6. Nontraumatic Deleterious Effects of Exercise in Children and Adolescents

Abnormal Response	*Underlying Disease or Condition*
Bronchoconstriction	Bronchial asthma, atopy, history of wheezy bronchitis
Chest pain	Aortic stenosis, bronchial asthma
Delayed menarche	Healthy*
Dehydration	Healthy
Dysrhythmia	A-V block, post-intracardiac surgery, healthy
Heatstroke, exhaustion	Dehydration, nonacclimatization, obesity, healthy
Hematuria	Healthy, glomerulonephritis, renal calculus
Hemoglobinuria	Healthy
High blood pressure	Aortic coarctation, hypertension, obesity
Hypoglycemia	Diabetes mellitus
Ischemic ST-T changes	Aortic stenosis and insufficiency, coarctation, mitral valve prolapse, familial hypercholesterolemia, sickle-cell anemia
Ketoacidosis	Diabetes mellitus (insulin deprivation)
Menstrual irregularities	Healthy*
Proteinuria	Healthy
Syncope	Aortic stenosis, A-V block
Stiffness spells	Hypoparathyroidism
Sudden death	Aortic stenosis, congenital coronary anomaly

* Resulting from chronic exercise. All other responses are to acute exercise.

The role of exertion in triggering sudden death among young cardiac patients was investigated in a cooperative international study.[31] As seen in Fig. 2.5, 58% of deaths occurred when the children were at complete physical rest. Thirty-two percent were playing games and 10% were engaged in sports. The authors concluded that "avoidance of strenuous activity seldom prevents this catastrophe" in the young cardiac patient. The underlying pathologies in those who die during, or immediately after, exercise are congenital coronary anomalies,[36] congenital aortic stenosis,[31,37] and hypertrophic cardiomyopathy.[35]

Most of the adverse reactions listed in Table 2.6 are related to a single bout of exercise. Some occur regularly in any given child, such as ST segmental depression in advanced aortic stenosis, ventricular dysrhythmia in congenital complete heart block, high arterial blood pressure in coarctation of the aorta, or exertional hematuria. Others will appear when several unfavorable conditions interact. For example, heatstroke can occur in a healthy, but unacclimatized, child who exerts for a prolonged period of time without sufficient fluid intake. Ketoacidosis may develop in an insulin-dependent child with diabetes who exercises with-

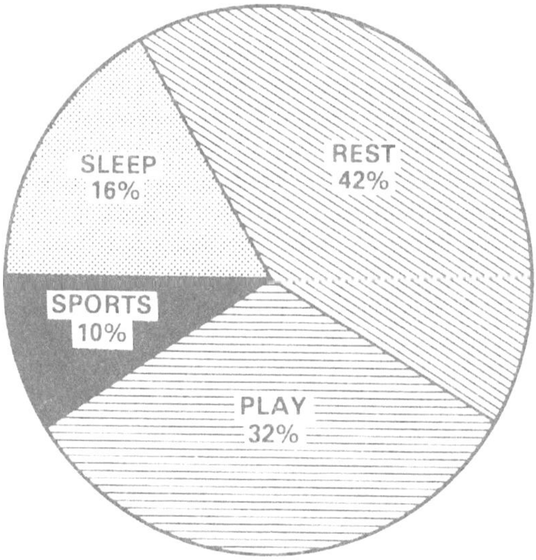

Figure 2.5. Physical activity prior to sudden death due to cardiovascular diseases, among 209 one to 21-year-old girls and boys who had been ambulatory and nonhospitalized. Results of an international cooperative study. Pooled data of 20 medical centers in nine countries. From Lambert et al.[31]

out having taken his scheduled insulin. Chronic exercise, in addition to causing overuse syndromes, can effect a delay in menarche[34,54] and secondary amenorrhea.[5,11]

Adverse physiologic responses to exercise are usually preventable, or their impact can be minimized by proper precautions. Exercise-induced hypoglycemia, for example, can be averted by ensuring sufficient carbohydrate intake prior to activity, by reducing the insulin dosage, and by selecting an appropriate site for insulin injection; heatstroke or heat exhaustion can be prevented by ensuring prior acclimatization to heat, fluid replenishment, and curtailment of activities when certain climatic conditions prevail; exercise-induced bronchoconstriction can be prevented, or ameliorated, by medication, by changing from running to swimming, and by reducing activities on dry or very cold days.

In some diseases parents or patients *assume* that exercise entails a risk which, in fact does not exist or is minimal and outweighed by the benefits of exercise. Such, for example, is the case with epilepsy, where in spite of popular belief convulsions are *not* triggered by exercise. Parents of a child with hemophilia sometimes keep him inactive to avoid bleeds. Current evidence suggests that such a risk is minimal. Children with hemophilia can assume normal activities, providing they take the appropriate replacement therapy. Similar overprotection is shown by parents of children with heart disease, real or assumed, for fear of a "heart attack."

By understanding the possible deleterious effects of exercise, the clinician can prevent them or minimize their impact. His role in alleviating fears of the *presumed* dangers of exercise is no less important.

References

1. American Academy of Pediatrics: Disqualifying conditions for contact sports. In: School Health: A Guide for Health Professionals. American Academy of Pediatrics, 1977, pp. 223–227.
2. Andersen KL, Masironi R, Rutenfranz J, Seliger V: Habitual physical activity and health. World Health Organization, Copenhagen, 1978.
3. Bergman AB, Stamm SJ: The morbidity of cardiac nondisease in schoolchildren. N Engl J Med 276:1008–1013, 1967.
4. Bodian M, Black JA, Kobayashi N, et al: Recurrent hematuria in childhood. Q J Med 34:359–382, 1965.
5. Bonen A, Belcastro AN, Simpson AA, Ling W: Comparison of LH and FSH concentrations in age group swimmers, moderately active girls, and adult women. In: Eriksson B, Furberg B (eds.) Swimming Medicine, IV. University Park Press, Baltimore, 1978, pp. 70–78.
6. Bradfield RB, Chan H, Bradfield NE, Payne PR: Energy expenditures and heart rates of Cambridge boys at school. Am J Clin Nutr 24:1461–1466, 1971.
7. Bruch H: Obesity in childhood. IV. Energy expenditure of obese children. Am J Dis Child 60:1082–1109, 1940.
8. Bullen BA, Reed, RB, Mayer J: Physical activity of obese and nonobese adolescent girls appraised by motion picture sampling. Am J Clin Nutr 14:211–223, 1964.
9. Chambers RB: Orthopaedic injuries in athletes (ages 6 to 17). Comparison of injuries occurring in six sports. Am J Sports Med 7:195–197, 1979.
10. Corbin CB, Pletcher P: Diet and physical activity patterns of obese and nonobese elementary school children. Res Q Am Assoc Health Phys Educ 39:922–928, 1968.
11. Dale E, Gerlach DH, Wilhite AL: Menstrual dysfunction in distance runners. Obstet Gynecol 54:47–53, 1979.
12. Dominguez RH: Shoulder pain in age group swimmers. In: Eriksson B, Furberg B (eds.) Swimming Medicine, IV. University Park Press, Baltimore, 1978, pp. 105–109.
13. Durnin JVGA: Physical activity by adolescents. Acta Paediatr Scand Suppl 217:133–135, 1971.
14. Durnin JVGA, Passmore R: Energy, work and leisure. Heineman, London, 1967.
15. Ellestad MH: Stress Testing. Principles and Practice, 2nd ed. F.A. Davis, Philadelphia, 1980.
16. Ellis MJ, Scholtz GJL: Activity and Play of Children. Prentice-Hall, Englewood Cliffs, N.J., 1978.
17. Engström L-M: Physical activity of children and youth. Acta Paediatr Scand Suppl 283:101–105, 1980.
18. Fox EL, Mathews DK, Kaufman WS, Bowers RW: Effects of football equipment on thermal balance and energy cost during exercise. Res Q Am Assoc Health Phys Educ 37:332–339, 1966.
19. Gandra YR, Bradfield RB: Energy expenditure and oxygen handling efficiency of anemic schoolchildren. Am J Clin Nutr 24:1451–1456, 1971.

20. Garrick JG: Sports medicine. The school-aged athlete. Pediatr Clin North Am 24:737–747, 1977.
21. Garrick JG, Smith NJ: Pre-participation sports assessment. Pediatrics 66:803–806, 1980.
22. Gilliam TB, Freedson PS, Geenen DL, Shartaray B: Physical activity patterns determined by heart rate in 6–7 year-old children. Med Sci Sports Exercise 13:65–67, 1981.
23. Gilliam TB, MacConnie SE, Geenen DL, et al: Exercise programs for children: a way to prevent heart disease? Physician Sportsmed 10:96–108, 1982.
24. Goldberg B, Witman PA, Gleim GW, Nicholas JA: Children's sports injuries: are they avoidable? Physician Sportsmed 7:93–101, 1979.
25. Goode RC: The child and physical activity. In: Goode RC, Volpe R (eds.) Proceedings of Workshop on the Child and Physical Activity. Ontario Heart Foundation, Toronto, 1979, pp. 20–35.
26. Griffiths M, Payne PR: Energy expenditure in small children of obese and non-obese parents. Nature 260:698–700, 1976.
27. Huber EG, Jani L, Keul J, et al: Sport im kindesalter. Munch Kinderheilkd 127:441–449, 1979.
28. Jackson DW, Wiltse LL: Low back pain in young athletes. Physician Sportsmed 2:53–60, 1974.
29. Jones NL, Campbell EJM: Clinical Exercise Testing, 2nd ed. W.B. Saunders, Philadelphia, 1982.
30. Kemper HC, Verschuur R: Validity and reliability of pedometers in habitual activity research. Eur J Appl Physiol 37:71–82, 1977.
31. Lambert EC, Menon VA, Wagner HA, Vlad P: Sudden unexpected death from cardiovascular disease in children. Am J Cardiol 34:89–96, 1974.
32. Larson RL, McMahan RO: The epiphyses and the childhood athlete. JAMA 169:99–104, 1966.
33. Mack RW, Kleinhenz ME: Growth, caloric intake and activity levels in early infancy: a preliminary report. Hum Biol 46:345–354, 1974.
34. Malina RM, Harper AB, Avent HH, Campbell DE: Age at menarche in athletes and non-athletes. Med Sci Sports 5:11–13, 1973.
35. Maron BJ, Roberts WC, McAllister HA, et al: Sudden death in young athletes. Circulation 62:218–229, 1980.
36. McClellan JT, Jokl E: Congenital anomalies of coronary arteries as cause of sudden death associated with physical exertion. Am J Clin Pathol 50:229–233, 1968.
37. Ongley PA, Nadas AS, Paul MY, et al: Aortic stenosis in infants and children. Pediatrics 21:207–221, 1958.
38. Orava S, Saarela J: Exertion injuries to young athletes. Am J Sports Med 6:68–76, 1978.
39. Oseid S, Bvjenth G, Evjenth D, et al: Lower back trouble in young female gymnasts—frequency, symptoms and possible causes. Federation Internationale d'Education Physique Bull 1:11–14, 1974.
40. ReFior HJ, Zenker: Wirbelsäule und Leistungsturnen. Wirbelkörper–und Bandscheibenveränderungen bei Kindern und Jugendlichen. Munch Med Wochenschr 11:463–467, 1970.

41. Rose HE, Mayer J: Activity, calorie intake, fat storage and the energy balance of infants. Pediatrics 41:18–29, 1968.
42. Rutenfranz J, Berndt I, Knauth P: Daily physical activity investigated by time budget studies and physical performance capacity of school boys. Acta Paediatr Belg 28[Suppl.]:79–86, 1974.
43. Saris WHM: Aerobic power and daily physical activity in children with special reference to methods and cardiovascular risk indicators. Doctoral dissertation, Catholic University, Krips Repro Meppal, Nijmegen, 1982.
44. Saris WHM, Binkhorst RA: The use of pedometer and actometer in studying daily physical activity in man. Part I: Reliability of pedometer and actometer. Eur J Appl Physiol 37:219–228, 1977.
45. Saris WHM, Binkhorst RA: The use of pedometer and actometer in studying daily physical activity in man. Part II: Validity of pedometer and actometer measuring the daily physical activity. Eur J Appl Physiol 37:229–235, 1977.
46. Saris WHM, Binkhorst RA, Cramwinckel AB, et al: The relationship between working performance, daily physical activity, fatness, blood lipids and nutrition in school children. In: Berg K, Eriksson BO (eds.) Children and Exercise IX. University Park Press, Baltimore, 1980, pp. 166–174.
47. Saris WHM, Binkhorst RA, Cramwinckel AB, et al: Evaluation of somatic effects of a health education program for schoolchildren. Bibl Nutr Diet 27:77–84, 1979.
48. Schulman JL, Reisman JM: An objective measure of hypoactivity. Am J Ment Defic 64:455–456, 1959.
49. Scotti P, Grisler R, Della Torre F, et al: Indagine sulla proteinuria da sforzo in un gruppo di giovani atleti. Med Lavoro 64:20–31, 1973.
50. Seliger V: Energy metabolism in selected physical exercises. Int Z Angew Physiol Enischl Arbeitphysiol 25:104–120, 1968.
51. Seliger V, Trefny Z, Bartunková S, Pauer M: The habitual activity and physical fitness of 12 year old boys. Acta Paediatr Belg 28[Suppl.]:54–59, 1974.
52. Shaffer TE: The adolescent athlete. Pediatr Clin North Am 20:837–849, 1973.
53. Shephard RJ: Physical Activity and Growth. Year Book Medical Publishers, Chicago, 1982.
54. Simri I, Hanne-Paparo N: Age of menarche and dysmenorrhea in female athletes and non-athletes in Israel (in Hebrew). In: Assif—The Wingate Scientific Book. Wingate Institute, Natanya, 1977.
55. Stefanik PA, Heald FP, Mayer J: Caloric intake in relation to energy output of obese and non-obese adolescent boys. Am J Clin Nutr 7:55–62, 1959.
56. Stunkard A, Pestka Y: The physical activity of obese girls. Am J Dis Child 103:116–121, 1962.
57. Taylor CM, Lamb MW, Robertson ME, MacLeod G: The energy expenditure for quiet play and cycling of boys seven to fourteen years of age. J Nutr 35:511–521, 1948.
58. Torg JS, Pollack H, Sweterlitsch P: The effect of competitive pitching on the shoulders and elbows of preadolescent baseball players. Pediatrics 49:267–272, 1972.

59. Wade MG, Ellis MJ: Measurement of free-range activity in children as modified by social and environmental complexity. Am J Clin Nutr 24:1457–1460, 1971.
60. Wasserman K, Whipp BJ: Exercise physiology in health and disease. Am Rev Respir Dis 112:219–249, 1975.
61. Wilkinson PW, Parkin JM, Pearlson G, et al: Energy intake and physical activity in obese children (abstract). Br Med J 1:756, 1977.
62. Winkler RB, Freed MD, Nadas AS: Exercise-induced ventricular ectopy in children and young adults with complete heart block. Am Heart J 99:87–92, 1980.
63. Wuellner LH: A method to investigate the movement patterns of children. Master's thesis, University of Illinois, 1969.

3

Pulmonary Diseases

Bronchial Asthma

The relevance of exercise to the health and well-being of the asthmatic child is fourfold:

1. Acute exertion triggers bronchoconstriction and an asthmatic attack.
2. Chronic exercise is of therapeutic value.
3. Exercise is an important diagnostic tool.
4. Research using exercise has been conducted to investigate the pathophysiology of asthma.

These topics are analyzed in this section.

Exercise-induced Bronchoconstriction (EIB)

Description. Post-exertional bronchoconstriction (also called exercise-induced asthma) is, clinically, the asthmatic child's most important response to exercise. It results in increased airway resistance, in lung hyperinflation, and, occasionally, in hypoxemia. Characteristically, bronchoconstriction starts some 2–4 min after exercise, peaking in children at 4–8 min post exercise (at 6–10 min in adults) and disappearing spontaneously within 20–40 min. Less frequently it may start *during* the activity; sometimes it may be sustained for more than one hour following exercise.

The functions commonly used for assessing bronchial patency and resistance to air flow are forced expiratory volume at the first second ($FEV_{1.0}$), peak expiratory flow, and maximum midexpiratory flow. Other functions, used mostly in research, include specific airway conductance and airway resistance.

Various criteria have been used to determine "abnormal" post-exertional bronchoconstriction. These are defined by a *percentage* drop (usually of $FEV_{1.0}$ or peak expiratory flow) from the pre-exercise value and range from 10 to 25%.[2,19,23,31,43,50,59,78,88,91,93,118,141,152] For details see section entitled The Exercise Challenge, below.

Some authors assigned importance to the *combined* bronchodilatory response during exercise and bronchoconstrictive response after exercise.[64,65,87] These combined changes presumably represent "airway lability," which was found to be greater in asthmatics than in healthy children. A Lability Index can be calculated that is considered abnormally high when it exceeds 20–22% of the pre-exercise function (usually peak expiratory flow). The use of this index has declined in recent years.

Epidemiology. When exercising in the laboratory, some 40–95% of asthmatics respond with EIB. Such diversity in reported response is due to the variety of protocols used by different investigators—especially the type, intensity, and duration of exercise; nonstandardized environmental temperature and humidity; different policies regarding drug withdrawal on the day of testing; and variability in the severity of disease. Furthermore, as discussed above, investigators have not agreed upon a universal criterion for a positive test. In spite of such variability, it is fair to state that about 70% of children with perennial asthma who are taken off medication some 6–8 hours before testing will respond with at least a 10–15% drop in $FEV_{1.0}$ following exercise.

Some authors refer to the above rates as "incidence"[42,93] or "prevalence"[134] of EIB. Their data, however, refer to a response to a standardized provocation test in the laboratory and do not reflect *real life* incidence and prevalence. There is only fair correlation between the response of an asthmatic child to exercise in the laboratory and his day-to-day response to spontaneous exercise.[93] It is therefore more appropriate to refer to the above percentages as "rate of response to a standardized exercise test."

In the absence of definitive epidemiologic data one can only speculate about the extent of EIB in real life. It is conceivable that *any* asthmatic child may, at some time, experience EIB when his or her activity happens to be intense and of sufficient duration. Potentially, therefore, all asthmatics are at risk. However, on the basis of their experience of such unpleasant episodes, many patients modify their activities in order to prevent triggering an attack. They do so spontaneously or on the advice of others, by avoiding intense exertion or by turning to less asthmogenic sports. Thus, the rate of occurrence of EIB during a given period of time may not be high.

Although EIB occurs typically in asthmatics, it is sometimes observed in patients with atopic disorders[23,93] or with cystic fibrosis.[35] A high Lability Index was found in first-degree relatives of asthmatics,[97] former asthmatics,[22,87] monozygotic twin siblings of asthmatics,[98] and patients with cystic fibrosis,[35] hay fever,[87] and past history of viral bronchiolitis.[140]

EIB—Nature of the Exercise Provocation

The degree of post-exertional bronchoconstriction varies with the nature of exercise. More specifically, it depends upon the type, intensity, and duration of the activity.

Type of Exercise. The information commonly given by the child or parent is that activities involving running, jumping, or cycling induce more discomfort than swimming. Indeed, a number of controlled studies have confirmed that swimming is the least asthmogenic type of exercise.[2,53,84] Fig. 3.1 shows an example comparing effects of running and swimming in an 8-year-old boy who performed both tasks at equal O_2 uptake and pulmonary ventilation.

What is the *most* asthmogenic type of exercise? Some authors have suggested that running induces greater and more consistent bronchoconstriction than walking or cycling.[4,49,64,85,89,137] It has further been suggested that free running provokes greater EIB than treadmill running.[4,133] Arm cranking was reported to be as potent a stimulus as running[2] and to cause greater EIB than leg cycling when both are done at the same absolute metabolic level.[147] As described in the section entitled Airway Cooling as a Trigger, below, the extent of EIB depends on the degree of airway cooling which, in turn, is a function of pulmonary ventilation and of climatic conditions. The main shortcoming of the above reports is that ventilation was not equated in the comparison between modes of exercise. Indeed, recent studies have indicated that when pulmonary ventilation, air temperature, and humidity are equal, walking, treadmill running, free running, cycling, and arm cranking induce similar degrees of bronchoconstriction.[37,95,110]

We can therefore conclude that, while swimming is the least asthmogenic exercise, there is no single type of activity that can be considered most asthmogenic. However, an individual child may find one specific "land activity" more asthmogenic than another.

Intensity of Exercise. The degree of EIB is related to the intensity of running[41,44,70,133,137,161] or cycling.[92] This relationship is presented in Fig. 3.2. As a rule, intense activities induce greater bronchoconstriction than light activities. During exercise that lasts 6–8 min, intensities of 70–85% of maximal aerobic power (heart rate = 160–180 beats/min) have the most asthmogenic effect.[44,66,161]

This can be explained by the associated high pulmonary ventilation and airway cooling. It is not clear, however, why *maximal* intensities

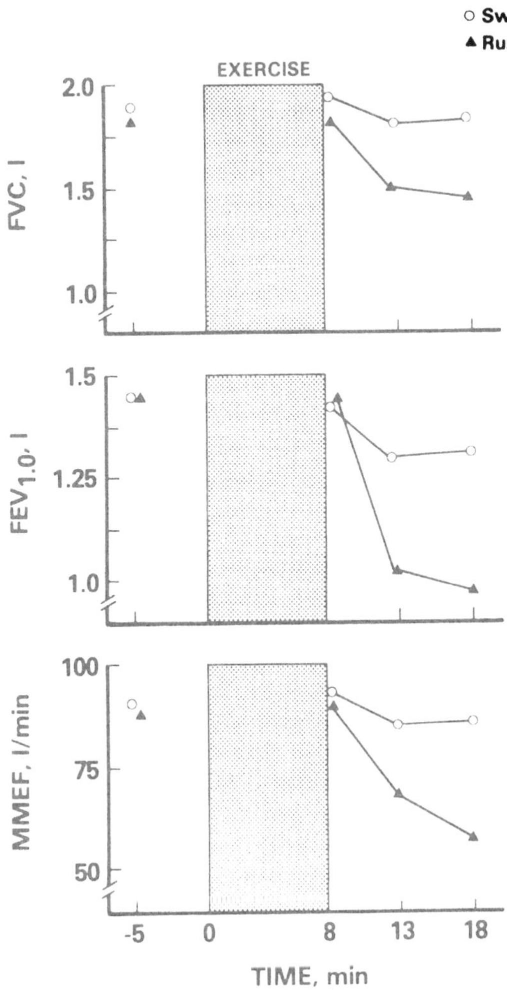

Figure 3.1. Swimming as a "non-provoker" of exercise-induced asthma. Pulmonary functions of an 8-year-old asthmatic boy before and after treadmill running and tethered swimming. FVC = forced vital capacity, $FEV_{1.0}$ = forced expiratory volume at the first second, MMEF = maximal mid-expiratory flow. Both tests lasted 8 min, during which time O_2 uptake was 29 ml/kg × min and minute ventilation 34 liters/min. Ambient temperature was 27°C and relative humidity, 30%. Data from the author's laboratory.

(which are accompanied by the highest pulmonary ventilation) do not cause the greatest EIB. One explanation is that such intense activities cannot be sustained long enough to induce sufficient airway cooling.

Duration of Exercise. The degree of EIB is related to the duration of exercise.[89,137] An exercise duration of 6–8 min was found to induce greater bronchoconstriction than shorter or longer protocols at exercise intensities equivalent to 60–85% of maximal aerobic power.[137] More research is needed, however, to find out whether this duration is optimal at other exercise intensities: for example, it may be less than 6–8 min for highly strenuous activities. A case in point is a study in which young

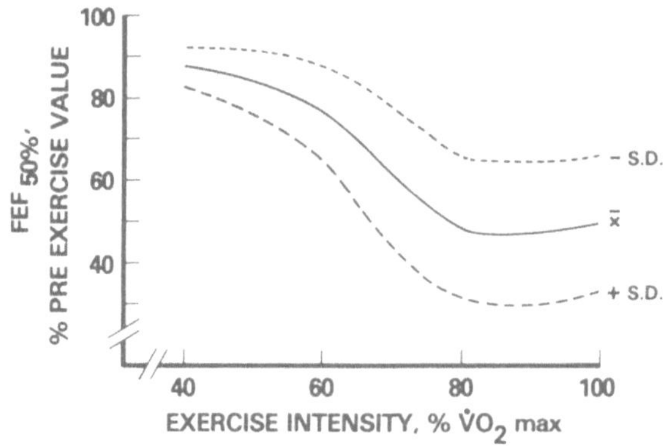

Figure 3.2. EIB and intensity of exercise. Nine asthmatic young adults ran on a treadmill at 40, 60, 80, and 100% of their maximal O_2 uptake ($\dot{V}O_2$ max). Each run lasted 6 min. Values are presented of post-exertional forced expiratory flow at 50% of vital capacity ($FEF_{50\%}$), expressed as percentage of pre-exercise value. Adapted from Wilson and Evans.[161]

adult asthmatics had a greater drop in maximal mid-expiratory flow following a supramaximal treadmill run, which exhausted them within 50 sec, than following a 7 min submaximal run.[83]

Prolonged activities (e.g., 20 min or more) are less asthmogenic than shorter ones. This is clinically important: an asthmatic who performs such prolonged activities may suffer no EIB and is said to have "run through" his asthma.

Time Since Previous Exercise. An asthmatic who responds to a bout of exercise with bronchoconstriction may react with little or no bronchoconstriction to a subsequent bout of exercise.[41,78,107] Such a patient is said to be in a "refractory period," which can last up to 2 hours following a prior episode of EIB.[41]

Patients and parents should be made aware of this phenomenon. A child who undergoes an episode of EIB can be reassured that subsequent activities (performed within the next 60–90 min) will trigger only a minor or no attack.

This refractoriness may be explained by depletion of mediators released from sensitized mast cells following the initial bout of exercise. The same mechanism may explain the phenomenon of "running through" asthma during a prolonged activity, in which the mediators are depleted during the first minutes of exercise.

Climatic Conditions and Asthma

Even when the type, intensity, and duration of exercise are standardized, EIB may not occur consistently in any given individual. The same child may respond to a certain task with marked EIB on one day and have little EIB on another. One possible cause for such inconsistency is the change in climatic conditions.

Climate and the Resting Asthmatic. Asthmatics often volunteer the information that their asthma is more severe on cold or dry days. Epidemiologic studies[40,74,75,153] and anecdotal data[24,146] have suggested a greater incidence of asthmatic attacks on cold days, especially during fall and winter. Controlled experiments have shown that in asthmatics, but not in healthy individuals, exposure to a cold shower[29] or inhalation of cold air[109,139] induced an increase in airway resistance and a drop in $FEV_{1.0}$. By contrast, inhalation of saturated air at 37°C can prevent bronchoconstriction even while the individual is exposed to a cold shower.[81]

It therefore seems that direct cooling of the airways, rather than generalized body cooling, triggers bronchoconstriction in the resting asthmatic. On the other hand, changes in air humidity do not seem to affect the asthmatic while at rest.[14,55]

Climate and EIB. It was first noted in the mid 1970s that air humidity and air temperature affect post-exertional bronchoconstriction in the following manner:

1. In neutral temperature (23–25°C), dry inspired air is more asthmogenic than humidified air (Fig. 3.3).[14,159]
2. Cold air is more asthmogenic than air at neutral temperature.[149]
3. Warming dry air to 30°C will partially protect against EIB.[28]
4. Saturated air at 37°C renders almost complete protection against EIB.[28]

Later studies have confirmed the above findings.[7,8,15,36,37,39,78,100,148] The clinical implications are twofold: first, a clinician wishing to use exercise in the evaluation of asthmatics should attempt to monitor and, if possible, control climatic conditions in his office—especially air humidity. Second, it is reasonable to assume that asthmatic children will be more susceptible to EIB on a cool day than during warm weather. Clinical experience supports this assumption.[124] Precautions should also be taken during spells of dry weather (a common phenomenon in some geographic regions—called Fohn, Sirocco, Chinook, or Hamsin). It is advisable to intensify drug therapy and, if this is of no avail, to curtail physical activity of asthmatic children whenever such weather is encountered.

Mechanisms Underlying EIB

The possible mechanisms by which exercise induces bronchoconstriction can be divided into: 1) those physical or chemical triggers that initiate a

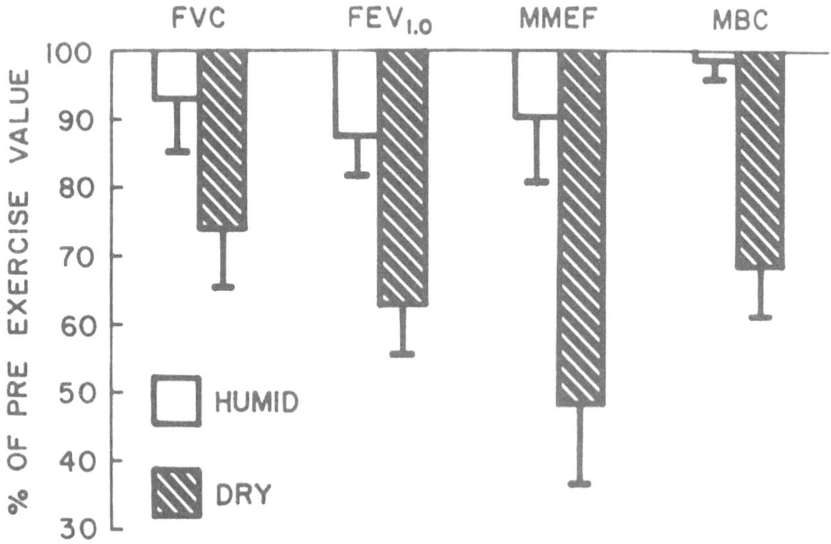

Figure 3.3. Air humidity and EIB. Pulmonary functions of 10 6- to 14-year-old asthmatic girls and boys measured 10 min after each of two treadmill walks. The children were free-breathing in a climatic chamber with air temperature of 25–26°C and humidity 25% ("dry") or 90% ("humid") relative humidity. FVC = forced vital capacity, $FEV_{1.0}$ = forced expiratory volume in first second, MMEF = maximal mid-expiratory flow, and MBC = maximal breathing capacity. Vertical lines denote 1 S.E.M. Based on data by Bar-Or et al.[14] Reproduced with permission from Bar-Or.[12]

physiologic response, and 2) the pathways by which such a response affects the bronchial smooth muscle.

Possible Triggering Stimuli. A variety of stimuli have been implicated as possible triggers of EIB. These include hypocapnia,[49] metabolic acidosis,[155] hypoxemia,[9] hyperpnea,[27,80] imbalance between α and β sympathetic receptors,[143] increased noradrenaline activity,[76] cooling of the facial skin,[90] and cooling of the airways.[122] Reviews of these possible mechanisms are available elsewhere.[66,104,134] The following discussion will focus on airway cooling, which currently is viewed by most investigators as the prime trigger of EIB.

Airway Cooling as a Trigger. As discussed above, either dry air or cool air aggravates EIB. The common denominator for these two air properties is that they both cool the respiratory mucosa—cool air by convection and dry air by evaporation. An example of such cooling during exercise is shown in Fig. 3.4. Note that the degree of cooling is similar in asthmatics and healthy controls who exercise in the same environmental conditions.

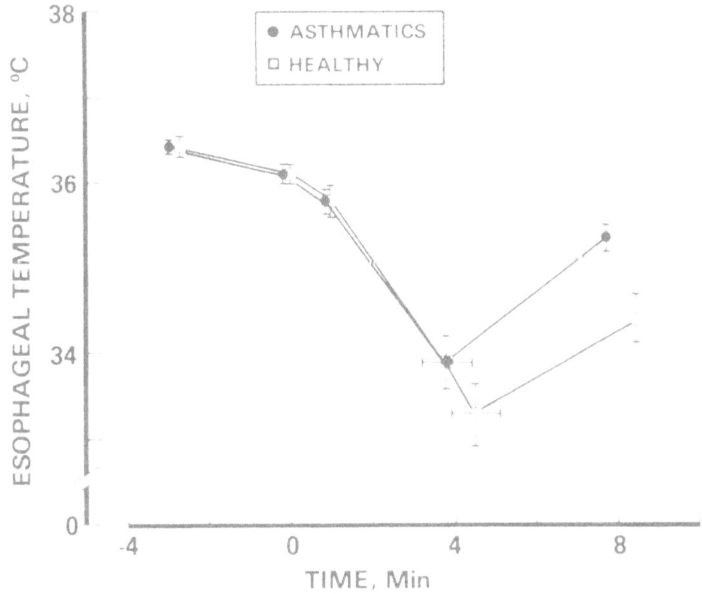

Figure 3.4. Changes in airway temperature due to exercise hyperpnea. Young adult asthmatics and healthy controls cycled at moderate intensity while inhaling dry air at −16 to −18°C. Esophageal temperature was measured by a thermistor at the level of the tracheal carina, to estimate airway temperature at that level. Zero time denotes start of exercise, which lasted for about 4 min. Vertical and horizontal lines are 1 S.E.M. Adapted from Deal et al.[38]

Air at a lower temperature than the airway mucosa will cool it by *convection* as per the equation:

$$HL_c = \dot{V}_E\, HC(T_E - T_I)$$

in which HL_c is heat loss in kiloJoules per minute; $\dot{V}_E$ = pulmonary ventilation (BTPS) in liters/min; HC = heat capacity (the product of specific heat and density) of air, which equals 0.00127 kJ/liter × °C; T_E and T_I = expiratory and inspiratory air temperature respectively in degrees centigrade, measured at the mouth. The factors that determine convective cooling are $\dot{V}_E$ and the temperature gradient between expired and inspired air, the former being close to body core temperature.

Inspired air that is not saturated by water vapor will cause mucosal cooling by *evaporation,* as per the equation:

$$HL_e = \dot{V}_E\, HV(WC_E - WC_I)$$

in which HL_e is evaporative heat loss in kiloJoules per minute, HV = latent heat loss of vaporization of water, which equals 2.43 kJ/g; WC_E and WC_I = expiratory and inspiratory water vapor content, respectively, in grams per liter. Evaporative cooling increases with the rise in gradient between the expired and inspired air and with increase in ventilation.

By combining the above two equations one obtains the overall respiratory heat loss (RHL) as follows:

$$\text{RHL} = \dot{V}_E\,[HC(T_E - T_I) + HV(WC_E - WC_I)]$$

The evaporative loss is ordinarily the dominant one in the overall RHL. For example, to induce cooling of 0.006 kJ/liter of air it is sufficient to dry the inspired air (at 37°C) from 100 to 50% relative humidity. To obtain the same RHL without drying the air, one will have to cool it from 37 to −10°C!

The increase in RHL during exercise is achieved primarily by the hyperpnea (increase in pulmonary ventilation) and, to a negligible extent, by an increase in T_E. Ventilation can rise as much as ten-fold in most active children, increasing RHL by that factor. Some investigators have gone as far as stating that the *only* role of exercise in the genesis of EIB is to increase pulmonary ventilation,[37] all other changes such as hypocapnia, acidosis, hormonal changes, or hypoxemia being immaterial.

There is a linear relationship between the degree of bronchoconstriction and respiratory heat loss,[28] as shown in Fig. 3.5. Such a relation-

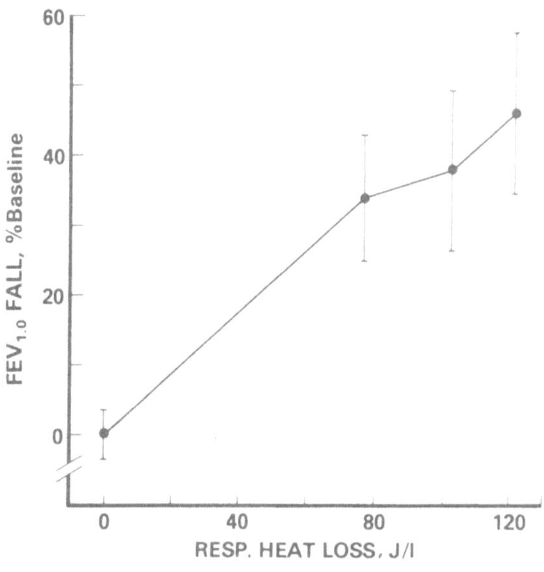

Figure 3.5. EIB as a function of respiratory heat loss. Eight asthmatic adolescents and young adults walked on a treadmill at four different combinations of air humidity and temperature. $FEV_{1.0}$ fall = the post-exercise decrease in forced expiratory volume at the first second, expressed in percentage of pre-exercise level. Mean ±1 S.E.M. Data by Chen and Horton.[28]

ship holds true for a wide range of inspired air temperatures (−10 to +80°C) and levels of humidity (0–100% relative humidity).[37,39] There is, however, a wide scatter in the degree of respiratory heat loss that causes a given level of EIB in asthmatic children.[7] There are also differences in response to a given heat loss between children and adults. At a given respiratory heat loss, children respond with greater EIB than do adults, as shown in Fig. 3.6. The reason may be that an identical absolute pulmonary ventilation is *relatively* greater for the child, who has a smaller area of respiratory mucosa.

It has been suggested that the protective nature of swimming is due to the highly humid air that the swimmer inhales at water level.[14] This possibility has been tested by letting asthmatic children swim while inhaling dried air.[15,84] As shown in Fig. 3.7, drying the air (i.e., increasing the respiratory heat loss) did generate bronchoconstriction in the swimming asthmatic, but the degree of EIB was lower than in running, even though equal levels of heat loss were achieved. It thus seems apparent that some unidentified mechanism, in addition to the protective humid air, renders swimming less asthmogenic than other sports.

In conclusion, the role of airway cooling in triggering EIB is now universally accepted. However, it may not be the *exclusive* trigger. Recent observations have raised the possibility that osmotic changes at the respiratory tract mucosa can also trigger EIB.[7]

Vagal Pathways vs. Chemical Mediators. While airway cooling by exercise hyperpnea is agreed upon as the major trigger of EIB, there are two schools of thought regarding the events that follow. One theory assumes propagation by the vagus from receptors located in the oropharynx, trachea, and large bronchi.[106,167] The other assumes a release from mucosal sensitized mast cells of chemical mediators (e.g., histamine, bradykinin, neutrophil chemotactic factor of anaphylaxis, prostaglandins, slow reacting substance of anaphylaxis).[41,66,107] Evidence for both theories is circumstantial, based on response to drugs, *in vitro* findings, or mechanisms found at rest; or on the presence of a refractory period to EIB following prior exercise. A suggestion has been made that *both* mechanisms may be in effect: obstruction of small airways may be a result of mediator release, whereas large airway obstruction is generated through vagal reflex pathways.[105]

Other questions, still unanswered, are why EIB appears *following* and not *during* exercise when cooling is at its peak (see Fig. 3.4) and why asthmatics respond to airway cooling with greater bronchoconstriction than do healthy individuals. The absence of bronchoconstriction during exercise could be due to the bronchodilatory effect of catecholamines and to the sympathetic drive, both of which are increased during exercise and subside during recovery.[66] The greater sensitivity of asthmatics to airway cooling represents their greater nonspecific airway reactivity,

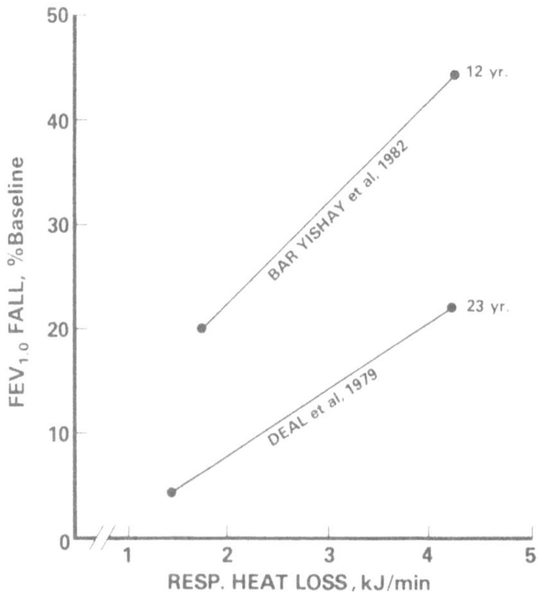

Figure 3.6. Effect of age or body size on the relationship between EIB and respiratory heat loss. A comparison of post-exertional fall in forced expiratory volume in the first second ($FEV_{1.0}$) between children and adults. Data from Bar Yishay et al,[15] and Deal et al.[39]

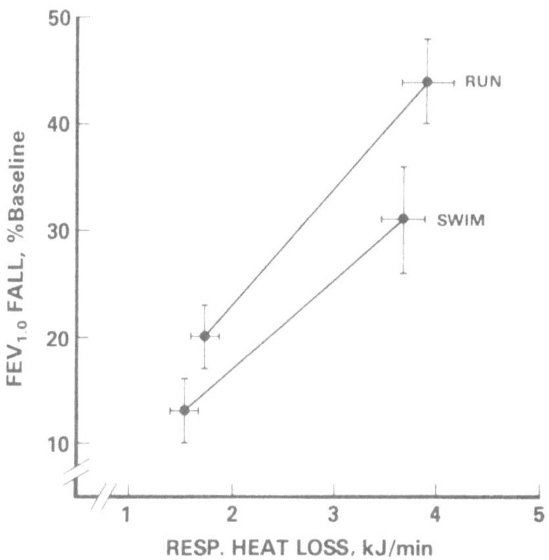

Figure 3.7. The protective nature of swimming against EIB as a function of respiratory heat loss. Thirteen adolescent asthmatics ran and swam at neutral temperature while inhaling dry or humid air. $FEV_{1.0}$ fall = post-exercise decrease as percentage of resting $FEV_{1.0}$. Mean ±1 S.E.M. Data of Bar Yishay et al.[15]

rather than an inability to condition the inspired air. It is possible, however, that those asthmatics who also suffer from allergic rhinitis must resort more to oral breathing (with a resultant greater airway cooling) than individuals with more patent nasal airways.

Other Responses of Asthmatics to Exercise

Most asthmatic children respond to exercise with adequate ventilatory and hemodynamic changes. In fact, *during* exercise they show greater bronchodilation than do nonasthmatics.[46,103] While pulmonary ventilation is appropriate to the metabolic level of the mildly and moderately affected patients,[46,48] it is excessively elevated in those severely affected.[17,61] Such high pulmonary ventilation is associated with high tidal volume. The work of breathing in these severely affected patients is abnormally high[16,61] since pleural pressure swings during the respiratory cycle are exceedingly high and the mean pleural pressure highly negative.[145] Alveolar gas exchange and lung diffusion capacity are usually normal in the exercising asthmatic.[48,57,61,71] However, children who develop marked post-exertional bronchoconstriction may have a rise in alveolar − arterial Po_2 difference and a mild O_2 desaturation.[92]

Heart volume and blood volume are normal in the asthmatic child,[16,72] as are cardiac output and stroke volume during submaximal and maximal exercise.[57,72] It has been suggested[16,157] that asthmatic children resort more than healthy ones to anaerobic metabolism during submaximal exercise, as manifested by higher levels of blood lactate. Neither the mechanism nor the implications of this phenomenon have been elucidated.

In conclusion, respiratory and circulatory functions alike are adequate in most exercising asthmatics. Only severely affected patients have respiratory deficiencies that affect their O_2 transport system.

Habitual Activity of the Asthmatic Child

Exercise is often equated with physical distress by the child with asthma—perhaps more than in any other chronic disease. An attack of EIB is alarming to the child and onlookers alike. Parents tend to become overprotective[120,130] and limit the physical activity of the child, who is progressively perceived as a "loser" by himself and his peers.

While systematic studies are not available on the habitual activity of asthmatics, some observations are noteworthy:

1. In the United States asthma is associated with limited physical activity in 30% of children and adolescents who have chronic lung disease (U.S. DHEW, 1973, quoted in Ref. 158).

2. Among asthmatic children who attended an exercise therapy clinic, only 30% were taking part regularly in physical education classes at school. Fifty-five percent had limited participation (avoiding mostly endurance-type and outdoor activities) and 15% did not participate at all.[82]
3. In a survey performed among 262 United States physical education instructors only 52% stated that asthmatic children consistently attended regular physical education classes. Forty-five percent reported that such pupils attended these classes intermittently.[158]
4. In the same survey, 76% of instructors allowed the asthmatic children to determine their own activity level in class.
5. Attendance at physical education classes is inversely related to the severity of asthma. In contrast, participation in leisure-time activity is independent of the severity of the disease.[71]

These findings show a high degree of absenteeism from physical education classes among asthmatics. Their teachers, perhaps due to insufficient knowledge or fear of medico-legal consequences, let the children select their own level of exercise rather than guide them to appropriate activities.

Physical Working Capacity of the Asthmatic Child

Fitness of asthmatics has ranged from low to Olympic championship levels. Factors that affect such variability include: the severity of the disease (often defined as the number of major attacks per year), the presence or absence of bronchoconstriction on the day of testing, and, especially, the level of habitual activity. Most studies in this area are of selected groups of volunteer patients and are subject to sample bias. Only seldom[135] are they conducted with a representative sample of the population.

In nonathletic asthmatics, maximal aerobic power ranges from normal[16,71,157] to moderately low.[62,119,135,156] Other fitness components such as muscle strength and sprinting ability are within the normal range.[135] The short stature of some children with severe asthma does not seem to interfere with their fitness.

There is a relationship between the sedentary lifestyle of many asthmatics and their low working capacity. The ability of such children to achieve normal maximal aerobic power through conditioning strongly suggests that low fitness is not inherent to their disease. An impressive number of asthmatics have reached world class level in sports, especially in swimming—where they have often won Olympic gold medals—but also in more "asthmogenic" sports such as long-distance running or cycling.[50]

Exercise as a Diagnostic Tool in Asthma

Rationale for Exercise Provocation Tests. Exercise testing of asthmatic children is indicated for the following:

Documentation of EIB. This is especially important for the child who presents with less common exercise-related complaints such as chest pain or cough. One should bear in mind that some asthmatics develop EIB during spontaneous activities but not in the laboratory, and others have the reverse pattern.

Evaluation of Medication for EIB. A standardized challenge test, repeated using different drug regimens, is highly informative as to the efficacy of a specific drug or adequacy of a certain dose. One must realize, however, that such information is specific to the effects of drugs on EIB and not on asthma in general.[66]

Diagnosis of Hyperreactive Airways. Exercise can be used as a nonspecific challenge test to determine asthma in patients with atypical respiratory symptoms, or in those who have long symptomless periods. This is analogous to challenge by cold or to inhalation of histamine or metacholine. The sensitivity of exercise challenge for the above purpose is somewhat lower than of histamine challenge.[108]

Comparison of Asthmogenicity of Different Activities. Some children report EIB following certain nonaquatic activities but not others. These activities can be simulated in the laboratory.

Determination of Exercise Tolerance. As in other diseases, this is an important aspect to consider before constructing an exercise program. A detrained child may hyperventilate at a relatively low power load, with a resultant high respiratory heat loss and EIB.

Instilling Confidence in Patient and Parent. This benefit of testing is often underestimated. Both parent and child can gain confidence from the *revelation* that the child can sustain high-level activity.

Assessment of the Emotional Component. It has been suggested[69] that asthmatic children who markedly benefit from the placebo effect of a drug on EIB have a major emotional component to their disease. This hypothesis needs further confirmation.

The Exercise Challenge. There are two conflicting considerations in administering an exercise challenge test to the asthmatic child: 1) optimizing the provocation so that maximal bronchial response is obtained, while 2) ensuring that the test is not detrimental to the child's health. The following guidelines take these considerations into account.

Pretest Preparation

1. Obtain medical history with emphasis on the cardiopulmonary system, habitual activity, and exercise-related symptoms.
2. Perform physical examination with emphasis on the cardiopulmonary system.
3. Define for yourself the rationale for exercise testing in each patient.
4. To alleviate their apprehension, explain to patient and parent the purpose and form of the test. Let the child become familiar with the procedures and equipment.
5. Obtain baseline pulmonary functions. For clinical purposes, sufficient information can be obtained with peak flowmeter or, preferably, a positive displacement one-breath spirometer.
6. *Do not proceed* with an exercise test if $FEV_{1.0}$ or peak expiratory flow are less than 60% of height-predicted level, or if the child is dyspneic at rest.[45] Pre-exercise wheezing *per se* is not a contraindication.

Withdrawal of Medication. Pretest withdrawal of anti-asthma drugs depends on the purpose of the test. If one wishes to maximize the response (e.g., to document EIB or airway hyperreactivity) all drugs, with the exception of corticosteroids, should be withdrawn. β_2-sympathomimetics, methylxanthines, and anticholinergics should be discontinued at least 8 hours prior to testing (12 hours for long-acting methylxanthines). Disodium cromoglycate should, ideally, be withdrawn 24 hours before exercise testing.[45]

If, on the other hand, the exercise test is conducted to assess efficacy of medication, withdrawal of drugs will be selective.

The Exercise Protocol. Unlike tests of physical working capacity that are of the multistage "progressive" type (see Appendix II, section entitled Prototypes of Exercise Tests), an exercise challenge test for EIB should comprise a single 8-min stage. The intensity should be such that heart rate reaches 85–90% of maximum (170–180 beats/min for children and adolescents).

When the child is tested for the first time, the exact power load that raises heart rate to this level is not known. The investigator must then select an arbitrary initial level and, based on interim heart rate, modify it within 2–3 min until the target heart rate is reached (see description of the single-stage with adjustments protocol in Appendix I). A test is technically successful if the required load is kept for *at least* 5 min. Heart rate should be monitored periodically by ECG or cardiotachometer. The former is preferable because it can disclose coincidental dysrhythmia or other electrocardiographic abnormalities.

The ergometer of choice is a treadmill, although a cycle ergometer or a step can also be used. The main advantage of treadmill running is for young children, who can master the technique and sustain the effort better than with other ergometers. Furthermore, test–retest reliability seems higher with treadmill than with cycle ergometer.[4] Some investigators advocate free running in or around the hospital or up and down a flight of stairs. Although such tasks are asthmogenic, they cannot be sufficiently standardized and are not recommended.

To safeguard standardization of an exercise challenge test, the following should be adhered to:

1. Avoid exertion for at least 2½–3 hours prior to testing. Failure to do so may lead to less than full bronchial response, as the child will be in a refractory period following prior activity.[41]
2. Climatic conditions, especially humidity, should be standardized. The use of commercially available air conditioners, humidifiers, and dehumidifiers is sufficient for clinical purposes.
3. When a test must be repeated, do not allow an interval shorter than a day or longer than a week. Test-retest variability at longer intervals is high.
4. To minimize diurnal variations attempt to retest any given child at the same time of the day.

Pulmonary Function Tests. Baseline testing is done just prior to exercise. If a certain drug is evaluated, a preliminary function test should precede its administration. Testing is repeated at 2, 5, and 10 min post-exercise and every 5 min thereafter until EIB subsides.

Interpretation of Post-exertional Bronchoconstriction. Assuming X = pre-exercise baseline and Y = the lowest post-exercise level, the degree of EIB can be expressed in a number of ways. The most common one is:

$$\text{EIB} = \frac{X - Y}{X} \times 100 \qquad (1)$$

i.e., the maximal drop in function as a percentage of pre-exercise function. Another expression of EIB is the degree of drop as a percentage of the height-predicted (P) function:

$$\text{EIB} = \frac{X - Y}{P} \times 100 \qquad (2)$$

This approach is more relevant to the *clinical* status of the patient prior to the test. He may have, for example, a small drop according to equation (1) but, due to low X, the drop in function may be sufficient to induce a major obstruction and discomfort. A third approach is to express EIB in absolute terms and not as a percentage of resting value:

$$\text{EIB} = X - Y \qquad (3)$$

In assessing the effect of a drug one should separately evaluate the pre-exercise bronchodilatory effect (see Drug Therapy in EIB, below) and the amelioration of EIB. The bronchodilatory effect is assessed by comparing the pre-drug and post-drug functions, both measured before exercise. To assess the affect on EIB some investigators use the pre-drug and others the post-drug resting value as a baseline.

Another consideration in evaluating anti-EIB drugs is their placebo effect, which is often very marked.[8,69,138,144] While relevant to research, this aspect also has clinical implications. To assess a placebo effect one must test the child twice, using drug and placebo in a blindfold fashion. Various mathematical approaches are available for such a comparison (for details see Ref. 86).

The criterion for a "positive exercise test" (i.e., EIB of a clinically significant degree) is not uniform among laboratories. Using equation (1) it ranges between 10 and 25%. Most workers in the field, however, advocate a 10 or 15% maximal drop as their criterion. Our own criterion is 15%.

Management of the Child with EIB

In 1970 the Committee on Children with Handicaps of the American Academy of Pediatrics[1] stated: "One must attain a balance between the needs of the asthmatic child to participate in the activities with as little restriction and emotional crippling as possible and the necessary limitations to living a full life." Today, with the variety of means at his disposal, the physician can all but eliminate these limitations.

Drug Therapy. With currently available anti-EIB medication, most asthmatic children can lead an active life. Treatment can be initiated as prophylaxis prior to exercise, or once EIB has begun. For obvious reasons the prophylactic approach is preferable. Children often do not plan their activities in advance, however, and they may resort to medication only after symptoms have begun.

Prophylactic Therapy. The most effective drugs in current use are the β_2-sympathomimetics, administered in aerosol. Another potent and widely used drug is disodium cromoglycate (DSCG), taken as inhaled powder. Also in use are some methylxanthines and anticholinergics.

β_2-sympathomimetics (e.g., salbutamol aerosol 0.2 mg per metered puff) are effective in preventing or ameliorating EIB in some 90–100%

of asthmatic children,[5,67] whether or not they are symptomatic prior to exercise. Such an aerosol is effective almost instantaneously, so it can be inhaled just prior to activity. Furthermore, due to its bronchodilatory effect and its direct action at the airways, it can relieve asthmatic symptoms *before* the start of the activity and potentiate the bronchodilatory changes *during* exercise.[143] A β_2-sympathomimetic aerosol (like other aerosols) has fewer systemic side effects than oral preparations, and its action does not depend on gastrointestinal absorption. The standard therapeutic dose for aerosol salbutamol is only 1/20 that of oral salbutamol.[5] For those children who do not get satisfactory protection, the dose can be increased (e.g., two puffs). The protective effect of the β_2-sympathomimetic aerosols lasts 2–6 hours, depending on the specific preparation.[6,132,143]

Although aerosol is preferable to the oral route, the latter is also useful if taken in high dose (salbutamol 0.15 mg/kg body weight). It is indicated for patients who, for some reason, cannot use aerosols.[56] In regular doses (salbutamol 0.10 mg/kg), however, the oral route is ineffective for children.[5]

Inhaled DSCG, unlike β_2-sympathomimetics or methylxanthines, does not cause bronchodilation at rest but is a specific EIB prophylactic agent.[67,113,142] Its use is effective in some 70–80% of asthmatic children, and it is especially indicated for patients with mild to moderate asthma who are asymptomatic prior to exercise. Although the regular pre-exercise dose is a capsule of 20 mg crushed into an inhaler, larger doses can be tried. The optimal time for administration is some 30 min before exercise, but it can also be taken effectively just prior to the activity.[138] DSCG is highly effective for up to 2 hours,[96] although some effect can still be detected for 4 hours. It is ineffective once exercise has begun.[138] To maximize its use in prevention of EIB, a dose of DSCG should be taken before exercise *in addition* to the regular daily dose (unless exercise happens to occur soon after the inhalation of a regular dose). DSCG is also administered in a nebulized solution, but the oral route is ineffective for prevention of EIB.[33] Children whose EIB cannot be prevented by either β_2-sympathomimetics or DSCG alone can sometimes benefit from a combination of both.

Based on *in vitro* data, DSCG apparently stabilizes mast cell membrane and prevents the release of mediators from the sensitized mast cells. There is, however, indirect *in vivo* evidence that DSCG may act through other mechanisms.[47]

Methylxanthines (especially theophylline), while commonly prescribed for the general control of asthma, are less often used for specific prevention of EIB. The major reason is their oral administration, which does not offer quick response. Like β_2-sympathomimetics they are potent bronchodilators at rest. They markedly reduce EIB in 65–80% of patients,[8,11,20,67] an effect which is dose-dependent,[21] minimal therapeutic concentration for theophylline being 10 g/ml.[125] There is a low correlation between the efficacy of methylxanthines in the general control of asthma and their prophylactic effect in EIB.[67] As with salbutamol[63] patients do not develop tolerance to theophylline following regular intake for a few weeks.[20]

Other drugs that have been tried for prevention of EIB include anticholinergics, such as ipratropium bromide aerosol;[8,162,163] calcium antagonists, such as nifedipine[26] or verapamil;[121] α adrenoreceptor blockers, such as indoramine;[18] prostaglandin antagonists, such as dextropropoxyphene;[8] or antihistamines such as ketotifen.[8]

Corticosteroids, whether oral or inhaled, are barely effective for the prevention of EIB,[96] even though they may potentiate the effect of other agents.[3]

Reversal of EIB. Once EIB has started, the drug of choice is a β_2-sympathomimetic aerosol such as salbutamol. While some children will respond to one puff, others may need two or three puffs, 3–4 min apart. Some patients respond better to nebulized salbutamol. The basic problem in reversing EIB is the inability of the distressed child to inhale properly. In addition, the narrowed airways may impede penetration of the drug even when inhalation is mechanically adequate. One way of inducing some bronchodilatation at that stage is to have the child exercise for about one minute and then inhale the drug.[3]

Nasal Breathing as Protection. The nose has long been recognized as a highly efficient conditioner of inspired air.[126] Nasal breathing can very effectively change the characteristics of the inspired air so that even in extremely low ambient humidity and temperature the air reaching the nasopharynx is humidified to near 100% relative humidity and its temperature is 32–37°C.[126,127] If nasal breathing were to be practiced by the asthmatic during exercise, the detrimental effect of cold air and dryness could be attenuated, or even abolished. Fig. 3.8 is a summary of a study showing that while oral breathing during a treadmill walk induces marked bronchoconstriction, an increase in airway resistance, and air trapping in the lungs, nasal breathing weakens these responses and, in practical terms, protects the child against EIB.[136] Similar effects have also been shown during free running[101] and during isocapnic hyperventilation at rest.[167]

The above data refer to mild or moderate exercise intensities. Although the feasibility of nasal breathing during strenuous exercise has yet to be determined experimentally, we do recommend that such a routine be practiced, at least during moderately intense activities. Patients with allergic rhinitis should be prescribed a nasal vasoconstrictor to increase nasal airway patency.

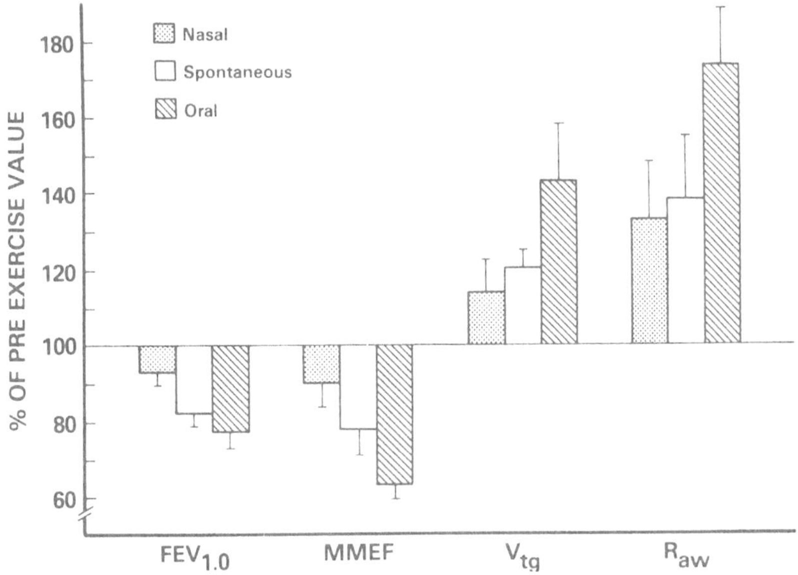

Figure 3.8. Post-exertional changes in pulmonary functions as related to the mode of breathing. Twelve 7- to 14-year-old asthmatics walked on three different days on a treadmill at 75–85% of maximal heart rate, using different inspiratory modes. Ambient conditions were 20–22°C and 25–30% relative humidity. Measurements were taken 7–9 min post-exercise. $FEV_{1.0}$ = forced expiratory volume at the first second, MMEF = maximal midexpiratory flow, V_{tg} = thoracic gas volume, R_{aw} = airway resistance. Vertical lines denote 1 S.E.M. Adapted from Shturman-Ellstein et al.[136]

Artificial Means for Warming and Humidifying Inspired Air. An alternative to nasal breathing is the use of a mouth mask to warm and humidify inspired air on cold days. This device, which is available commercially, causes mixing of inspired air with air from the preceding expiration, thus increasing its temperature and water vapor content. A disadvantage of this procedure is some increase in dead space. The efficacy of such a mask as protection against EIB has been shown experimentally.[129] Air temperature inside the mask was over 30°C during exercise, even when outside temperature was only 4°C. In the absence of a mask one can cover the mouth and nose with a scarf and obtain similar results.

Choosing the Right Activities. As described above, some activities are more asthmogenic than others. For most asthmatics swimming is the exercise of choice and should be strongly recommended, whenever feasible. Among land activities, preference should be given to the intermittent ones (1–3 min exercise bouts, interspersed by rest or, preferably, by mild exertion),[120] which are less asthmogenic than prolonged, continuous activities. Examples are circuit training, gymnastics, diving, or downhill skiing. Among team games water polo, American football, and baseball are suitable.

The importance of warming up as a means of preventing EIB has not been established experimentally. While a 3-min walk or jog was not found beneficial,[112] repeated 30-sec runs 10 min prior to an exercise task did reduce the fall in peak expiratory flow by 36%.[131] In spite of the lack of scientific proof, we and others[120] recommend a warm-up at the beginning of each exercise session. It should last 6–10 min, raising heart rate to not more than 150 beats/min.

When EIB starts, exercise can be used to reverse or even stop the attack. This is based on the *bronchodilatation* that takes place *during* exercise. A short bout (up to 1 min) of activity has been found effective in reversing EIB,[89] and it is especially effective in conjunction with aerosol therapy.[3]

Physical Conditioning and the Asthmatic Child

Two aspects of the effects of conditioning and training on asthmatic children are clinically relevant: improvement in exercise performance and reduction in asthmatic complaints, especially EIB.

Conditioning and Improvement of Fitness. Numerous studies have shown that most asthmatics are trainable and, through proper programs, can increase their maximal aerobic power, muscle strength, and other fitness components.[51,54,62,73,99,118,119,123] As in healthy children, the degree of improvement in work performance is related to the intensity of conditioning.[73] Patients with severe asthma may not improve their fitness by a conditioning program because they are often not able to withstand intense activity without triggering EIB.

The improved ability of asthmatics to participate in play and sports yields important psychosocial dividends. These include improved school attendance, increased acceptance by peers, improved self-confidence and sense of accomplishment, decrease in frequency and intensity of emotional upsets, and above all, the recognition by the asthmatic that sickness need not be a way of life.[1,10,82,120,130,150] A case of a young asthmatic child is instructive. This Australian girl decided to start swimming in order "to lick asthma." Some years later she became an Olympic champion and one of the most notable figures in swimming history.[52]

Conditioning and EIB. To the clinician, the key question is whether physical conditioning can improve asthmatic control. More specifically, can it prevent or reduce EIB? Early studies suggested that, in spite of improving fitness, conditioning does not reduce the risk of EIB.[54] Recent evidence shows, however, that once conditioned, the asthmatic re-

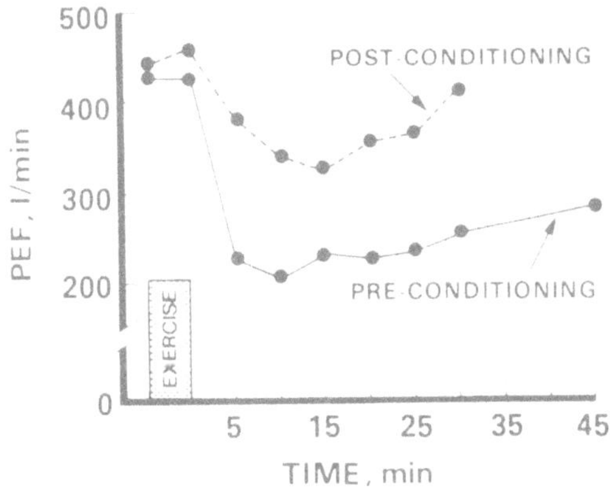

Figure 3.9. Physical conditioning and exercise-induced bronchoconstriction. Post-exercise peak expiratory flow (PEF) in a 13.5-year-old girl before and after a 14-day conditioning program. The exercise provocation was cycling at 87 Watts for 6 min. Reproduced with permission from Oseid and Haaland.[119]

sponds to a given power load with less bronchoconstriction.[77,79,111,119,151] Such an improvement is demonstrated in Fig. 3.9. Although the mechanism for such a change is not clear, the reduced EIB could be related to the reduction in submaximal pulmonary ventilation and in respiratory heat loss. Whatever the mechanism, physical conditioning seems to raise the threshold for the triggering of EIB in some patients. There is, however, no proof of a relationship between the *degree* of improvement in EIB and the degree of improvement in fitness. In fact, an asthmatic may reach top world standards in some sports, such as swimming, without any appreciable alleviation of EIB.[51] Nor do we know whether conditioning at a young age has any bearing on the long-range prognosis in asthma.

Ideally, asthmatic children should take part in regular physical activities at school. There are, however, those who may first need a short (4–6 weeks) specialized "initiation program" during which they can improve their fitness and regain self-confidence. As part of their own education, teachers should be instructed to allow asthmatic children to bring their medication to school and self-administer it, whenever needed.

Cystic Fibrosis (CF)

Causes of Deficient Physical Working Capacity

Exertional or post-exertional dyspnea is common among CF children with advanced lung damage. The exercise performance of such children is limited and is in direct relation to the clinical severity of their disease. Table 3.1 summarizes those pathophysiologic changes that are relevant to the exercise performance of the child with CF.

Airflow limitation results in reduced maximal breathing capacity at rest[68] and maximal pulmonary ventilation during exercise in the child

Table 3.1. Pathophysiologic Changes in Children with Advanced Cystic Fibrosis Which Affect Their Physical Working Capacity

1. Airflow limitation	Airway obstruction Low lung recoil "Hyperreactive" airways
2. Cardiovascular damage	Destruction and remodeling of pulmonary vascular bed Pulmonary induced heart disease—right and left

Based on Chipps et al,[30] Day and Mearns,[35] Godfrey and Mearns,[68] Mansell et al,[102] Mellis and Levison,[108] and Ryland and Reid.[128]

with advanced lung disease.[25] Healthy children use only 60–70% of their resting maximal breathing capacity during maximal exercise. In contrast, a child with advanced CF may reach, and even exceed, 100% of his maximal breathing capacity during such exercise, which suggests a deficient ventilatory reserve.[116] In healthy people the O_2 transport system during maximal exercise is limited by maximal cardiac output and by the energy turnover capability of the exercising muscles. In contrast, the patient with advanced CF is limited by his ventilatory capacity.

Airflow limitation induces an increase in dead space and a reduction in alveolar ventilation, as manifested by CO_2 retention.[25,60,68] Pulmonary ventilation during submaximal exercise is excessive, probably to compensate for the increased dead space. The disadvantage of such excessive ventilation is that a high proportion of O_2 uptake is diverted to the respiratory muscles rather than to those muscles that perform the exercise. This may lead to easy fatigability of the latter.

The high ventilatory demand of the child with CF acts as a training stimulus to the respiratory muscles, such that their endurance (i.e., ability to sustain a high level of ventilation for prolonged periods) is increased. In fact, ventilatory endurance in patients with CF is *higher* than in healthy controls.[94]

Destruction of lung parenchyma and of the capillary network is reflected by a reduced lung diffusion capacity at rest and, especially, by an inability to increase diffusion during exercise.[68,168] This pattern is shown in Fig. 3.10. The compromised lung performance results also in "wasted" ventilation and increased physiologic dead space.[68]

Another characteristic of the severely affected patient is arterial desaturation during maximal exercise.[25,32] It is unrelated to symptoms or signs other than cyanosis. As seen in Fig. 3.11, a mild deficiency in O_2 saturation is associated with a marked deficiency in exercise performance. It is unclear, however, to what extent O_2 desaturation is the *cause* of decreased performance in CF.

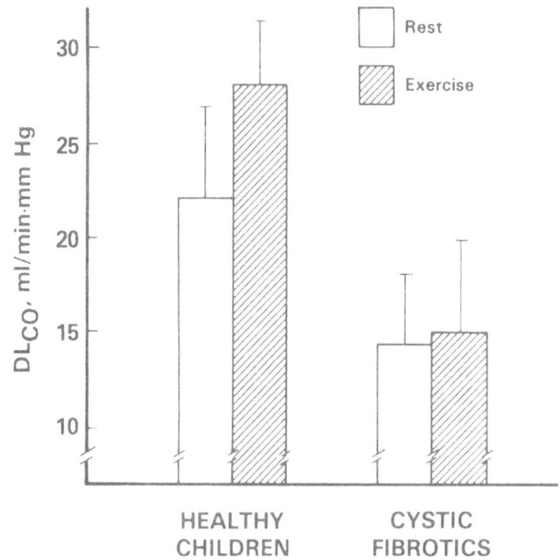

Figure 3.10. Lung diffusing capacity for carbon monoxide (DL_{CO}) in children with cystic fibrosis (n = 18) and in healthy controls (n = 8), 5–15 years old. A comparison between resting values in the sitting position and values obtained immediately after a maximal cycle ergometer exercise. Vertical lines denote 1 S.D. Adapted from Zelkowitz and Giamonna.[168]

Cystic fibrosis often leads to *cor pulmonale*, but can also impair left ventricular function. Exercise-induced dyspnea in advanced CF may not be due only to the lung disease but also to left ventricular dysfunction.[30]

In conclusion, exercise performance of the child with CF, although normal in the mildly affected patient, is limited when the disease is severe. Primarily affected arc those fitness components that tax the O_2 transport system. In contrast, performance of brief, intense "anaerobic"-type tasks such as jumps, sit-ups, or strength-related activities is not deficient[34] unless the child is inactive. The physiologic limiting factor in high-level exercise is not circulatory, but ventilatory, both airway and parenchymal functions being affected.

Exercise in Assessment of the Child

Exercise is not relevant for the actual diagnosis of CF. It is useful, though, for the following:

1. Identification of patients who respond with bronchoconstriction.
2. Evaluation of alveolar hypoventilation and O_2 desaturation during intense exercise.

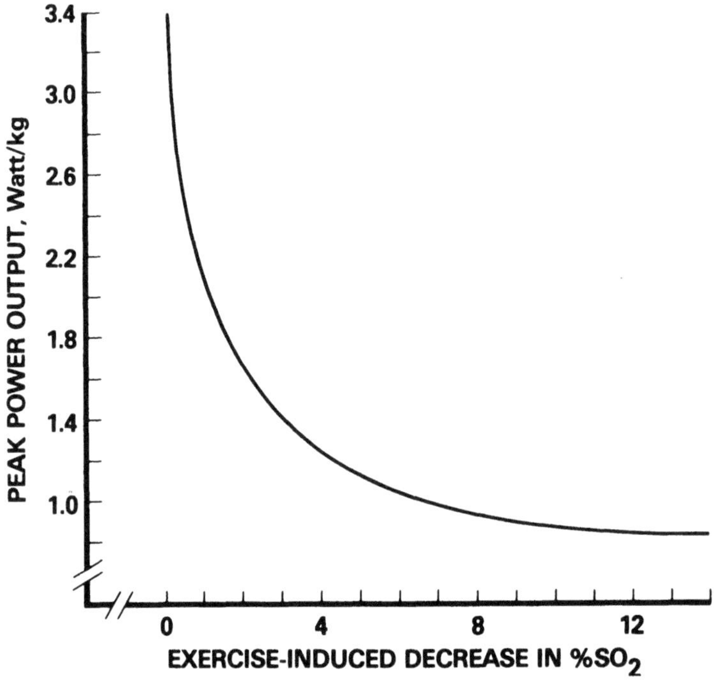

Figure 3.11. Arterial O_2 saturation (%So_2) as a limiting factor in physical working capacity (peak power output) of children with cystic fibrosis. Twenty adolescent patients underwent a progressive continuous maximal cycling task. %So_2 was measured at rest and continuously during exercise. Schematic adaptation from Cropp et al.[32]

Between 30 and 50% of CF patients respond to walking or running by a drop of more than 15% in peak expiratory flow.[35,166] Such a response cannot be predicted from resting pulmonary functions. Therefore, an exercise provocation test is warranted whenever the patient complains of exertional or post-exertional dyspnea.

Centers where exercise prescription is given as a therapy advocate an exercise-provocation test prior to the commencement of the program. Children who respond to the test with alveolar hypoventilation or with arterial O_2 desaturation are identified and given special attention during the ensuing program.[25] Such children should, at least initially, be given only mild activities.

An attempt has been made to use exercise provocation for identifying those CF patients who also have an atopic disease, by comparing pulmonary functions before and after exercise. Response to a treadmill test, however, was noncontributory,[166] and it seems that an exercise provocation test alone is of no differential diagnostic value in separating atopic from nonatopic patients with CF.

Some patients with moderate CF respond to exercise with a decrease in the left ventricular ejection fraction. At rest, the same patients do not show any impairment of left ventricular function.[30] Thus, an exercise test might be used to diagnose early stages of left cardiac involvement in CF. This finding needs further confirmation.

Beneficial Effects of Conditioning

The use of sports in the therapy of children with CF has only recently been evaluated and the following beneficial effects documented:

Improved clearance of mucus
Increased endurance of respiratory muscles
Reduced airway resistance
Improved exercise performance

During a swimming program, sputum volume was 15% higher on those days when the patients exercised.[165] A similar benefit was reported during a walk/jog program.[117,160] In fact, during a 17-day intense activity program, sputum clearance was adequate even though aerosol inhalation and chest physiotherapy had been discontinued.[164] In adults with chronic bronchitis, directly measured mucus clearance from the bronchi is greater during exercise than during postural drainage.[115] The mechanism of such enhanced clearance is unknown. More data are needed on the specific activities that help clear the bronchi, before exercise can be considered as a substitute for chest physiotherapy.

The endurance of the respiratory muscles of patients with CF (see above) was found to improve following programs of walking and jogging,[116] swimming and canoeing,[94] and specific training of the respiratory muscles.[94] As seen in Fig. 3.12, this improvement was greater than that found in healthy controls. Increased endurance of the respiratory muscles can help improve the exercise performance of children with CF. Whether such conditioning can help the child function better during exacerbation of the disease is not known.

Airway resistance decreases with conditioning, particularly with swimming.[164,165] There is no change, however, in lung volumes or in the residual volume : total lung capacity ratio.[116,164] The improved airway flow is short-lived. As seen from Fig. 3.13, improvement in $FEV_{1.0}$ and maximal midexpiratory flow following a 7-week swimming program disappeared 10 weeks after swimming was discontinued.

Children and adolescents with CF are trainable. Peak power output and maximal O_2 uptake increase with conditioning, especially in those with very low fitness at the start of the program.[116] Some can improve markedly in their athletic performance and even reach championship levels.[117]

An effective conditioning program does not have to push the child to

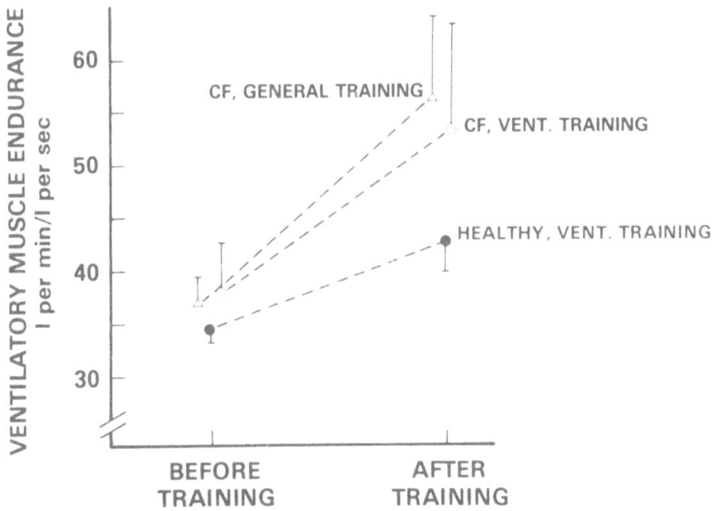

Figure 3.12. Effect of training on endurance of ventilatory muscles. CF adolescents underwent a 4-week training program that consisted of either specific ventilatory muscle training (n = 4) or swimming and canoeing (n = 7). Four healthy adults served as controls. The criterion for ventilatory muscle endurance was the maximal level of normocapnic ventilation that could be sustained for 15 min, divided by $FEV_{1.0}$. Data by Keens et al.[94]

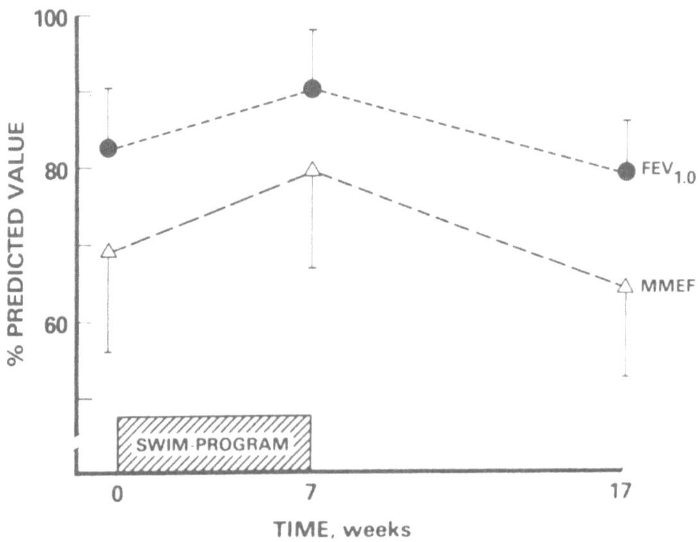

Figure 3.13. Durability of conditioning-induced changes in pulmonary functions of children with cystic fibrosis. Ten 6- to 18-year-old patients participated in a 7-week swimming program, two to three sessions per week. Forced expiratory volume at the first second ($FEV_{1.0}$) and maximal mid-expiratory flow (MMEF) are compared before the program, at its conclusion, and again 10 weeks later. Vertical lines denote 1 S.E.M. Data by Zach et al.[165]

his limit: exercise intensities that raise heart rate to 70–80% of maximum are sufficient.[116] Emphasis should be given to an individualized approach, keeping in mind the wide variability in pulmonary function, exercise capacity, nutritional status, and level of apprehension in these children.

The main, as yet unanswered question is whether improvement in exercise tolerance, in endurance of the respiratory muscles, and in general well-being can slow down deterioriation in clinical status.

Precautions to Be Taken During Exercise Programs

Some children with clinically advanced disease (e.g., 60 or less on the Taussig-NIH score) respond to intense exercise with arterial O_2 desaturation.[25,60] It has been suggested that in a child with an already compromised pulmonary function, such a response might be deleterious, inducing or aggravating pulmonary hypertension and right ventricular hypertrophy. With the concomitant hypoxemia, exercise might induce cardiac dysrhythmia.[32] Although the above detrimental effects are yet to be documented, one should practice caution when prescribing exercise programs to severely affected CF patients. Their activities should be mild and progression in intensity must be very gradual. Periodic monitoring by an exercise test is a recommended precaution for such children while they participate in exercise programs.

The increased salt loss through sweat in patients with CF does not seem to impair their exercise tolerance. However, as discussed in Chapter 10, these patients are highly prone to heat-related illness when the climate is warm and humid. Even though children with CF can thermoregulate efficiently during short bouts of exercise in the heat,[117] the effect of salt loss (and possibly fluid deficit) during *prolonged* exercise has not been evaluated.

We recommend that patients who exercise in a warm or humid climate be encouraged to drink above and beyond their thirst and to supplement their water intake with salt. Until experimental data are available, we recommend that salt intake be some 30% higher than that shown in the guidelines in Chapter 9, section entitled Guidelines for Conduct of Athletic Events in the Heat.

Interstitial Lung Disease

Response to exercise was tested in children and adolescents with interstitial lung disease of various etiologies, such as sarcoidosis, hemosiderosis, scleroderma, Hodgkin's disease,[58] or lymphoid interstitial pneumonia.[114] The degree of exercise-induced drop in arterial O_2 pressure was taken as the criterion for efficacy of the respiratory system. This drop was attributed to deficient lung-diffusing capacity and to decreased dynamic compliance of the lungs. This further suggests that respiration may be a limiting factor in the exercise capacity of children with diseased lungs, which is not the case in the healthy child.

Pulmonary Tuberculosis

Maximal aerobic power ($\dot{W}_{170}$) of children and adolescents with pulmonary tuberculosis who were treated in a sanatorium was lower than the height-predicted values, especially among the older adolescents.[154] The deficit in maximal aerobic power was related neither to the severity of the disease (X-ray criteria) nor to its duration. Apparently, the low habitual activity of these hospitalized patients was a major cause of their low physical working capacity.

References

1. American Academy of Pediatrics (Committee on Children with Handicaps). The asthmatic child and his participation in sports and physical education. Pediatrics 45:150–151, 1970.
2. Anderson SD: Physiological aspects of exercise-induced bronchoconstriction. PhD Thesis, University of London, 1972.
3. Anderson SD: Exercise-induced asthma: current views. Patient Management 15:43–55, 1982.
4. Anderson SD, Connolly NM, Godfrey S: Comparison of bronchoconstriction induced by cycling and running. Thorax 26:396–401, 1971.
5. Anderson SD, Pojer R, Smith ID, Temple D: Exercise-related changes in plasma levels of 15-keto-13, 14-dihydro-prostaglandin F_2 and noradrenaline in asthmatic and normal subjects. Scand J Resp Dis 57:41–48, 1976.
6. Anderson SD, Rozea PJ, Dolton R, Lindsay DA: Inhaled and oral bronchodilator therapy in exercise-induced asthma. Aust NZ J Med 5:544–550, 1975.
7. Anderson SD, Schoeffel RE, Follet R, et al: Sensitivity to heat and water loss at rest and during exercise in asthmatic patients. Eur J Respir Dis 93–105, 1982.
8. Anderson S, Seale JP, Ferris L, Schoeffel R, Lindsay DA: An evaluation of pharmacotherapy for exercise-induced asthma. J Allergy Clin Immunol 64:612–614, 1979.
9. Astin TW, Penman RWB: Airway obstruction due to hypoxemia in patients with chronic lung disease. Am Rev Respir Dis 95:567–575, 1967.
10. Backman A: Physiological and psychological aspects of the training of asthmatic children. In: Oseid S (ed.) The Asthmatic in Play and Sports. Pitman, London, in press.
11. Badiei B, Faciane J, Sly RM: Effect of theophylline, ephedrine and their combination upon exercise-induced airway obstruction. Ann Allergy 35:32–35, 1975.

12. Bar-Or O: Climate and the exercising child—a review. Int J Sports Med 1:53–65, 1980.
13. Bar-Or O: Climatic conditions and their effect on exercise-induced asthma. In: Oseid S (ed.) The Asthmatic Child in Play and Sport. Pitman, London, in press.
14. Bar-Or, O, Neuman I, Dotan R: Effects of dry and humid climates on exercise-induced asthma in children and preadolescents. J Allergy Clin Immunol 60:163–168, 1977.
15. Bar-Yishay E, Gur I, Inbar O, et al: Differences between swimming and running as stimuli for exercise-induced asthma. Eur J Appl Physiol, in press.
16. Bevegård S, Eriksson BO, Graff-Lonnevig V, et al: Circulatory and respiratory dimensions and functional capacity in boys aged 8 to 13 years with bronchial asthma. Acta Paediatr Scand Suppl 217:86–89, 1971.
17. Bevegård S, Eriksson BO, Graff-Lonnevig V, et al: Respiratory function, cardiovascular dimensions and work capacity in boys with bronchial asthma. Acta Paediatr Scand 65:289–296, 1976.
18. Bianco S, Griffin JP, Kamburoff PL, Prime FJ: Prevention of exercise-induced asthma by indoramin. Br Med J 4:18–20, 1974.
19. Bierman CW, Pierson WE: Summary—Symposium on exercise and asthma. Pediatrics 56:950–952, 1975.
20. Bierman CW, Shapiro GG, Pierson WE, Dorsett CS: Acute and chronic theophylline therapy in exercise-induced bronchospasm. Pediatrics 60:845–849, 1977.
21. Bierman CW, Shapiro GG, Pierson WE, Cho YW: Exercise-induced bronchospasm in asthmatic children as a dose-response model for theophylline. Int J Clin Pharmacol Biopharm 16:245–248, 1978.
22. Blackhall MI: Ventilatory function in subjects with childhood asthma who have become symptom free. Arch Dis Child 45:363–366, 1970.
23. Burr ML, Eldridge BA, Borysiewicz LK: Peak expiratory flow rates before and after exercise in schoolchildren. Arch Dis Child 49:923–926, 1974.
24. Bury JD: Climate and chest disorders (letter). Br Med J 4:613, 1972.
25. Cerny FJ, Pullano TP, Cropp GJA: Cardiorespiratory adaptations to exercise in cystic fibrosis. Am Rev Respir Dis 126:217–220, 1982.
26. Cerrina J, Denjean A, Alexandre G, et al: Inhibition of exercise-induced asthma by a calcium antagonist, nifedipine[1–3]. Am Rev Respir Dis 123:156–160, 1981.
27. Chang-Yeung MMW, Vyas MN, Grybowski S: Exercise-induced asthma. Am Rev Respir Dis 104:915–923, 1971.
28. Chen WY, Horton DJ: Heat and water loss from the airways and exercise-induced asthma. Respiration 34:305–313, 1977.
29. Chen WY, Horton DJ: Airways obstruction in asthmatics induced by body cooling. Scand J Respir Dis 59:13–20, 1978.
30. Chipps BE, Alderson PO, Roland J-MA, et al: Non-invasive evaluation of ventricular function in cystic fibrosis. J Pediatr 95:379–384, 1979.
31. Cropp GJA: The exercise bronchoprovocation test: standardization of procedures and evaluation of response. J Allergy Clin Immunol 64:627–633, 1979.

32. Cropp GJ, Pullano TP, Cerny FJ, Nathanson IT: Exercise tolerance and cardiorespiratory adjustments at peak work capacity in cystic fibrosis. Am Rev Respir Dis 126:211–216, 1982.
33. Dahl R, Henriksen JM: Effect of oral and inhaled sodium cromoglycate in exercise-induced asthma. Allergy 35:363–365, 1980.
34. Darby CW, Davidson AG, Desai ID: Muscular performance in cystic fibrosis patients and its relation to vitamin E. Arch Dis Child 48:72–75, 1973.
35. Day G, Mearns MB: Bronchial lability in cystic fibrosis. Arch Dis Child 48:355–359, 1973.
36. Deal EC Jr, McFadden ER Jr, Ingram RH Jr, Jaeger JJ: Effects of atropine on the potentiation of exercise-induced bronchospasm by cold air. J Appl Physiol: Respir Environ Exercise Physiol 45:238–243, 1978.
37. Deal EC Jr, McFadden ER Jr, Ingram RH Jr, Jaeger JJ: Hyperpnea and heat flux: initial reaction sequence in exercise-induced asthma. J Appl Physiol: Respir Environ Exercise Physiol 46:476–483, 1979.
38. Deal EC Jr, McFadden ER Jr, Ingram RH Jr, Jaeger JJ: Esophageal temperature during exercise in asthmatic and non-asthmatic subjects. J Appl Physiol: Respir Environ Exercise Physiol 46:484–490, 1979.
39. Deal EC Jr, McFadden ER Jr, Ingram RH Jr, et al: Role of respiratory heat exchange in production of exercise-induced asthma. J Appl Physiol: Respir Environ Exercise Physiol 46:467–475, 1979.
40. Derrick EH: The seasonal variation of asthma in Brisbane: its relation to temperature and humidity. Int J Biometeorol 9:239–251, 1965.
41. Edmunds AT, Tooley M, Godfrey S: The refractory period after EIA: its duration and relation to the severity of exercise. Am Rev Respir Dis 117:247–254, 1978.
42. Eggleston PA: Exercise-induced asthma in children with intrinsic and extrinsic asthma. Pediatrics 56:856–859, 1975.
43. Eggleston PA: Laboratory evaluation of exercise-induced asthma: methodologic considerations. J Allergy Clin Immunol 64:604–608, 1979.
44. Eggleston PA, Guerrant JL: A standardized method of evaluating exercise-induced asthma. J Allerg Clin Immunol 58:414–425, 1976.
45. Eggleston PA, Rosenthal RR, Anderson SA, et al: Guidelines for the methodology of exercise challenge testing of asthmatics (Study Group on Exercise Challenge, Broncho-Provocation Committee, American Academy on Allergy). J Allergy Clin Immunol 64:642–645, 1979.
46. Engström I, Karlberg P, Kraepelien S, Wengler G: Respiratory studies in children. VIII. Respiratory adaptation during exercise tolerance test with special reference to mechanical properties of the lungs in asthmatic and healthy children. Acta Paediatr 49:850–858, 1960.
47. Fanta CH, McFadden ER Jr, Ingram RH Jr: Effects of cromolyn sodium on the response to respiratory heat loss in normal subjects. Am Rev Respir Dis 123:161–164, 1981.
48. Feisal KA, Fuleihan FJD: Pulmonary gas exchange during exercise in young asthmatic patients. Thorax 34:393–396, 1979.
49. Fisher HK, Hatton P, Buxton RStJ, Nudel JA: Resistance to breathing during exercise-induced asthma attacks. Am Rev Respir Dis 101:885–896, 1970.

50. Fitch KD: Exercise-induced asthma and competitive athletics. Pediatrics 56:942–943, 1975.
51. Fitch K: Swimming medicine and asthma. In: Eriksson B, Furberg B (eds.) Swimming Medicine IV. University Park Press, Baltimore, 1978, pp. 16–31.
52. Fitch KD, Godfrey S: Asthma and athletic performance. JAMA 236:152–157, 1976.
53. Fitch KD, Morton AR: Specificity of exercise in exercise-induced asthma. Br Med J 4:577–581, 1971.
54. Fitch KD, Morton AR, Blanksby BA: Effects of swimming training on children with asthma. Arch Dis Child 51:190–194, 1976.
55. Fontana VJ, Fost A, Rappaport I: Effects of rapid change in humidity on pulmonary function studies in normal and asthmatic children in a controlled environment. J Allergy 43:16–21, 1969.
56. Francis PWJ, Krastins IRB, Levison H: Oral and inhaled salbutamol in the prevention of exercise-induced bronchospasm. Pediatrics 66:103–108, 1980.
57. Friedman M, Kovitz KL, Miller SD, et al: Hemodynamics in teenagers and asthmatic children during exercise. J Appl Physiol: Respir Environ Exercise Physiol 46:293–297, 1979.
58. Gaultier CL, Buvry A, Boule Y, et al: Epreuve d'effort chez 16 enfants atteints d'une maladie interstitielle. Bull Eur Physiopath Resp 15:409–411, 1979.
59. Gerhard H, Schachter EN: Exercise-induced asthma. Postgrad Med 67:91–102, 1980.
60. Germann K, Orenstein D, Horowitz J: Changes in oxygenation during exercise in cystic fibrosis (abstract). Med Sci Sports Exercise 12:105, 1980.
61. Geubelle F, Dechange J, Louis I, Beyer M: Respiratory function, energetic metabolism and work capacity in boys with asthma syndrome. Acta Paediatr Belg 31:79–86, 1978.
62. Geubelle F, Ernould C, Jovanovich M: Working capacity and physical training in asthmatic children, at 1800 m altitude. Acta Paediatr Scand Suppl 217:93–98, 1971.
63. Gibson GJ, Greenacre JK, König P, et al: Use of exercise challenge to investigate possible tolerance to beta-adrenoreceptor stimulation in asthma. Br J Dis Chest 72:199–206, 1978.
64. Godfrey S: Exercise Testing in Children. Applications in Health and Disease. W.B. Saunders, Philadelphia, 1974.
65. Godfrey S: Exercise-induced bronchial lability in wheezy children and their families. Pediatrics 56:851–855, 1975.
66. Godfrey S: Exercise-induced asthma. Review article. Allergy 33:229–237, 1978.
67. Godfrey S, König P: Inhibition of exercise-induced asthma by different pharmacological pathways. Thorax 31:137–143, 1976.
68. Godfrey S, Mearns M: Pulmonary function and response to exercise in cystic fibrosis. Arch Dis Child 46:144–151, 1971.
69. Godfrey S, Silverman M: Demonstration by placebo response in asthma by means of exercise testing. J Psychosom Res 17:293–297, 1973.

70. Godfrey S, Silverman M, Anderson S: The use of treadmill for assessing EIA and the effect of varying the severity and duration of exercise. Pediatrics 56:893–899, 1975.
71. Graff-Lonnevig V: Cardio-respiratory function, aerobic capacity and effect of physical activity in asthmatic boys. MD Thesis, Karolinska Institute, Stockholm, 1978.
72. Graff-Lonnevig V, Bevegård S, Eriksson BO: Cardiac output and blood pressure at rest and during exercise in boys with bronchial asthma. Scand J Respir Dis 60:36–43, 1979.
73. Graff-Lonnevig V, Bevegård S, Eriksson BO, et al: Two years' follow-up of asthmatic boys participating in a physical activity program. Acta Paediatr Scand 69:347–352, 1980.
74. Greenberg L, Field F, Reed JI, Erhardt CL: Asthma and temperature change. Arch Environ Health 8:642–647, 1964.
75. Greenberg L, Field F, Reed JI, Erhardt CL: Asthma and temperature change. II—1964 and 1965 epidemiological studies of emergency clinic visits for asthma in three large New York City Hospitals. Arch Environ Health 12:561–563, 1966.
76. Griffiths J, Leung FY, Grzybowski, Chan-Yeung MMW: Sequential estimation of plasma catecholamines in exercise-induced asthma. Chest 62:527–533, 1972.
77. Grilliat JP, Viniaker H, Vaillandet M, Ohlsson MG: Réadaptation des asthmatiques a l'effort. Rev Fr Mal Respir 5:431–440, 1977.
78. Henriksen JM, Dahl R, Lundquist GR: Influence of relative humidity and repeated exercise on exercise-induced bronchoconstriction. Allergy 36:463–470, 1981.
79. Henriksen JM, Nielsen TT: Effects of physical training on exercise-induced bronchoconstriction. In: Oseid S (ed.) The Asthmatic Child in Play and Sports. Pitman, London, in press.
80. Herxheimer H: Hyperventilation asthma. Lancet 1:83–87, 1946.
81. Horton DJ, Chen WY: Effects of breathing warm humidified air on bronchoconstriction induced by body cooling and by inhalation of metacholine. Chest 75:24–28, 1979.
82. Hyde JS, Swarts CL: Effect of an exercise program on the perennially asthmatic child. Am J Dis Child 116:383–396, 1968.
83. Inbar O, Alvarez DX, Lyons HA: Exercise-induced asthma—a comparison between two modes of exercise stress. Eur J Respir Dis 62:160–167, 1981.
84. Inbar O, Dotan R, Dlin RA, et al: Breathing dry or humid air and exercise-induced asthma during swimming. Eur J Appl Physiol 44:43–50, 1980.
85. James L, Faciane J, Sly RM: Effect of treadmill exercise on asthmatic children. J Allergy Clin Immunol 57:408–416, 1976.
86. Johnson JD: Statistical considerations in studies of exercise-induced bronchospasm. J Allergy Clin Immunol 64:634–641, 1979.
87. Jones RHT, Jones RS: Ventilatory capacity in young adults with a history of asthma in childhood. Br Med J 2:976–978, 1966.
88. Jones RS: Assessment of respiratory function in the asthmatic child. Br Med J 2:972–975, 1966.

89. Jones RS, Wharton MJ, Buston MH: The place of physical exercise and bronchodilator drugs in the assessment of the asthmatic child. Arch Dis Child 38:539–545, 1963.
90. Josenhans WT, Melville GN, Ulmer WT: The effect of facial cold stimulation on airway conductance in man. Can J Physiol Pharmacol 47:453–457, 1969.
91. Kattan M, Keens TG, Mellis CM, Levison H: The response to exercise in normal and asthmatic children. J Pediatr 92:718–721, 1978.
92. Katz RM, Siegel SC, Rachelefsky GS: Blood gas in exercise-induced bronchospasm: a review. Pediatrics 56[Suppl.]:880–882, 1975.
93. Kawabori I, Pierson WE, Conquest LL, Bierman DW: Incidence of exercise-induced asthma in children. J Allergy Clin Immunol 58:447–455, 1976.
94. Keens TG, Krastins IRB, Wannamaker EM, et al: Ventilatory muscle endurance training in normal subjects and patients with cystic fibrosis. Am Rev Respir Dis 116:853–860, 1977.
95. Kilham H, Tooley M, Silverman M: Running, walking and hyperventilation causing asthma in children. Thorax 34:582–586, 1979.
96. König P: Clinical implications of bronchial lability in relation to asthma. PhD Thesis, University of London, 1974.
97. König P, Godfrey S: Exercise-induced bronchial lability and atopic status of families of infants with wheezy bronchitis. Arch Dis Child 48:942–946, 1973.
98. König P, Godfrey S: Exercise-induced bronchial lability in monozygotic (identical) and dizygotic (non-identical) twins. J Allergy Clin Immunol 54:280–287, 1974.
99. Leisti S, Finnila M-J, Kiura E: Effects of physical training on hormonal responses to exercise in asthmatic children. Arch Dis Child 54:524–528, 1979.
100. Malo JL, Filiatrault S, Martin RR: Combined effects of exercise and exposure to outside cold air on lung functions of asthmatics. Bull Eur Physiopathol Respir 16:623–635, 1980.
101. Mangla PK, Menon MPS: Effect of nasal and oral breathing on exercise-induced asthma. Clin Allergy 11:433–439, 1981.
102. Mansell A, Dubrawsky C, Levison H, et al: Lung elastic recoil in cystic fibrosis. Am Rev Respir Dis 109:190–197, 1974.
103. Mansfield L, McDonnell J, Morgan W, Souhrada JF: Airway response in asthmatic children during and after exercise. Respiration 38:135–143, 1979.
104. McFadden ER Jr, Ingram RH Jr: Exercise-induced asthma. Observations on the initiating stimulus. N Engl J Med 301:763–769, 1979.
105. McFadden ER Jr, Ingram RH Jr, Haynes RL, Wellman JJ: Predominant site of flow limitation and mechanisms of postexertional asthma. J Appl Physiol: Respir Environ Exercise Physiol 42:746–752, 1977.
106. McNally JF Jr, Enright P, Souhrada JF: The role of the oropharynx in exercise-induced bronchoconstriction (abstract). Am Rev Respir Dis 117[Suppl.]:372, 1978.
107. McNeill RS, Nairn JR, Millar JS, Ingram CG: Exercise-induced asthma. Q J Med 35:55–67, 1966.

108. Mellis CM, Levison H: Bronchial reactivity in cystic fibrosis. Pediatrics 61:446–450, 1978.
109. Millar JS, Nairn JR, Unkles RD, McNeill RS: Cold air and ventilatory function. Br J Dis Chest 59:23–27, 1965.
110. Miller GJ, Davies BH, Cole TJ, Seaton H: Comparison of the bronchial response to running and cycling in asthma using an improved definition of the response to work. Thorax 30:306–311, 1975.
111. Miller LC, Miller WW, Johnson RL Jr, Schneider M: The effect of physical training on exercise-induced asthma (abstract). Clin Res 24:588A, 1976.
112. Morton AR, Fitch KD, Davis T: The effect of "warm-up" on exercise-induced asthma. Ann Allergy 42:257–260, 1979.
113. Morton AR, Turner KJ, Fitch KD: Protection from exercise-induced asthma by pre-exercise cromolyn sodium and its relationship to serum IgE levels. Ann Allergy 31:265–271, 1973.
114. O'Brodovich HM, Moser MM, Lee L: Familial lymphoid interstitial pneumonia: a long-term follow-up. Pediatrics 65:523–528, 1980.
115. Oldenburg FA, Dolovich MB, Montgomery JM, Newhouse MT: Effects of postural drainage, exercise and cough on mucus clearance in chronic bronchitis. Am Rev Respir Dis 120:739–745, 1979.
116. Orenstein DM, Franklin BA, Doershuk CF, et al: Exercise conditioning and cardiopulmonary fitness in cystic fibrosis. The effects of a three-month supervised running program. Chest 80:392–398, 1981.
117. Orenstein DM, Henks KG, Cerny FJ: Exercise and cystic fibrosis. Physican Sportsmed 11:57–63, 1983.
118. Oseid S: Exercise-induced asthma: A review. In: Berg K, Eriksson BO (eds.) Children and Exercise IX. University Park Press, Baltimore, 1980, pp. 277–288.
119. Oseid S, Haaland K: Exercise studies on asthmatic children before and after regular physical training. In: Eriksson B, Furberg B (eds.) Swimming Medicine IV. University Park Press, Baltimore, 1978, pp. 32–41.
120. Oseid S, Kendall M, Larsen RB, Selbekk R: Physical activity programs for children with exercise-induced asthma. In: Eriksson B, Furberg B (eds.) Swimming Medicine IV. University Park Press, Baltimore, 1978, pp. 42–51.
121. Patel KR: The effect of calcium antagonist, nifedipine, in exercise-induced asthma. Clin Allergy 5:429–432, 1981.
122. Pearson RB: The effect of exercise in asthma. Acta Allergol (Kbh) 5:310–311, 1952.
123. Petersen KH, McElhenney TR: Effects of a physical fitness program upon asthmatic boys. Pediatrics 35:295–299, 1965.
124. Pierson WE, Bierman CW: Free running test for exercise-induced bronchospasm. Pediatrics 56[Suppl.]:890–892, 1975.
125. Pollock J, Kiechel F, Cooper D, Weinberger M: Relationship of serum theophylline concentration to inhibition of exercise-induced bronchospasm and comparison with cromolyn. Pediatrics 60:840–844, 1977.
126. Proctor DF: The upper airways. I. Nasal physiology and defense of the lungs. Am Rev Respir Dis 115:97–129, 1977.

127. Proctor DF, Andersen I, Lundqvist GR: Human nasal mucosal function at controlled temperatures. Respir Physiol 30:109–124, 1977.
128. Ryland D, Reid L: The pulmonary circulation in cystic fibrosis. Thorax 30:285–292, 1975.
129. Schachter EN, Lach E, Lee M: The protective effect of a cold weather mask on exercise-induced asthma. Ann Allergy 46:12–16, 1981.
130. Scherr MS, Frankel L: Physical conditioning program for asthmatic children. JAMA 168:1996–2000, 1958.
131. Schnall RP, Landau LI, Phelan PD: The use of short periods of exercise in the prevention and reversal of exercise-induced bronchoconstriction. In: Oseid S (ed.) The Asthmatic Child in Play and Sports. Pitman, London, in press.
132. Schoeffel RE, Anderson SD, Seale JP: The protective effect and duration of action of metaproterenol aerosol on exercise-induced asthma. Ann Allergy 46:273–275, 1981.
133. Shapiro GG, Pierson WE, Furukawa CT, Bierman CW: A comparison of the effectiveness of free-running and treadmill exercise for assessing exercise-induced bronchospasm in clinical practice. J Allergy Clin Immunol 64:609–611, 1979.
134. Shephard RJ: Exercise-induced bronchospasm—a review. Med Sci Sports 9:1–10, 1977.
135. Shephard RJ: Physical Activity and Growth. Year Book Medical Publishers, Chicago, 1982.
136. Shturman-Ellstein R, Zeballos RJ, Buckley JM, Souhrada JF: The beneficial effect of nasal breathing on exercise-induced bronchoconstriction. Am Rev Respir Dis 118:65–73, 1978.
137. Silverman M, Anderson SD: Standardization of exercise tests in asthmatic children. Arch Dis Child 47:882–889, 1972.
138. Silverman M, Andrea T: Time course of effect of disodium cromoglycate on exercise-induced asthma. Arch Dis Child 47:419–422, 1972.
139. Simonsson BG, Jacobs FM, Nadel JA: Role of autonomic nervous system and the cough reflex in the increased responsiveness of airways in patients with obstructive airway disease. J Clin Invest 46:1812–1818, 1967.
140. Sims DG, Downham MAPS, Gardner PS, et al: Study of 8-year-old children with a history of respiratory syncytial virus bronchiolitis in infancy. Br Med J 1:11–14, 1978.
141. Sly RM: Exercise-related changes in airway obstruction: Frequency and clinical correlates in asthmatic children. Ann Allergy 28:1–16, 1970.
142. Sly RM: Effect of cromolyn sodium on exercise-induced airway obstruction in asthmatic children. Ann Allergy 29:362–366, 1971.
143. Sly RM: Effect of β-adrenoreceptor stimulants on exercise-induced asthma. Pediatrics 56[Suppl.]:910–915, 1975.
144. Sprenkle AC, Van Arsdel PP, Bierman CW: New drug evaluation using exercise-induced bronchospasm. Pediatrics 56[Suppl.]:937–939, 1975.
145. Stalcup SA, Mellins RB: Mechanical forces producing pulmonary edema in acute asthma. N Engl J Med 297:529–595, 1977.
146. Steer RG: Asthma and the weather (letter). Med J Aust 7:38, 1976.

147. Strauss RH, Haynes RL, Ingram RH Jr, McFadden ER Jr: Comparison of arm vs leg work in induction of acute episodes of asthma. J Appl Physiol 42:565–570, 1977.
148. Strauss RH, McFadden ER Jr, Ingram RH Jr, et al: Influence of heat and humidity on the airway obstruction induced by exercise in asthma. J Clin Invest 61:433–440, 1978.
149. Strauss RH, McFadden ER Jr, Ingram RH Jr, Jaeger JJ: Enhancement of exercise-induced asthma by cold air. N Engl J Med 297:743–747, 1977.
150. Strunk RC, Kelly LJ: The recreation therapy programme at National Jewish Hospital/National Asthma Center. In: Oseid S (ed.) The Asthmatic Child in Play and Sport. Pitman, London, in press.
151. Swann IL: Improvement in asthmatic children as a result of physical training. In: Oseid S (ed.) The Asthmatic Child in Play and Sport. Pitman, London, in press.
152. Tower J: Office testing for exercise-induced asthma. Alaska Med 20:70–72, 1978.
153. Tromp SW, Bouma J: Effect of weather on asthmatic children in the eastern part of the Netherlands. Int J Biometeorol 9:233–238, 1965.
154. Välimäki I, Liuko L, Peltonen T, Hirvonen L: Physical working capacities of children with pulmonary tuberculosis. Scand J Respir Dis 49:260–263, 1968.
155. Vassallo CL, Gee JBL, Domm BM: Exercise-induced asthma. Observations regarding hypocapnia and acidosis. Ann Rev Respir Dis 105:42–49, 1972.
156. Vávra J, Máček M, Mrzena B, Spicak V: Intensive physical training in children with bronchial asthma. Acta Paediatr Scand Suppl 217:90–92, 1971.
157. Vávra J, Máček M, Spicak V: Working capacity of asthmatic children (in French). Rev Pediatr 5:3–7, 1969.
158. Verma S, Hyde JS: Physical education programs and exercise-induced asthma. Clin Pediatr 15:697–699, 1976.
159. Weinstein RE, Anderson JA, Kvale P, Sweet LC: Effects of humidification on exercise-induced asthma (EIA) (abstract). J Allergy Clin Immunol 57:250–251, 1976.
160. Wilbourn K: The lung distance runner. Runner's World 13:62–65, 1978.
161. Wilson BA, Evans JN: Standardization of work intensity for evaluation of exercise-induced bronchoconstriction. Eur J Appl Physiol 47:289–294, 1981.
162. Wolkove N, Kreisman H, Frank H, Gent M: The effect of ipratropium on exercise-induced bronchoconstriction. Ann Allergy 47:311–315, 1981.
163. Yeung R, Nolan GM, Levison H: Comparison of the effects of inhaled SCH 1000 and Fenoterol on exercise-induced bronchospasm in children. Pediatrics 66:109–114, 1980.
164. Zach M, Oberwaldner B, Hausler F: Cystic fibrosis; physical exercise vs. chest physiotherapy. Arch Dis Child 57:587–589, 1982.
165. Zach MS, Purrer B, Oberwalder B: Effect of swimming on forced expiration and sputum clearance in cystic fibrosis. Lancet II:1201–1203, 1981.
166. Zambie MF, Gupta S, Lemen RJ, et al: Relationships between response to exercise and allergy in patients with cystic fibrosis. Ann Allergy 42:290–294, 1979.

167. Zeballos RJ, Shturman-Ellstein R, McNally JF Jr, et al: The role of hyperventilation in exercise-induced bronchoconstriction. Am Rev Respir Dis 118:877–884, 1978.
168. Zelkowitz PS, Giammona ST: Effects of gravity and exercise on the pulmonary diffusing capacity in children with cystic fibrosis. J Pediatr 74:393–398, 1969.

4
Cardiovascular Diseases

Aortic Stenosis (AS)

Physiologic Responses to Acute Exercise

The child with congenital AS often responds to exercise with abnormal hemodynamic function and with myocardial ischemia.

Hemodynamic Abnormality. The underlying lesion causes increased resistance at the left ventricular outflow tract, which results in an excessively high systolic (mean and peak) pressure in the left ventricle and a reduced post-stenotic pressure in the aorta. Hemodynamic responses to acute exercise are summarized in Table 4.1. Whereas healthy children respond with elevation of systolic arterial blood pressure and an increase in pulse pressure, the child with AS will have smaller increases of systolic arterial pressure, absolute levels being as much as 50 mm Hg lower than expected (see Fig. 4.1).[1,13,51,59,95] Occasionally, no change or even a *drop* in systolic arterial pressure may occur. Diastolic arterial pressure is normal or somewhat elevated so that pulse pressure is well below the expected. Such pressure abnormalities may exist in the resting patient, but are more apparent during exertion.

Depressed arterial blood pressure in the exercising patient is in part caused by the outflow obstruction. However, a paradoxical vasodilatation may take place in the nonexercising muscles of patients with AS (instead of the normal vasoconstriction), with a concomitant lowering of vascular resistance in these muscles and a lower than expected arterial pressure.[73] (See also Clinically Detrimental Effects of Acute Exercise, below, for discussion of syncope.)

While systolic arterial pressure is abnormally low, left ventricular systolic pressure is exceedingly high. Peak ventricular pressures as high as

Table 4.1. Abnormal Hemodynamic Responses to Exercise of the Child with Aortic Stenosis

Variable	*Comparison with a Healthy Child*
Arterial systolic blood pressure	Low
Pulse pressure	Low
Peak left ventricular systolic pressure	High
Left outflow pressure gradient	High
Left ventricular end-diastolic pressure	High
Stroke volume	Low
Cardiac output	Low
Left ventricular stroke work index	High
Myocardial O_2 supply : demand	Low

Comment: *Vasodilatation* in nonexercising muscle was found during exercise in adults with AS.

230–250 mm Hg have often been recorded during exercise, compared with 160–180 mm Hg at rest. Successful surgical repair will lower these values to 150–180 and 120–150, respectively.[91]

It is evident from the above that the left outflow pressure gradient ("aortic gradient"), high in the child at rest, is even higher during exer-

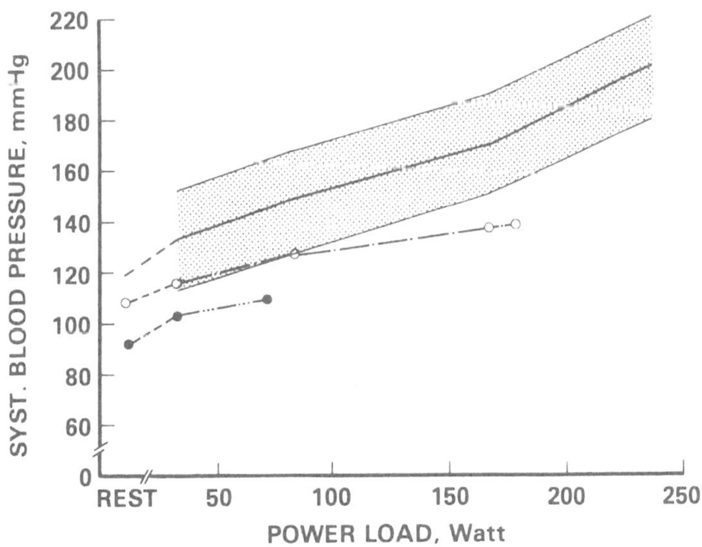

Figure 4.1. Systolic arterial blood pressure of children with aortic stenosis at rest and during various levels of exercise. Data on nine patients were reported by James[56] (○) and on 20 patients by Cueto and Moller[13] (●). The shaded area represents healthy children (mean ± 1 S.D.) as reported by James.[56]

cise. This exercise-induced gradient change is important to the clinician because, more than any other physiologic function, it reflects the *degree* of stenosis. Exercise aortic gradients are often 20–25 mm Hg higher than values at rest.[13,91] Following aortic valvotomy this difference is decreased to about 5–10 mm Hg.[91]

Left ventricular end-diastolic pressure is a hemodynamic function which, according to Starling's principle, affects the contractile forces in the subsequent systole and hence the stroke volume. Left ventricular end-diastolic pressure does not normally change from rest to exercise. It may even decrease by as much as 5 mm Hg. In contrast, left ventricular end-diastolic pressure of a child with AS may increase with exercise by up to 6 mm Hg.[91] In some patients this rise in pressure is not accompanied by the normal increase in stroke volume, which suggests decreased contractile function.

The child with AS has a low resting stroke volume, which barely increases with exercise.[13] Therefore, any increase in cardiac output depends almost entirely on the elevation of heart rate. The end result is an abnormally low cardiac output during submaximal[108] and maximal exercise. This will limit the O_2 transport capability and be a major cause for the low maximal aerobic power of children with AS.

An additional abnormal function is the left ventricular stroke-work index, which is proportional to a product of stroke-index and the difference between peak-systolic and end-diastolic pressures. This index is excessively elevated on transition from rest to exercise in either leg pedaling[91] or static hand-grip.[97]

Myocardial Ischemia. Normally, exercise is associated with an increase in coronary blood flow. This is not the case in some children with AS. Rather than receiving more blood, the myocardium of these children may indeed be confronted with *reduced* blood flow. The mechanisms for ischemic changes in the child with AS have not been fully elucidated, but some observations have been made: in children with severe AS the ratio O_2 supply : O_2 demand in the myocardium drops to levels compatible with subendocardial ischemia.[70,97] Such ischemia may result from a decreased aortic pressure on the one hand and an increased left ventricular end-diastolic pressure on the other. Corrective surgery will reduce but not eliminate the exercise-induced deficit in myocardial O_2 supply.[91] In some children improvement is gradual and may not be noticeable until more than a year after the operation.[124] The drop in O_2 supply : demand ratio is apparently heart rate-dependent. Tachycardia is accompanied by a reduction in the relative O_2 supply, perhaps due to shortening of the left ventricular diastolic time and to a simultaneous rise in the relative myocardial contraction period, which increases the metabolic demand.

An insufficient myocardial O_2 supply is often manifested by "ischemic" ECG changes, mostly ST segmental depression.[11,44,51,53,59,72,93,115] The diagnostic value of such changes is discussed below under Evaluation Based on Electrocardiography.

Physical Working Capacity

Many children with AS complain of excessive fatigue and marked exertional dyspnea. Such patients may have decreased maximal aerobic power.[24,39,41,59,81,122,124] Others, especially when the obstruction is mild, have practically normal exercise performance.[14,41,108] Subnormal maximal aerobic power may be caused by the low myocardial O_2 supply and maximal cardiac output. However, low fitness is not entirely due to the disease. In all likelihood such children are detrained because of overprotection and fear of chest pain, syncope, or a "heart attack." In those patients with no contraindication for exercise and sports (see Permissible Activities, below), increased activity may improve exercise performance and alleviate complaints such as fatigue and exertional dyspnea.

Clinically Detrimental Effects of Acute Exercise

There are only a few pediatric diseases where physical exertion is detrimental to health. AS is one of these. The major clinical effects of exercise are: syncope, chest pain, or sudden death with or without prior left heart failure. Their occurrence is related to the hemodynamic severity of the disease (e.g., left outflow pressure gradient),[23,120] and they often appear in patients who also complain of high fatigability and excessive exertional dyspnea.

The pathogenesis of *effort-induced syncope* is not completely understood. It seems to occur in a patient with AS who can raise his cardiac output and systolic arterial pressure at low exertional levels but who, at a critical level of exercise, suffers an abrupt fall in both.[31] This drop could be triggered by sudden peripheral vasodilatation in the nonexercising muscles, which is a reflex response to activation of baroreceptors at the left ventricle.[73] Aortic valvotomy in adults with a preoperative history of exertional syncope can reverse such peripheral vasodilatation.[73]

Exertional chest pain is evident in some 5–10% of children with AS. It results from an inadequate myocardial O_2 supply in the face of increasing demands during exercise, as discussed above.

Sudden death occurs in about 2–7% of children with AS. As stated by Nadas and Fyler,[86] the question confronting the pediatric cardiologist is: which patient will safely reach middle age and which will suffer sudden death in early adolescence following minimal physical exertion? Most cases of sudden death occur in the resting patient. In some, however, a temporal connection has been shown between exertion and sudden death.[23,64] Almost invariably, such patients had been symptomatic, complaining of high fatigability, excessive dyspnea, syncope, or exertional chest pain.

Exercise as a Diagnostic Tool

The diagnosis of congenital AS will have been confirmed well before the child arrives at the exercise laboratory. There are, however, specific questions of clinical importance that can be answered by use of an exercise test:

Should the child undergo cardiac catheterization?
Is surgical correction indicated?
How severe is the narrowing?
Is there myocardial ischemia?
Should athletic activity be curtailed?
Is the disease progressing?

For some of the answers, catheterization of the left heart and aorta may be required in conjunction with the exercise test. For others a noninvasive exercise protocol is sufficient to monitor arterial blood pressure and ECG.[57,124] Although the physical working capacity of children with AS is often subnormal, its assessment is of no diagnostic or prognostic value. It does, however, inform the clinician of the level of fitness of the child and the possible need for modification of his habitual activity.

It is sometimes inconvenient to administer an exercise test during cardiac catheterization. This applies mostly to very young children or to those who are apprehensive, uncoordinated, or ill-motivated. An alternative stress test has been suggested for children with AS, which simulates the inotropic and, especially, the chronotropic effects of exercise. This is the intravenous infusion of isoproterenol, which, like exercise, can be titrated to induce a certain level of heart rate.[85,114]

Evaluation Based on Hemodynamic Variables. The main variables measured in an exercise test which are of diagnostic or prognostic importance are the following:

Systolic Arterial Blood Pressure. A drop in systolic pressure with exercise, a non-rise in systolic arterial pressure, or a rise that is inappropriately low are suggestive of a marked obstruction or of a low myocardial contractility. Such a finding (even if the patient is asymptomatic) is in itself an indication for catheterization. In contrast, an asymptomatic patient with a normal exercise ECG and a normal or near normal systolic arterial pressure response to exercise will require only supervision and periodic check-ups. The above reasoning is valid for pre- and postoperative children alike.[51,57,59,95]

Left Ventricular End-Diastolic Pressure and Stroke Volume. In some laboratories a rise of more than 3 mm Hg in left ventricular end-diastolic pressure during exercise, especially if combined with no rise or a decrease in stroke volume, is an indication for surgery (M.D. Freed, J.H. Moller, personal communication).

Peak Left Ventricular Systolic Pressure. Values of 200 mm Hg or more, when measured during moderate exercise, have been considered an indication for surgery (J.H. Moller, personal communication). This figure, however, is not absolute and depends, among other factors, on the age of the child and the intensity of exercise.

Myocardial O_2 Supply : Demand Ratio. This index is assessed by the product of the diastolic pressure–time interval and arterial O_2 content, divided by systolic pressure–time interval. When this ratio falls during exercise to less than 10, subendocardial ischemia can be assumed.[70,91] This is an indication for surgery. The index can also be used as a guide to permissible levels of exercise in AS patients.

Evaluation Based on Electrocardiography. Exercise ECG is a highly informative noninvasive procedure for evaluation of the child with AS. In such a child, an ECG taken during or *immediately following* exercise is more sensitive for identifying ischemic ST depression (>1 mm) than the resting ECG.[11,44,51,53] Furthermore, the *magnitude* of the ST depression during exercise reflects the severity of the outflow tract obstruction, i.e., the gradient across the aortic valve (see Fig. 4.2). When the gradient at rest is below 50 mm Hg, there is little or no ST depression during exercise. However, gradients of 50 mm Hg or more are often accompanied by segmental ST depression of 1–3 mm. In practical terms, exercise ECG can be used as a screening test in AS. Although negative findings cannot rule out a hemodynamic deficit, an ST depression of more than 1 mm is highly suggestive of major stenosis, even if the child is asymptomatic. Such a child, as a rule, should be catheterized. Repeated periodically at least once a year, exercise ECG testing can gauge a possible aggravation of the obstruction in a patient hitherto asymptomatic.[11,51,93] The exercise ST changes were found to diminish following surgical removal of the left ventricular tract obstruction.[59] There are, however, cases where the obstruction has been removed and the child still responds to exercise with ischemic ST depression.[115]

As with diagnosis of coronary heart disease in adults, ST depression in the AS child is greater and lasts longer with an increase in exercise intensity. Indeed, one of the main reasons for a false negative result (i.e.,

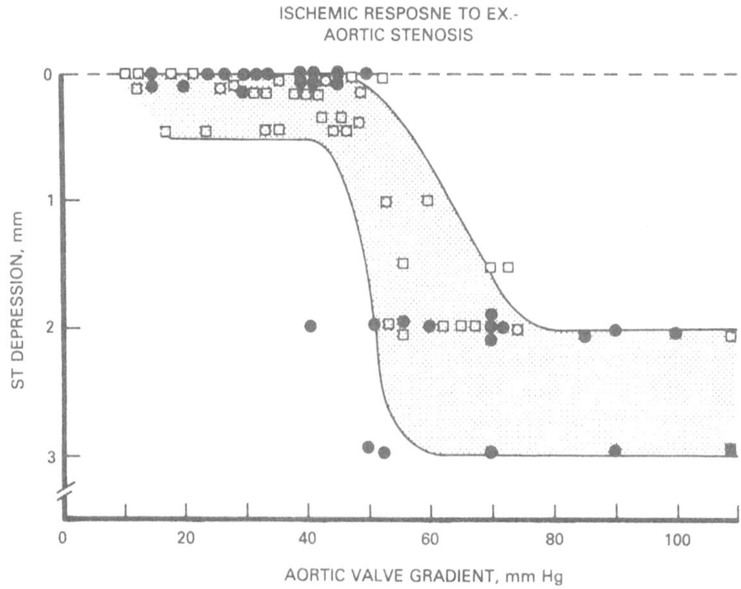

Figure 4.2. Segmental ST depression during exercise as a function of pressure gradient across the aortic valve in children with aortic stenosis. Data from Chandramouli et al[11] (□), Halloran[44] (●), and James[57] (■).

no ischemic changes in spite of marked narrowing) is an exercise test of insufficient intensity.[93] If intensity is such that heart rate is only 150 beats/min, the ST depression may disappear within 30 sec after exercise. If, on the other hand, heart rate is raised to 180 or more, the ST depression may still be evident 5 min post-exercise.[44]

We can conclude that an exercise ECG of the child with AS is a potent tool for: 1) selection of children for catheterization, 2) periodic testing to determine the natural history of the disease, and 3) management of the child regarding permissible levels of habitual activity.

Criteria for Successful Surgical Repair. Not all children who undergo surgical repair for AS achieve complete normalization of function. Some residual hemodynamic or electrocardiographic abnormality can occur. Nevertheless, certain improvements in response to exercise can be expected:

1. Arterial systolic blood pressure will assume a normal rise with power load.
2. Peak systolic ventricular pressure will become normal (e.g., will not exceed 160–170 mm Hg during moderate exercise).
3. There will be no rise in left ventricular end-diastolic pressure.
4. The aortic pressure gradient will be less than 40 mm Hg.

5. Stroke volume will rise from rest to moderate exercise.
6. The myocardial O_2 supply : demand ratio will be greater than 10.
7. There will be no ST segmental depression of 1 mm or more during strenuous exercise.

Permissible Activities

What policy should govern the participation of AS children in sports? The answer depends on the type of activity and on the severity of the lesion. First and foremost, physicians and parents alike must be made aware of the difference between a high-intensity activity regimen for competition and less intense recreational activities. The latter can be permitted the asymptomatic child unless ECG findings indicate left ventricular strain at rest, or segmental ST depression of more than 1 mm, or ventricular dysrhythmia during exercise.

Any youngster with AS who wishes to enter competitive sports should first be assessed by exercise ECG and catheterization, even if he is asymptomatic. If the left outflow pressure gradient is less than 20 mm Hg, the child can be allowed any activity. When gradients are 20–40 mm Hg, recreational activities only should be allowed. A higher pressure gradient is a definite contraindication for physical exertion and the child should be reevaluated after surgical correction (G.R. Cumming, M.D. Freed, personal communication). Because of the progressive nature of this disease, a periodic exercise ECG (e.g., twice yearly) should be performed on any youngster with AS who pursues athletic activities.[96]

What about the symptomatic child? Although no clear-cut guidelines can be given, it is sound policy to curtail the activity of those with a history of syncope and chest pain. On the other hand, if the complaints are limited to exertional dyspnea and "fatigue," and there is no electrocardiographic or other objective evidence of an advanced obstruction, it is likely that the child merely has a low maximal aerobic power due to imposed inactivity. Moderate levels of conditioning may, indeed, be beneficial to him.

Coarctation of the Aorta (CA)

Clinical and hemodynamic findings in this anomaly vary with the location of the narrowing along the aorta and the patency of the ductus arteriosus. It is beyond the scope of this book to analyze the various nuances of this defect. We shall therefore approach CA as a single entity insofar as responses to exercise are concerned. Exertion-related complaints are pain in the calves during intense jumping or running and, especially in those children who also have a left-to-right shunt, exertional dyspnea and fatigue.

Hemodynamic Abnormalities

Hemodynamic functions of patients with CA and healthy youths, at rest and during exercise, are compared in Table 4.2. Systolic blood pressure, as measured in or over the brachial artery, is extremely high during exercise. Values of 250–300 mm Hg are not uncommon in unoperated patients who perform strenuous activities (see Fig. 4.3).[20,58,111] Even months or years following corrective surgery that is considered anatomically successful, one can find systolic arterial pressure of 200–250 mm Hg.[58,111] In contrast, diastolic arterial pressure in exercising patients with CA is usually normal. Thus both mean arterial pressure and pulse pressure are elevated.

As a result of narrowing, the pressure gradient across the lesion is high. It may exceed 40 mm Hg at rest and 120 mm Hg during exercise.[20] Both the volume load and the pressure load will raise the work done by the left myocardium in the unoperated child.[111] Successful operation should abolish the high gradient and the excessive myocardial work.

Urinary prostaglandin E-like material is abnormally elevated following exercise in patients with CA.[107] The significance of this finding is unclear.

Exercise as a Diagnostic Tool

Hemodynamic Response to Exercise. Although most children with CA should undergo surgical correction, the *urgency* of the operation can be indicated through their blood pressure response to exercise[111] (also

Table 4.2. Hemodynamic and Other Variables Taken at Rest and During Exercise in Patients with Coarctation of the Aorta Before and After Surgical Correction (Comparison with Healthy Individuals)

Function	*Preoperative*		*Postoperative*	
	Rest	*Exercise*	*Rest*	*Exercise*
Systolic arterial blood pressure	High	Very high	Normal/high	High
Diastolic arterial blood pressure	Normal	Normal	Normal	Normal
Pulse pressure	High	Very high	Normal	High
Pressure gradient across narrowing	High	Very high	—	—
Left ventricular work	High	Very high	Normal	Normal/high
"Ischemic" ECG changes	Common	Very common	Common	Very common

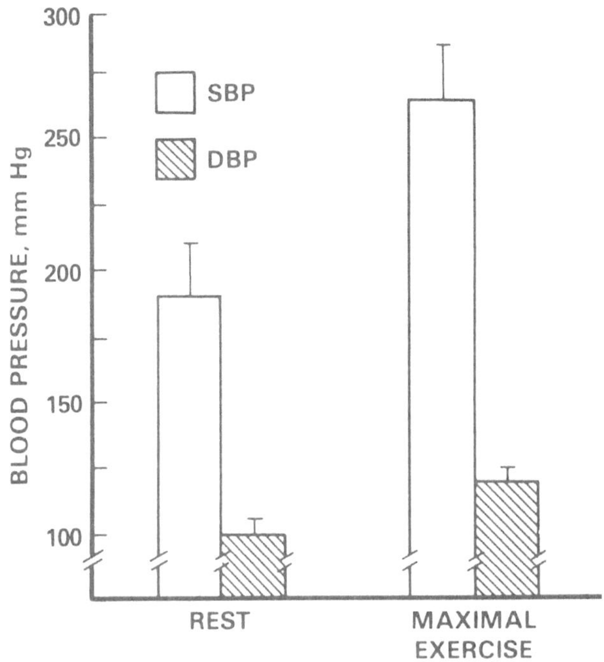

Figure 4.3. Arterial blood pressure during rest and maximal exercise in eight adolescent girls with coarctation of the aorta. Vertical lines denote 1 S.E.M. Data from Taylor and Donald.[111]

M.D. Freed, personal communication). Some clinicians feel that when maximal systolic arterial pressure is less than 200 mm Hg there is no urgency for surgical correction. Others base their decision upon the pressure gradient across the narrowing. When this gradient exceeds 40 mm Hg at rest or 80 mm Hg during exercise, surgical correction is indicated.

High and persistent postoperative exercise aortic pressure can serve as an indication for hypotensive medication.[20,57,59] This is an important point to consider, for even successful surgical correction is no guarantee against complications related to hypertension, or even sudden death, years later.

Ischemic Changes During Exercise. Early coronary heart disease has been found in adolescents and young adults who had undergone coarctectomy during childhood.[74] There is an obvious need to identify those youngsters who may develop coronary heart disease later in life. Little information is so far available on the use of exercise in this context but its importance warrants some attention. Forty-seven post-coarctectomy patients, 6–27 years old, underwent maximal exercise ECG testing. Two-thirds of the females and more than one-third of the males had a segmental ST depression of 1–4 mm. This rate was markedly greater than that found in a control group (17% for females and 7% for males).[61] Whether those individuals who respond with ischemic changes have a greater risk for future coronary heart disease has yet to be shown. Neither is any comparison available between the sensitivity of exercise ECG and that of other methods in the detection of potential coronary heart disease in post-coarctectomy patients.

Congenital Complete Heart Block (CCHB)

In this lesion, ventricular rate is independent of the sino-atrial pacemaker. It is low and not always responsive to such stimuli as emotional stress, increased metabolic demand, or administration of adrenergic agents. Among symptomatic children the complaints relevant to exercise are fatigue and exertional dyspnea. Syncope of the Adams-Stokes type is rare and may or may not be related to exertion. A summary of the physiologic responses to exercise in CCHB is presented in Table 4.3.

Hemodynamic Response to Exercise

Whereas atrial rate is normal at rest and during different levels of exercise, ventricular rate is low at rest and often rises very little with exertion. In some patients, this rise is proportional to the power load, or to the rise in atrial rate. In most patients, however, the degree of ventricular acceleration is independent of atrial acceleration.[55,112] Such interindividual variability in response is shown in Fig. 4.4. In most children ventricular rate does not rise beyond 100–120 beats/min, even during all-out exercise. With the same exercise, atrial rates of 180 and more are easily

Table 4.3. Response to Exercise of Children with Complete Congenital Heart Block

Variable	*Submaximal*	*Maximal*
Atrial rate	Normal	Normal
Ventricular rate	Low	Very low
Stroke volume	High	Normal/high
Cardiac output	Normal/low	Low
Arteriovenous O_2 difference	High	Normal/high
O_2 uptake	Normal	Low
Premature ventricular contraction	Often	Often

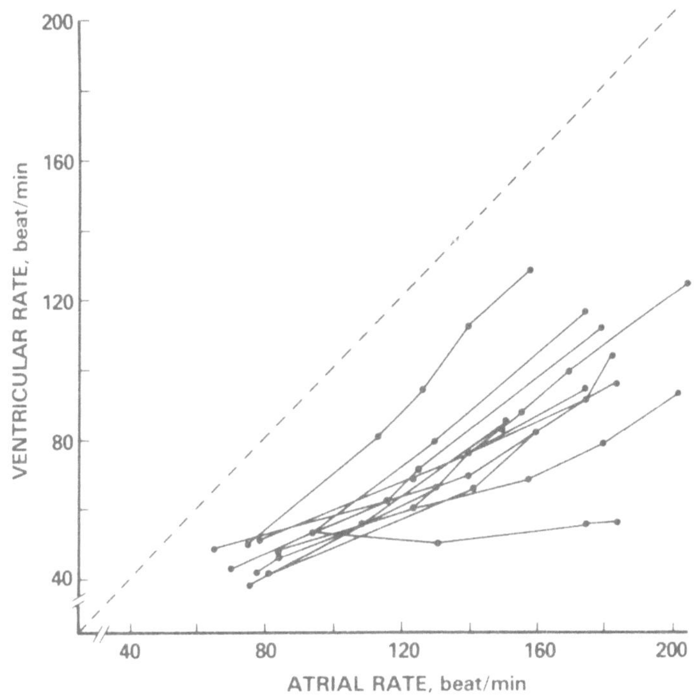

Figure 4.4. Complete congenital heart block. Relationship between ventricular and atrial rate at rest, submaximal, and maximal exercise. Individual values of 11 children 7–15 years old. Broken line denotes identity (Ventricular Rate = Atrial Rate). Data from Thorén et al.[112]

reached.[55,78,110,112,121] Unpaced adults with complete A-V block respond similarly.[50,55,110]

Although stroke volume is usually higher than normal both at rest and during exercise,[78,110] cardiac output is low and sometimes hardly rises with exertion.[110,121] Based on the Fick principle the ability of many patients with CCHB to raise their O_2 uptake is dependent, therefore, on an increased O_2 extraction from the capillary blood, which results in low mixed-venous O_2 content.[110] Those children who can raise neither arteriovenous O_2 difference nor cardiac output become highly fatigued with the slightest exercise provocation and their effort tolerance is extremely low.

Ventricular Dysrhythmia During Exercise

A puzzling finding in CCHB is the frequent occurrence during exercise of premature ventricular contractions and other types of ventricular ectopy. These are very infrequent in resting patients but appear in about

30–70% of patients who exercise.[12,55,110,112,121,125] In a study systematically designed to analyze ventricular ectopy (Fig. 4.5), 57% of children and young adults with CCHB responded to maximal exercise with multifocal, or paired, premature beats or with ventricular tachycardia. The prevalence of dysrhythmia was positively related to age and to exercise level, but not to the working capacity of the patients. Nor was it different between those with isolated A-V block and those who had multiple lesions.[125] It is tempting to speculate that premature ventricular contractions appear as a compensatory mechanism to enhance ventricular pace and cardiac output.[121] Continuous measurements of intra-arterial pressure in an exercising young adult patient,[50] however, demonstrated a drop in blood pressure simultaneous with a bout of premature ventricular contractions, suggesting inadequate ventricular filling prior to the premature contraction; and there was no correlation between the appearance of premature contractions and a rise in cardiac output.

Exercise-induced Syncope

Exercise-induced syncope in CCHB is quite rare and occurs inconsistently in any given patient. Some with a history of Adams-Stokes syncope

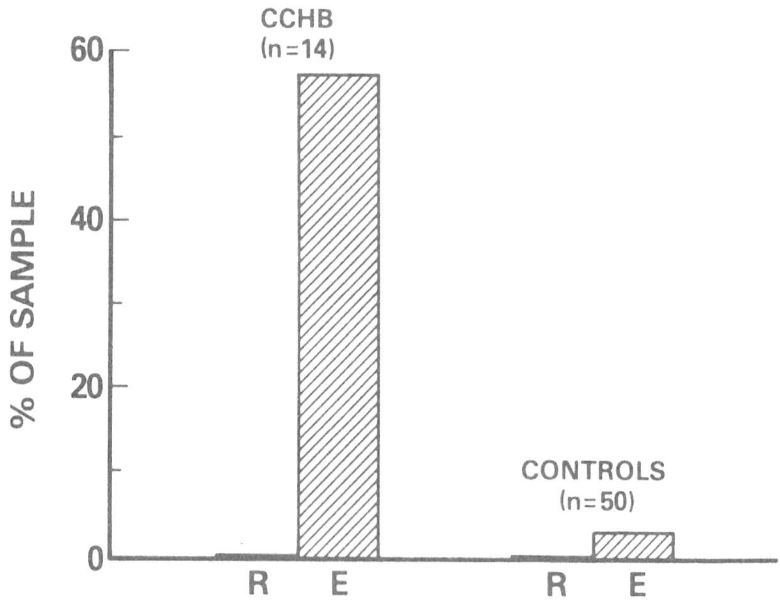

Figure 4.5. Frequency of ventricular ectopy at rest (R) and during, or immediately after, maximal exercise (E) in 4- to 24-year-old patients with complete congenital heart block and in controls. One of the patients had exercise-induced ventricular tachycardia; the others, multifocal or paired premature ventricular contractions. (CCHB = Congenital Complete Heart Block.) Adapted by permission from Winkler et al.[125]

at rest do not have adverse reactions to exercise.[10] Others[121] respond to exercise with dizziness, pallor, and exhaustion but without a loss of consciousness. In some cases syncope appears *after* the conclusion of exercise.[75] An overshoot of vagal response has been implicated as the possible mechanism.

Physical Working Capacity

While some patients with CCHB are extremely unfit and cannot perform strenuous activities, others have a normal physical working capacity.[86,110,112,118,121,125] Children with a low fitness level cannot sufficiently raise their cardiac output or peripheral O_2 extraction during exercise.[37,50,112,121] Some may have additional cardiac defects. As a rule of thumb, doubling of ventricular rate from rest to maximal exercise is needed if the child is to participate successfully in normal daily activities. If such an increment is not reached, it is unlikely that the child will have a normal working capacity (unless his stroke volume or O_2 extraction are exceptionally high). Maximal O_2 uptake as low as 15–20 ml/kg × min has been reported for such patients, compared with 40–55 ml/kg × min in healthy individuals.

At the other end of the spectrum are those patients with a normal working capacity who perform well as athletes. Ikkos and Hanson[55] describe a 16-year-old boy who regularly took part in ice-hockey training. He could almost triple his ventricular rate with exercise (from 46 to 120 beats/min). Four other patients in that study were described as capable of "usual athletics." One boy, 16 years of age, who had been diagnosed at the age of 4 to have CCHB, could run 5 miles daily. His cardiopulmonary and metabolic response to submaximal exercise was normal.[121] Maximal O_2 uptake as high as 52 ml/kg × min has been reported for a patient who, since the age of 15, played league soccer with no functional limitation.[45]

Habitual Activity

Due to their low working capacity, many young patients do not participate in activities with their peers. There are, however, others whose sedentary life is objectively not justified but imposed by their parents or physicians. A case in point[110] is a 23-year-old woman who, since the age of 6—when diagnosed as suffering from CCHB—was instructed "never to exert herself." As a result, she never played at school and always led a sedentary life. Detailed testing of this patient did not reveal any hemodynamic deficiency. It is quite likely that her low effort tolerance merely reflected detraining. Another example is that of a male patient, diagnosed at 2 years of age to have CCHB. This boy was not allowed to play games at school due to an enlarged heart and a slight ventricular septal defect. At the age of 20 he applied to the air force, participated in arduous training, and passed the tests for flying duties.[9]

The compatibility of CCHB with a long and active life has been shown in several reports.[9,10,28] Of interest is the story of a man who had frequent syncopal attacks since the age of 2 years, continuing until his fourth decade. These, however, did not deter him from an active life. At 25 he was engaged in hard manual jobs, including a voluntary role in the fire brigade. In addition, he took part in soccer, cycling, and cricket and felt well. At the age of 50 he was asymptomatic and still quite active although his resting ventricular rate was only 37 beats/min, rising to 55 with moderate exercise.[10]

Exercise in the Assessment of CCHB

A diagnostic exercise test can help answer the following:

What is the *degree* of ventricular defect?
Should the child be paced? given drugs?
What is the likelihood of ventricular dysrhythmia?
Is the functional ability of the child compatible with an active life?

The degree of rise in ventricular rate determines the potential of the child to increase his cardiac output and to cope with the metabolic demands of daily activities. It is therefore of importance to identify those children who cannot sufficiently raise their ventricular rate. As shown in Fig. 4.6, there is a very low correlation between ventricular bradycardia at rest and the inability to raise ventricular rate during maximal exercise. It is, therefore, impossible to predict the response of the child to exertion by monitoring only his resting ventricular rate (see also Fig. 4.4).

Advanced cases of incapacity, with or without episodes of syncope or of heart failure, will be artificially paced, with no need for exercise testing. An exercise test is warranted, however, for the child with only *mild* complaints of fatigue, dyspnea, or dizziness. In some medical centers[86] such patients are given a submaximal exercise task that, in healthy children, will raise heart rate to about 160–170 beats/min. Failure of the ventricular rate to rise to at least 50% above the resting value is an indication for adrenergic therapy. If the latter fails, exogenous pacing should be considered.

An additional value of exercise testing is the chance of discovering individuals with ventricular ectopy.[12,125] The appearance with exercise of multiple premature ventricular beats or ventricular tachycardia suggests risk for Adams-Stokes episodes and life-threatening dysrhythmias.[125] Here, the protocol of choice is an all-out progressive (continuous or interrupted) test with ECG monitoring.

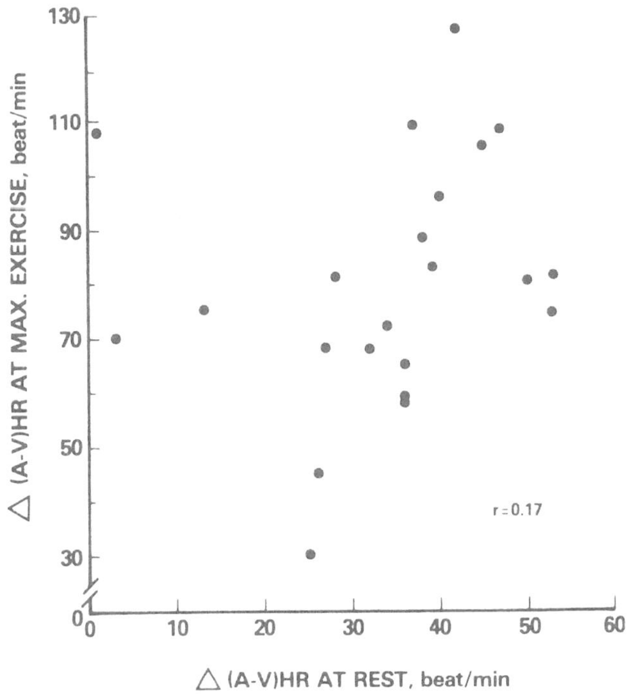

Figure 4.6. The deficit of ventricular rate [Δ(A-V)HR] during maximal exercise plotted against Δ(A-V)HR at rest. Individual values of 7- to 15-year-old patients with complete congenital atrioventricular block. Based on data by Ikkos and Hanson,[55] Mocellin and Bastanier,[78] Taylor and Godfrey,[110] and Thorén et al.[112]

In assessing maximal aerobic power, the investigator must avoid a pitfall: he cannot use methods which depend on heart rate response to standardized exercise, such as $\dot{W}_{170}$ or *predicted* maximal O_2 uptake (see Appendix II). These methods assume a normal maximal ventricular rate and a linear response of heart rate to power load, neither of which occurs in CCHB. The use of a prediction equation in a child whose maximal heart rate is, for example, 100 instead of 200 beats/min, will lead to drastic overestimation of his maximal aerobic power.[77] An alternative approach can be the use of *atrial* rate for the calculation of $\dot{W}_{170}$.[50,118]

A more appropriate criterion is one that does not depend on heart rate response, such as the actual peak mechanical power output, or the total time for which the child can sustain a progressive task (e.g., the Bruce treadmill test). In a well equipped exercise physiology laboratory, a direct measurement of maximal O_2 uptake is the method of choice.[55,112]

Coronary Heart Disease

Physical Exercise in Childhood: Does It Prevent Coronary Heart Disease?

Coronary heart disease as a pediatric issue has been gaining recognition in recent years.[36,88,119] The possible relationship between lifestyle in early years and eventual development of coronary heart disease is of obvious public health importance. It is especially useful to know whether an intervention program at a young age will modify "tracking" of risk factors.

Can an increase in activity during childhood help prevent coronary heart disease? A definitive study of this issue would require a prospective randomized follow-up from childhood to middle age. Such an ambitious project has yet to be launched. A compromise approach is to evaluate the *short-term* impact of conditioning on coronary risk factors. Conditioning programs of a few weeks' duration have shown the following changes in risk factors:

1. Decrease in adiposity level.
2. Mild decrease in resting blood pressure.
3. Increase in serum high-density lipoprotein cholesterol (HDL-C) and a decrease in serum triglycerides (TG).[34,82] In cross-sectional comparisons, active children had higher HDL-C and lower TG than had less active children.[98,113,117]
4. Increase in habitual physical activity (following an 8-month conditioning program).[35]

Smoking is a major coronary risk factor. Surveys have shown[6] that adolescents who belong to sports clubs smoke less than those who are not active in sports. While such comparisons cannot prove a causal relationship between smoking and habitual activity, it is likely that, during their formative years, youths who participate in sports adopt better health-related attitudes and habits than do nonathletes.

Hypertension (HT)

Hemodynamic Response to Rhythmic Exercise

Children and adolescents with HT respond to rhythmic exercise with above normal systolic and diastolic arterial pressure at any given metabolic level.[26,30,42,56,89,103] In most patients, however, the *increments* in arterial pressure from rest to maximal exercise are not out of line with the normal response. The above pattern is shown in Fig. 4.7. During recovery from maximal exertion, normotensives and adolescents with HT have a similar rate of decline in arterial pressure.[26]

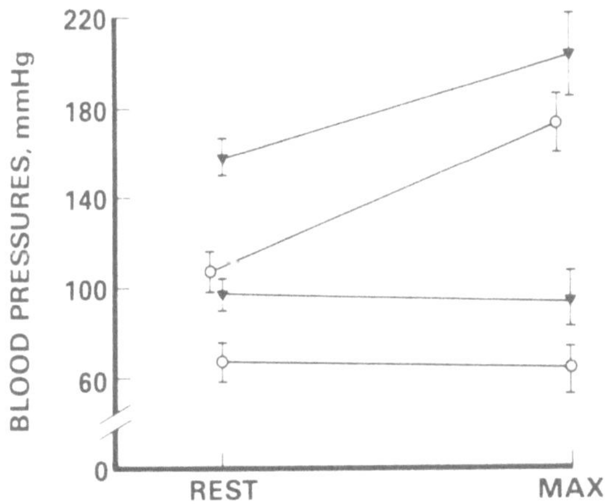

Figure 4.7. Juvenile hypertension and arterial blood pressure response to maximal exercise. Hypertensive adolescents (▼) and controls (O) were tested at rest and during an all-out cycling test. Vertical lines denote 1 S.D. Based on Nudel et al.[90]

In some juveniles with HT, the exercise-induced increase in blood pressure *is* abnormally high. To understand their hemodynamic characteristics one should bear in mind that the pressure gradient along a vessel is a product of blood flow and the vessel's resistance. A high systemic arterial pressure reflects, therefore, either a high cardiac output, a high total peripheral resistance, or a combination thereof. Healthy youths respond to exercise with a rise in cardiac output and a drop in peripheral resistance. Some hypertensive youngsters, on the other hand, have either an inappropriately high cardiac output during submaximal exercise or an insufficient decrease of total peripheral resistance.[43,62,90] The latter type of response suggests an insufficient compliance of the "resistance vessels" (arterioles and capillaries) during exercise.

Hemodynamic Response to Static Exercise

Static exercise has been given special consideration in the analysis of blood pressure response in hypertension. As discussed in Chapter 1, static exercise induces in adults a blood pressure rise above and beyond the response expected from mere metabolic demands.[71] On the other hand, in children—healthy or hypertensive—a blood pressure rise with static exercise is less dramatic than in adults.[29,99,104] There are no studies, however, comparing arterial pressure response of hypertensive children to dynamic and static exercise at equal metabolic levels. As in dynamic exercise, the *increments* in arterial pressure between rest and static exercise are similar in children with HT and in healthy children.

Physical Working Capacity and Habitual Activity

To what extent does their high blood pressure affect the exercise ability of youngsters with hypertension? Although some reports[29,43,90] suggest that maximal aerobic power of adolescents with hypertension is below normal, control groups tested in these studies[29,90] also had a low maximal aerobic power. Nor was there a difference in the anaerobic threshold of the two groups.[90]

The normal exercise performance in juvenile HT reflects a normal level of habitual activity. In one survey 41% of 14- to 17-year-old mildly hypertensive girls and boys reported participation in organized competitive sports or in vigorous conditioning. The respective participation among controls was 43%. It is, most probably, the asymptomatic nature of the disease that helps the patients keep normally active.

Exercise in the Assessment of Juvenile HT

Specific abnormalities to watch for are:

Inappropriate rise of systolic blood pressure (beyond 230 mm Hg in maximal power load).
Ventricular dysrhythmia.
"Ischemic" ST-T changes

In addition to the above proven usefulness of exercise testing, there are several *potential* benefits, still under research. Exercise stressing, for example, has unmasked renal hemodynamic deficiency (exaggerated reduction in renal plasma volume and glomerular filtration rate) in young adults with hypertension who showed no detectable renal damage at rest.[92] Whether this applies also to children has yet to be shown. It has been suggested that exercise tests might be useful to evaluate the effectiveness of antihypertensive drugs.[103] No data are available to substantiate this approach.

Another promising use of exercise is the screening of healthy children or adolescents for eventual development of HT. Such early identification will become of immense public health importance once means are available for prevention of the disease. Ideally, to isolate a valid predictor of future hypertension, one should conduct a prolonged longitudinal follow-up, the "end point" of which is hypertension. An alternative, more feasible approach is to take matched groups of adults or adolescents with or without hypertension and retrospectively evaluate their documented response to exercise in early years. In a study completed in the author's laboratory,[22] 11% of normotensive adolescents and young men who had an inappropriately high blood pressure response to exercise became hypertensive 3–14 years later. None of the matched individuals, with normal blood pressure response to exercise, developed HT. When normotensive children of hypertensive parents were studied,[47] a relationship was found between the level of hypertension of the mothers and the blood pressure response to maximal rhythmic exercise of their children. There was no such relationship between the mother's hypertension and the resting blood pressure of her children.

It has been suggested[5,104] that arterial pressure response to *static* exercise may be used as a predictor of future HT. This notion is based on the assumption that an arterial pressure response to mild static exercise primarily reflects changes in peripheral vascular resistance and, to a lesser degree, in cardiac output. While a healthy child responds to exercise by a dilatation of "resistance vessels" in the exercising muscle, a child with risk of future HT may have some early loss of vasodilatory capacity. This abnormality may not be manifested at rest, but only during exercise provocation—especially during static exercise. Other advantages of a static stress test are its simplicity and low cost, which make it more feasible for mass screening than rhythmic ergometry.

Meticulous evaluation is needed before exercise can be suggested as a screening test for future hypertension. Without a doubt, the vast importance of such screening is a challenge for further research.

Beneficial Effects of Chronic Exercise

Endurance-type conditioning may induce a mild reduction in the resting systolic and diastolic arterial pressure of adults, whether normotensive or with HT.[94] Similar results have been reported for adolescents with HT.[43] Although the biologic benefit of such a mild decrease—less than 10 mm Hg—is not clear, conditioning may be beneficial to borderline cases. Other effects of aerobic conditioning include a decrease in resting peripheral resistance, submaximal systolic arterial pressure, and heart rate and an increase in maximal O_2 uptake.[43]

Hypoactivity of children, on the other hand, may cause an *increase* in resting blood pressure. Such was the case among 12-year-old children who were immobilized in a cast or in traction. These patients had a fourfold greater incidence of above-normal blood pressure than had a control group of hospitalized but not immobilized children.[110]

Little information is available on the effects on HT of nonendurance conditioning. A 2-month weight-training program did not induce changes in blood pressure nor in the dimensions of the myocardium of 15- to 16-year-old HT patients.[68]

Is Exercise in Hypertension Detrimental to Health?

Should children and adolescents with HT be allowed intense exertion? In 1977, a Task Force on Blood Pressure Control in Children[106] concluded that:

> It seems appropriate to recommend supervised dynamic exercises, such as calisthenics, swimming, running, baseball, or basketball playing which improve cardiovascular fitness. On the other hand, static exercise such as weight-lifting, wrestling and isometrics raises both systolic and diastolic pressure and markedly stresses the heart; it may, therefore, impose an unacceptable load on hypertensive subjects, particularly those with evidence of left ventricular hypertrophy.

This cautionary statement against static exercise was based on extrapolation of findings from healthy adult individuals rather than direct evidence on juveniles with hypertension. Since then, more specific data have been gathered on the effects of static exercise on children and adolescents with HT. When 109 mildly hypertensive adolescents performed static hand grip (25% of maximal strength, sustained for 4 min), their diastolic arterial pressure increased by 12 mm Hg. In contrast, the same patients had a decrease of 21 mm Hg in diastolic pressure as a result of an all-out rhythmic exercise.[29,30] These findings are summarized in Fig. 4.8. Clinically hypertensive youths who were exposed to moderate static[29] or maximal rhythmic[29,89,90] exercise had no chest pain, ST depression, ventricular dysrhythmia, or hemodynamic abnormalities. Nor did a 2-month weight-training program induce deleterious effects on the resting arterial pressure of adolescent hypertensives or on their cardiac dimensions, as assessed by echocardiography.[68]

The above data pertain to mild and moderate HT only, and to specific types of static exercise. It seems justifiable not to restrict the physical activities of such patients, unless ST-T changes, left ventricular strain, or inappropriate blood pressure rise (maximal systolic pressure beyond 230 mm Hg, as recommended by the American Academy of Pediatrics[2]) appear during exercise testing. Until proven otherwise, a prudent approach should be taken regarding sports participation by *severe* hypertensives or those who respond with abnormal ECG or excessive blood pressure rise during exercise. For such patients precautions should be

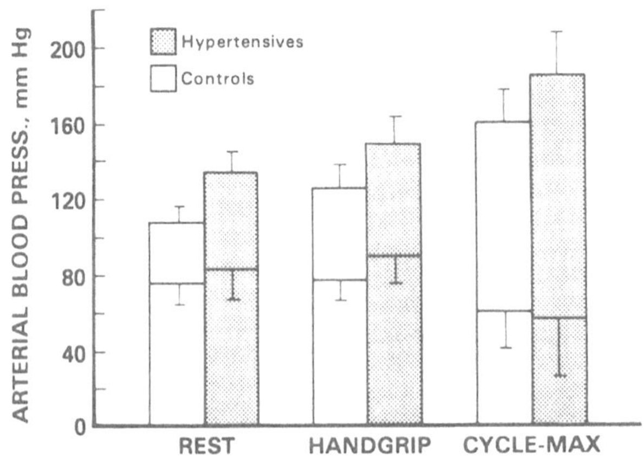

Figure 4.8. Hypertension and blood pressure response to static (handgrip) and maximal dynamic (cycle) exercise. Subjects were 109 hypertensive and 74 normotensive 14- to 17-year-old girls and boys. The handgrip was at 25% maximum voluntary force and lasted 4 min. An all-out progressive protocol was used for cycling. Upper horizontal lines represent systolic blood pressure, lower horizontal lines, diastolic blood pressure. Vertical lines denote 1 S.D. Data by Fixler et al.[29]

taken with sports such as wrestling, football line play, gymnastics, or some types of sailing, in which intense isometric muscle contraction is required.

Neurocirculatory Asthenia (NCA)

Response to Exercise and Physical Working Capacity

This syndrome (also called vasoregulatory asthenia, irritable heart, Da Costa's syndrome, and soldier's heart) appears mostly in young adults but is also seen among adolescents. Affected individuals complain of palpitations, chest pain, breathlessness, and fatigue. In some patients these complaints are manifested mostly during exercise. In others they appear during rest and exercise alike. In both groups, especially the latter, there is evidence of neurosis, with no signs of organic heart disease.

The main sign is a high heart rate at rest, during mild exercise, and following orthostatic changes. During strenuous activities (which the patient is often reluctant to perform), heart rate is still high but not as excessively as in milder loads. Maximal heart rate is normal for age. Hemodynamic functions such as stroke volume and intracardiac pressures are all normal.[48] Characteristically, patients have high pulmonary ventilation and ventilatory equivalent, suggesting wasteful ventilation. Nonspecific ST-T changes and ventricular dysrhythmia may appear at rest and disappear with exercise.[100]

NCA patients often have a low physical working capacity, which is a combined manifestation of their anxiety, fear of exertion, and detraining. Although the pathogenesis of this syndrome is not clear, it has been assumed to result from vasoregulatory instability and sympathetic (β-receptor) overactivity.

Exercise as a Diagnostic Tool

Testing in NCA is indicated primarily for assessment of working capacity. However, a *diagnosis* of this syndrome may sometimes be made through exercise. This occurs when a patient appears at the exercise laboratory for assessment of exertional chest pain or excessive dyspnea. A high heart rate at mild exercise levels without evidence of organic heart disease can yield the diagnosis. Further questioning will often reveal apprehension, reluctance to take part in sports, and, occasionally, anxiety.

One should be cautious in interpreting the results of a submaximal test in which only heart rate is measured. If such a procedure is adopted (e.g., the $\dot{W}_{170}$ test or predicted maximal O_2 uptake), high heart rate at mild power loads will lead to underestimation of maximal aerobic power. Such patients should therefore be given higher-intensity tasks, preferably all-out ones. It has been our experience that some NCA patients who have exceedingly high heart rate at a low exercise level can still reach a close to normal peak power output.

NCA should be differentiated from the "hyperkinetic heart syndrome." Here, as in NCA, resting heart rate is high, reaching 100–120 beats/min. Unlike NCA, however, it is high at *all* levels of exercise, with maximal values reaching 15–20 beats/min above the age-predicted maximal heart rate.[105] Adolescents with this syndrome are often active athletes, with above normal maximal aerobic power. An example of heart rate response to exercise is shown in Fig. 4.9 for three adolescents—one with NCA, one with hyperkinetic heart syndrome, and one a healthy control.

Beneficial Effects of Conditioning

NCA patients can markedly benefit from physical conditioning. Regimens of various durations and intensities have been found effective. They induce a relief in subjective complaints, disappearance of ECG

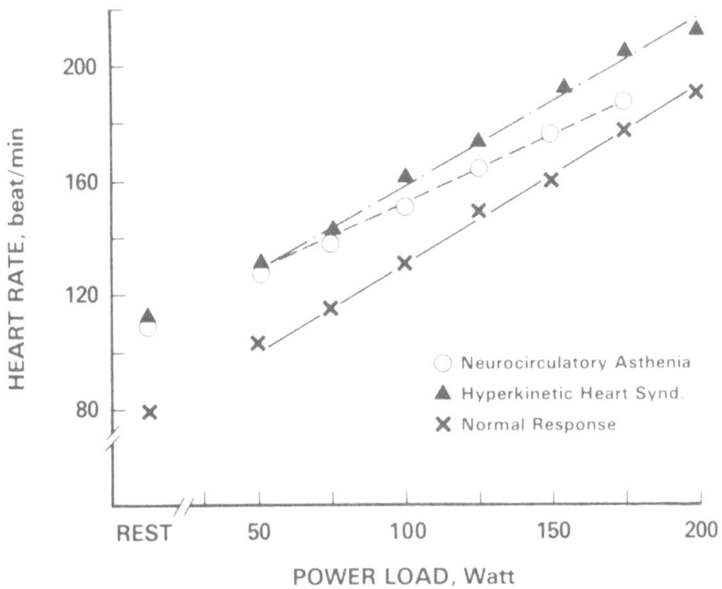

Figure 4.9. Comparison of heart rate response to exercise in three nonathletic 15-year-old boys, one with neurocirculatory asthenia, one with hyperkinetic heart syndrome, and one healthy child.

abnormalities, improvement in orthostatic response, and a decrease in resting and exercise heart rate. These are far beyond changes expected from mere improvement in cardiovascular dimensions or the O_2 carrying capacity of the blood.[49,67,100] The end result is a better subjective tolerance of exertion and an objective improvement in maximal aerobic power.

Those NCA patients who can be persuaded to participate in exercise rehabilitation programs are primarily the non-neurotic. Patients who have a major neurotic component are harder to motivate.[100] The latter often have a home environment that is not conducive to such participation.

Pulmonary Stenosis (PS)

The main hemodynamic abnormality in PS stems from an increased resistance at the right ventricular outflow tract, which results in a high pressure gradient across the stenotic site. The most common disabilities suffered by children with PS are exertional dypsnea and easy fatigability.

Hemodynamic Abnormalities in the Preoperative Child

Table 4.4 is a summary of hemodynamic responses to exercise in the child with PS. The most consistent pressure-related abnormality is an elevation at rest, and especially during exercise, of peak right ventricular systolic pressure. This pressure is high due to valvar narrowing and a compensatory ventricular hypertrophy. To date only a weak relationship has been found between the *degree* of narrowing and the exercise-induced rise in right ventricular systolic pressure.

As in aortic stenosis, there is a post-stenotic decrease of pressure in PS. The result is a high right outflow pressure gradient. Values as high as 50–80 mm Hg have been described during exercise, compared with 5–

Table 4.4. Abnormal Hemodynamic Responses to Exercise of the Child with Pulmonary Stenosis

Variable	*Comparison with a Healthy Child*
Peak right ventricular systolic pressure	High
Right outflow pressure gradient	High
Right ventricular end-diastolic pressure	High
Stroke volume	Low
Cardiac output	Low

30 mm Hg at rest.[19,27,114] The gradient is related to, and reflects, the cross-sectional area of the pulmonary valve.

Whereas the normal response to exercise is no change, or a slight drop, in right ventricular end-diastolic pressure,[83,101] this pressure rises with exercise in children with PS.[83,87,101,114] Furthermore, a relationship has been found between the degree of stenosis and the elevation with exercise of right ventricular end-diastolic pressure, as depicted in Fig. 4.10.

Whereas the stroke volume of healthy children increases with exercise, the child with PS often does not elevate his stroke volume on transition from rest to exercise. This is especially so when stenosis is severe.[52,83] Figure 4.11 represents a relationship between the degree of valvar narrowing and the difference between resting and exercising stroke index. While the children with mild narrowing had a normal stroke index response to exercise, those with extreme stenosis (valve area less than 0.5 cm^2/m^2) did not raise their stroke index at all.[51,83]

An insufficient rise of stroke volume during exercise results in subnormal cardiac output, especially at high levels of exertion when heart rate can no longer rise. Elevation of right ventricular end-diastolic pressure with no concomitant increase in stroke volume reflects a reduced compliance of the myocardium which, in children with PS, could be due to right ventricular hypertrophy.[83]

Hemodynamic Abnormalities Following Pulmonary Valvotomy

From a hemodynamic point of view, successful pulmonary valvotomy should result in a markedly decreased pulmonary pressure gradient, in normal right ventricular systolic and end-diastolic pressures, and in an adequate increase in stroke volume and cardiac output during exercise. Such a return to normal has not always been demonstrated in those operated upon as adults. The mechanisms causing persistent subpar function in adults are not clear, but decreased right ventricular compliance due to myocardial fibrosis[63] and long-standing peripheral vascular or left myocardial adaptation[65] have been implicated. Surgical results with children are more encouraging: stroke volume rises with exercise,[3,27,101] and right ventricular end-diastolic pressure assumes a normal pattern.[27,101]

Physical Working Capacity

The ability to exert strenuously is somewhat reduced in children with PS.[3,17] Tested by the Bruce treadmill test, for example, very few children with mild or moderate stenosis scored above the 75th percentile. Some

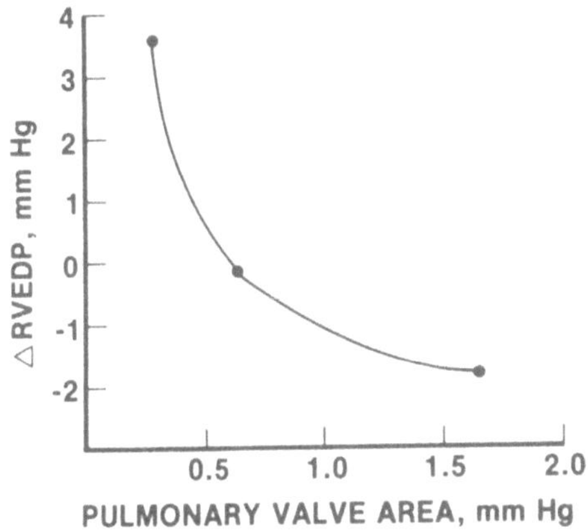

Figure 4.10. Exercise minus resting right ventricular end-diastolic pressure (ΔRVEDP) as a function of pulmonary valve area in 64 children with pulmonary stenosis. Data from Moller et al.[83]

70% of all patients scored below the 50th percentile. Those with severe stenosis had, as a group, the lowest scores.[17]

Perhaps most relevant to the low exercise performance of patients with PS are their subnormal stroke volume and cardiac output. In submaximal exercise, the child can compensate for low cardiac output by a greater extraction of oxygen from the blood, as manifested by an increased arterial mixed-venous O_2 difference. Such compensation is not possible, however, during maximal exercise, when O_2 extraction can no longer rise. Thus the low maximal cardiac output results in reduced maximal aerobic power. After valvotomy, the ability to raise stroke volume determines the improved exercise performance, those children with subnormal stroke volume still having a deficient maximal O_2 uptake.[3]

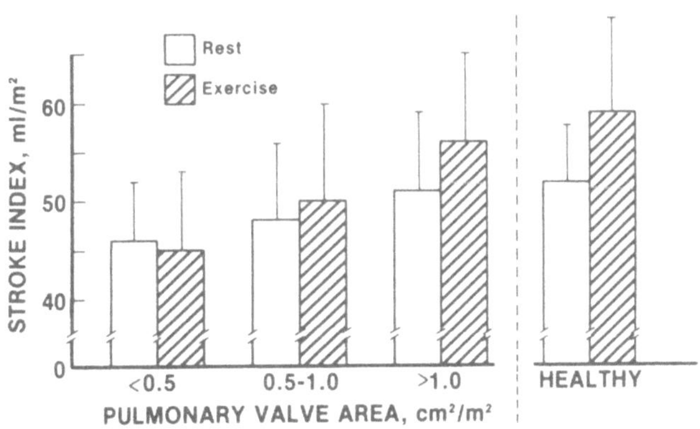

Figure 4.11. Stroke index at rest and during moderate supine exercise as a function of pulmonary valve area in 64 children with pulmonary stenosis (adapted from Ref. 83). A comparison is made with 31 boys with no cardiac disease.[15] Mean ± 1 S.D.

Exercise as a Diagnostic Tool

Although decisions regarding clinical management are based on evaluation of the resting patient, exercise testing has become a valuable auxiliary diagnostic tool, for the following:

Indications for pulmonary valvotomy.
Functional evaluation of surgical repair.

Right ventricular systolic pressure of 75 mm Hg or more at rest is a sufficient indication for surgical intervention. The exercise-related indication is a rise in right ventricular end-diastolic pressure, especially when accompanied by a decline in stroke volume and a cardiac output of less than 6 liters/min at an O_2 uptake of 600 ml/min (J.H. Miller, M.D. Freed, personal communication).

Evaluation by exercise of surgical repair is based on changes in stroke volume. As shown in Fig. 4.12, the effects of the operation on stroke volume are hardly noticeable at rest. In contrast, it becomes distinctly higher during exercise. No rise in stroke volume would suggest residual myocardial damage. The use of stroke volume for diagnosis is especially important because it can be determined by noninvasive techniques.

Isoproterenol as a Substitute for Exercise Testing. The recognized value of exercise stress testing in PS has led pediatric cardiologists to seek alternative stressors that may simulate the chronotropic and inotropic effects of exercise.[8,85] Such a substitute is often needed for the very young, the feeble, or the ill-motivated child who cannot perform a standardized exercise test. Two studies[87,114] directly compared the effects of exercise and isoproterenol on PS patients. Both stimuli caused an increase in heart rate to about 150 beats/min. However, exercise induced a

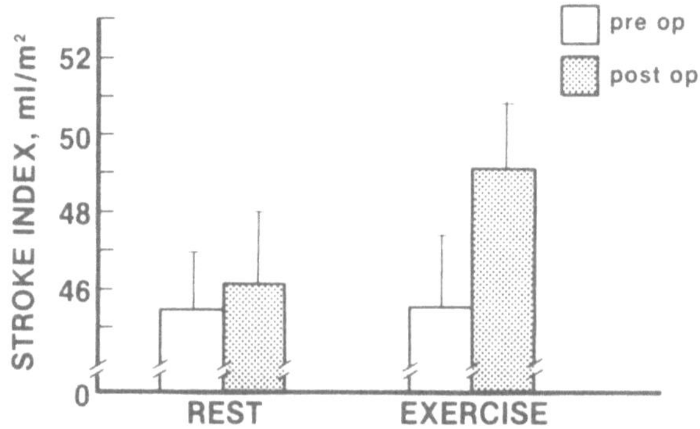

Figure 4.12. Pre- and postoperative stroke index at rest and during exercise in 20 children with pulmonary stenosis. Vertical lines denote 1 S.D. Data from Stone et al.[101]

higher increase in cardiac output than did isoproterenol. The latter may cause an overestimation of the severity of the stenosis. For a complete hemodynamic evaluation of the pulmonary stenotic child, the use of exercise is therefore preferable. Whenever an exercise test is not feasible, isoproterenol provocation can be used as a substitute.

Septal Defect—Atrial (ASD)

While most patients with this common defect are asymptomatic, some may complain of excessive fatigue during exercise and play.

Hemodynamic Response to Exercise

Two interdependent hemodynamic functions that change with exercise are of interest: one is the extent of the left-to-right shunt; and the other, an increase in pulmonary arterial pressure. There are conflicting data on the exercise-induced change in the shunt. In some studies[54] it has been shown to increase, while in others[4] it has not changed, or has even decreased from rest to mild or moderate exercise. The main reason for such a variable response is the difficulty in obtaining accurate values of mixed-venous O_2 content. These values are needed for the calculation of cardiac output by the Fick equation. Pulmonary arterial pressure increases with exercise, especially in older age-groups. In children, on the other hand, the rise is mild and systolic values seldom exceed 40–50 mm Hg.[4]

Physical Working Capacity

The exercise performance of children with mild to moderate ASD of the secundum type is quite normal or mildly deficient.[17,24,40] In patients with more severe lesions the deficiency is more pronounced. For example, 33% of children with severe ASD scored at the 10th percentile or lower in the Bruce treadmill test.[17] Especially handicapped are those children with pulmonary hypertension. However, the extent of the shunt *per se* does not seem to appreciably affect exercise tolerance.[24,40] Successful operation will reverse the deficiency so that the patient may have normal working capacity.[17] If surgical correction is postponed until adulthood, however, cardiac output remains abnormally low during exercise and the working capacity is deficient. Sparse data available[76] indicate that some children with corrected ASD are trainable and can increase their maximal aerobic power in response to conditioning. The degree of trainability is not related to the extent of the preoperative shunt.

Exercise in the Assessment of the Child with ASD

The main use of exercise testing is in post-surgical evaluation. Abnormalities that may appear during exercise—but not at rest—are:

Inappropriately low cardiac output.
Rare dysrhythmias.

When catheterization is not feasible, one can measure maximal aerobic power, which, with successful surgical correction, should improve. If this is not reached, residual myocardial or pulmonary vascular deficiency should be suspected.

Septal Defect—Ventricular (VSD)

In this lesion, functional severity depends on the size of the septal defect and on the pressure developing in the pulmonary circulation. These two factors determine the extent of the left-to-right (or the less common right-to-left) shunt. Whereas children with a small shunt and no pulmonary hypertension are virtually asymptomatic, those with a large shunt may fatigue easily and have excessive exercise dyspnea.

Hemodynamic Response to Exercise

Because most VSD children undergo corrective surgery before the age of 2 years, there is little information on the response to exercise of the preoperative patient or the one who is treated by pulmonary banding. In 7- to 15-year-old children who had VSD with pulmonary hypertension (with or without pulmonary banding), the main hemodynamic abnormality during intense exercise was a marked rise in mean pulmonary arterial pressure. Values ranged between 65 and 148 mm Hg, which in some cases equaled or exceeded the systemic pressure. The end result was diminution of the left-to-right shunt and a marked increase in the ratio of pulmonary vascular resistance to systemic resistance. In some children, the shunt may even be reversed during intense exercise.[80] In contrast, VSD patients without pulmonary hypertension will respond to exercise with an *increase* of their left-to-right shunt.[54]

Surgical correction will improve the hemodynamic function of most children with VSD. There is an inverse relationship between age at operation and cardiovascular function: those operated on before the age of 10 have better myocardial function (as manifested by low cardiac output) and respond to intense exercise with less elevation of pulmonary arterial pressure than those operated on in later years.[74] The mechanism for such an age-related response is not clear.

Physical Working Capacity

The maximal aerobic power of the child with VSD can vary from normal to highly deficient.[17,40] It is directly related to the extent of the lesion: patients with a severe septal defect seldom score above the 25th percentile.[17] Surgical correction, especially if performed before school age, can normalize the child's working capacity.

Exercise in the Assessment of the Child with VSD

Indications for operation or signs of inoperability are based on the resting values of pressures and blood flow. Exercise is valuable, however, in post-surgical evaluation. Abnormalities that appear with exercise—but not at rest—are:

Inappropriately low cardiac output.
High pulmonary blood pressure.
Rare dysrhythmias.

Tetralogy of Fallot (TF)

There are two abnormal hemodynamic functions in TF that cause circulatory and respiratory deficits at rest and, particularly, during exercise: high right ventricular outflow-tract resistance and a right-to-left shunt. The first results in an elevated pressure gradient across the stenotic area (infundibulum, pulmonary valve, or major pulmonary artery), reduced pulmonary blood flow, and, eventually, right ventricular hypertrophy. The shunt results in venous admixture and peripheral hypoxia. Subjectively, many of these patients complain of excessive fatigability and exertional dyspnea. The major hemodynamic and pulmonary changes that occur with exercise are summarized in Table 4.5.

Hemodynamic Characteristics—The Preoperative Child

As in isolated pulmonary stenosis, exercise induces an increase in the pressure gradient across the stenotic area in right ventricular systolic and end-diastolic pressures. A typical response is the reversal of a left-to-right shunt (or a small right-to-left shunt) in a noncyanotic child with TF into a considerable right-to-left shunt with cyanosis.[37,108] Such a reversal occurs when the pulmonary stenosis is mild and, at rest, does not cause much obstruction to flow. With exercise, myocardial contraction rises and the narrowed pulmonary valve can no longer accommodate the increased flow of blood. The end result is back-pressure in the right

Table 4.5. Hemodynamic and Pulmonary Response to Exercise of Children with Uncorrected Tetralogy of Fallot

Variable	*Response to Exercise (compared with normal)*
Peak right ventricular systolic pressure	High
Right outflow pressure gradient	High
Right ventricular end-diastolic pressure	High
Pulmonary blood flow	Low
Right-to-left shunt	High
Right "forward" cardiac output	Low
Arterial O_2 saturation	Low
Minute ventilation	High
Physiologic dead space	High

ventricle, which reverses the trans-septal shunt. Exercise can induce up to a four-fold increase in the right-to-left shunt, with a marked venous admixture and cyanosis.[37] In some children the very first occurrence of cyanosis may be related to a bout of exercise.

Respiratory Characteristics—The Preoperative Child

A diminished ventilatory response to hypoxia has been described for the resting child with TF. In contrast, most exercising patients show marked hyperventilation, the degree of which is related to arterial O_2 desaturation,[109] as depicted in Fig. 4.13. In extreme cases, exercise can induce a fall of 40% in arterial O_2 saturation and a concomitant increase in pulmonary ventilation of 150% above the expected value for a given metabolic level. The reduced pulmonary blood flow and increased ventilation result in a large physiologic dead space.

Response to Exercise after Surgical Repair

When only a palliative operation is performed (i.e., aorto-pulmonary shunting), the hemodynamic and ventilatory responses to exercise remain abnormal,[38] despite some improvement in lung perfusion.

Following total surgical correction, some 90% of patients will survive to adulthood and 80% of these will lead a normal, active life.[33] The most obvious hemodynamic improvements are the lower pulmonary pressure gradient and right ventricular systolic pressure, the higher pulmonary blood flow, and the diminution or disappearance of the right-to-left shunt. In many children there is also a decrease in right ventricular end-diastolic pressure.

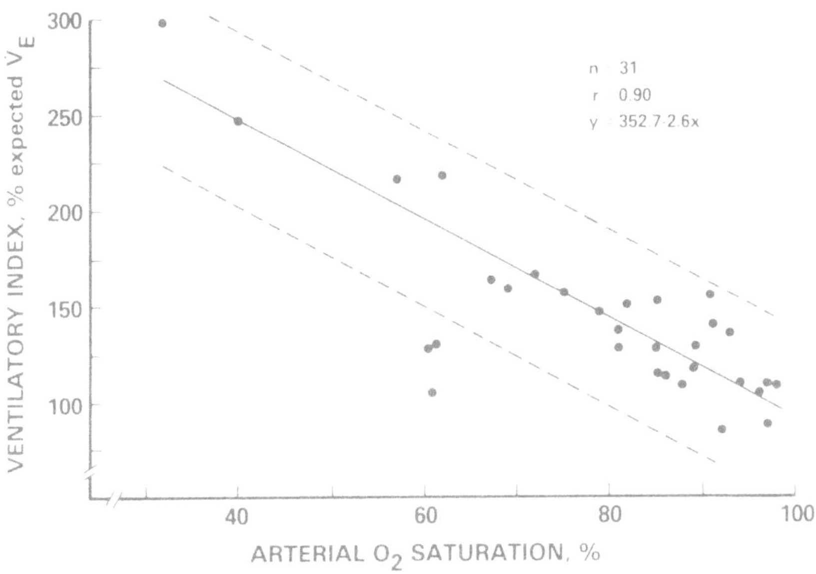

Figure 4.13. Ventilatory response to exercise in cyanotic congenital heart disease. Ventilatory index (observed ventilation as percentage of expected ventilation, $\dot{V}_E$) as a function of resting arterial O_2 saturation in 17 patients, 6- to 14.5-years old. Dotted lines are ±2 S.E.E. Data from Taylor.[109]

There may be residual deficiencies, however, which are more apparent during exercise than at rest. These include high right ventricular systolic pressure,[16,18,25] with values exceeding 100 mm Hg during maximal exercise; elevated right ventricular end-diastolic pressure,[7,18,25,46,66] with values reaching 20 mm Hg; arterial desaturation and increased physiologic dead space;[102] appearance of ventricular dysrhythmia;[32] and reduced stroke volume and cardiac output during submaximal and maximal exercise.[7,18,66,79] During submaximal exercise, such a "hypokinetic" circulatory response can be compensated for by an increase in arterial mixed-venous O_2 difference. In maximal exercise, however, the low stroke volume and cardiac output result in a reduced maximal aerobic power.[79]

The mechanisms for the deficient right ventricular function after surgery are not clear. They could result from the ventricular incision, residual outflow-tract obstruction, pulmonary valve insufficiency, an increase in right ventricular systolic work, or reduced myocardial compliance due to hypertrophy and scarring.[18,79] The residual O_2 desaturation may reflect pulmonary vascular disease, especially in those individuals who have undergone palliative aorto-pulmonary shunting.[102]

The degree of hemodynamic improvement following total correction depends on the residual right outflow-tract resistance and the closure of the septal defect. In Fig. 4.14 are shown right ventricular functions of two groups of children with TF, both having undergone total correction.[18] One group had a residual high pulmonary pressure gradient, with or without a residual septal defect. The other had neither a high gradient nor a residual septal defect. As seen in Fig. 4.14, the right ventricular function is distinctly better in the group without residual obstruction, at rest and, especially, during exercise.

Habitual Activity

The sparse data available on the activity of children with TF refer only to those patients who have undergone total surgical repair. Using a 6-

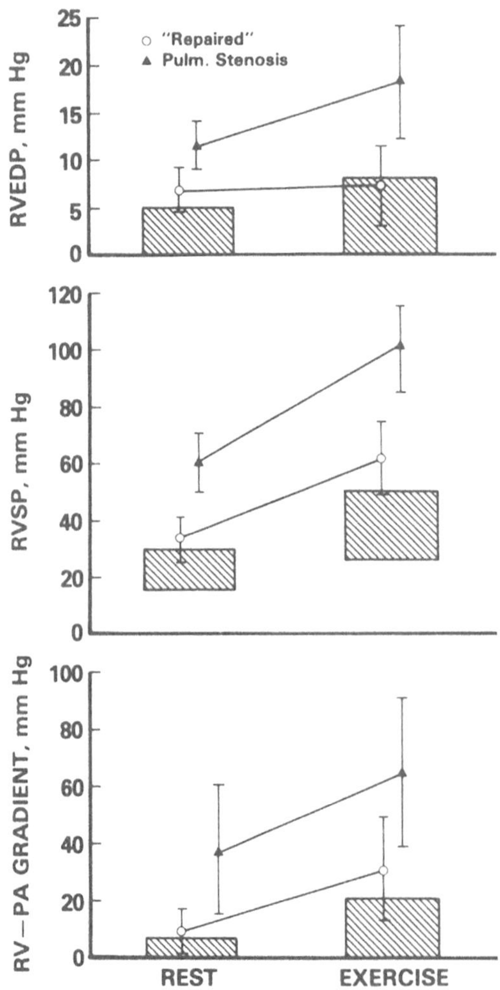

Figure 4.14. Right ventricular function in tetralogy of Fallot, following surgical correction. Patients were 5- to 14-year-old girls and boys who had undergone surgical correction 1–5 years prior to testing. In one group ("Repaired," $n = 10$) the septal defect was fully repaired and there was a resting right outflow gradient of 20 or less mm Hg. The other group ("Pulmonary stenosis," $n = 12$) had a right outflow gradient of more than 22 mm Hg, with or without a residual septal defect. Values at rest and following maximal supine exercise are shown for right ventricular end-diastolic pressure (RVEDP), right ventricular systolic pressure (RVSP), and pressure gradient between the right ventricle and the pulmonary artery (RV-PA gradient). Shaded areas denote normal range. Vertical lines = 1 S.D. Adapted by permission from Cumming.[18]

month recall questionnaire, Lambert et al[69] did not find any difference between the habitual activity of male adolescents with TF and that of healthy controls. Among the girls, however, TF patients attended fewer physical education classes than did the controls. A significant correlation between maximal O_2 uptake and level of activity was found among the females. In another study, only 50% of postoperative patients with TF took part in physical activities at school.[79] An overprotective attitude of the parents, which may persist for years after the operation, has been cited as the major cause of hypoactivity in these patients.[69,79]

Physical Working Capacity

Unlike most other congenital heart defects, TF is accompanied by a distinctly low maximal aerobic power. Successful surgical correction causes an improvement in maximal aerobic power of most patients such that performance approaches normal levels.[17,18,21,32,46,69,79,84,102,123] One example is a study by Cumming[17]: performing the Bruce treadmill test, some 90% of unoperated children with TF scored below the 25th percentile. Palliative surgery did not seem to improve the score, but total correction induced a near normalization of performance.

Why do some patients still have a low maximal aerobic power after total correction? Possible causes of such a residual abnormality are outlined in Table 4.6. Among these, the most important are residual pulmonary stenosis or insufficiency, disturbed myocardial contractility, and a sedentary lifestyle. Disturbed contractility is manifested by low maximal stroke volume, which is strongly related to the decreased maximal O_2 uptake. An inverse relationship has been found between maximal aerobic power and the age at which surgical correction took place. As shown schematically in Fig. 4.15, a patient who undergoes correction at the age

Table 4.6. Possible Causes of Low Physical Working Capacity in Tetralogy of Fallot Patients Who Have Undergone Total Surgical Correction

Hemodynamic	Pulmonary valve insufficiency Residual pulmonary valve stenosis Low stroke volume and cardiac output Right outflow aneurism Residual right-to-left shunt
Pulmonary	Pulmonary vascular disease Deficient gas exchange
History and habits	Operation performed late in life Sedentary lifestyle

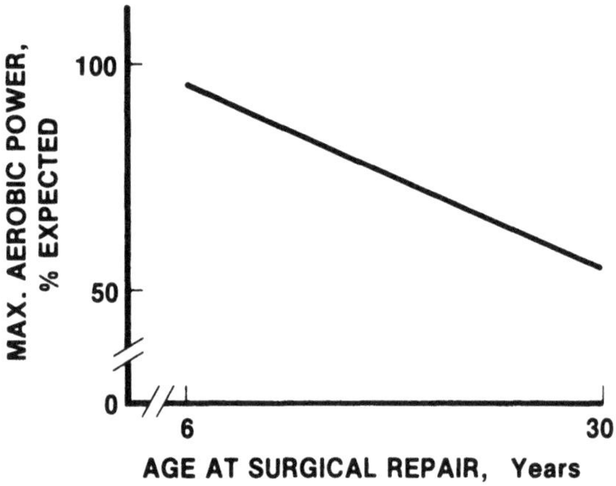

Figure 4.15. Maximal aerobic power as a function of the age at which surgical repair of tetralogy of Fallot was performed. A schematic presentation of data reported by Strieder et al[102] and Mocellin et al.[79]

of 6 years will have a good chance of reaching normal maximal aerobic power. Any delay in surgical intervention may result in reduced fitness.[18,79,84,102]

Exercise as a Diagnostic Tool

Preoperative assessment by exercise is useful in all patients because it can serve as a baseline for postoperative comparisons. Specifically, one would like to answer the following questions:

Has operation improved hemodynamic function?
Has operation improved physical working capacity?
Is there a post-surgical ventricular dysrhythmia?

Postoperative changes in hemodynamic and respiratory function and in physical working capacity have been described in previous sections. Exercise-induced ventricular dysrhythmia may appear after surgical correction and is of prognostic value. Any child who responds to exercise with ventricular dysrhythmia, especially if multifocal, is at some risk of sudden death.[32,60] We therefore strongly recommend that an exercise ECG be periodically determined (e.g., twice yearly) in any child with TF who has undergone total surgical correction. Moreover, a fairly strong relationship has been shown between exercise-induced ventricular ectopy and abnormal hemodynamic status.[32] An exercise ECG can therefore be used as a simple noninvasive tool to screen those patients with a residual hemodynamic deficiency.

Permissible Activities

The preoperative child will spontaneously limit his or her activity due to dyspnea and fatigue. Restrictions need not be imposed, therefore, at that stage. With functional improvement following surgery, the child may wish to engage in more strenuous activity. As a rule, he or she may enjoy normal participation in physical education classes 6 months after the operation (G.R. Cumming, personal communication). Participation in competitive sports should be permitted on a gradual basis, barring any appearance of dysrhythmia or pulmonary hypertension.

The main concern for the postoperative child is ventricular dysrhythmia. This conduction defect (single premature ventricular beat or more severe dysrhythmia) may be elicited by strenuous exercise in 20–30% of patients.[32]

Sudden death may occur in patients with TF soon after, or even many years following, "successful" surgical correction. Although no cause-and-effect relationship has been proven between sudden death and exertion, some patients die while exerting.[60] Such a possibility must be borne in mind whenever advising a child with TF about permissible physical activity. A prudent policy is to determine periodically the ECG of each patient both at rest and during exercise. Those who respond with ventricular ectopy should be instructed to avoid strenuous exertion and should be given antidysrhythmic therapy.

References

1. Alpert BS, Kartodihardjo W, Harp P, et al: Exercise blood pressure response—a predictor of severity of aortic stenosis in children. J Pediatr 98:763–765, 1981.
2. American Academy of Pediatrics Policy Statement: Cardiac evaluation for participation in sports. American Academy of Pediatrics, Evanston Ill., 1977.
3. Bastanier C, Kaltwasser B, Mocellin R: Postoperative Belastungsuntersuchungen bei Kindern und Jugendlichen mit valvulärer Pulmonalstenose. Z Kardiol 66:587–593, 1977.
4. Bay G, Abrahamsen AM, Muller C: Left-to-right shunt in atrial septal defect at rest and during exercise. Acta Med Scand 190:205–209, 1971.
5. Berenson GS, Voors AW, Webber LS, et al: Racial differences of parameters associated with blood pressure levels in children. The Bogalusa heart study. Metabiologica 28:1218–1228, 1979.
6. Biener K: Tabakkonsum und Sportverhaltung Jugendlicher. Schweiz Rundchau Med 65:78–81, 1976.
7. Bristow JD, Kloster FE, Lees MH, et al: Serial cardiac catheterization and exercise hemodynamics after correction of tetralogy of Fallot. Circulation 41:1057–1066, 1970.
8. Brodsky SJ, Krovetz LJ, Schiebler GL: Assessment of severity of isolated valvular pulmonic stenosis using isoproterenol. Am Heart J 80:660–670, 1970.
9. Campbell M: Congenital complete heart block. Br Heart J 5:15–18, 1943.

10. Campbell M, Emanuel R: Six cases of congenital complete heart block followed for 34 to 40 years. Br Heart J 29:577–587, 1967.
11. Chandramouli B, Ehmke DA, Lauer RM: Exercise-induced electrocardiographic changes in children with congenital aortic stenosis. J Pediatr 87:725–730, 1975.
12. Chawla K, Serratto M, Cruz J, et al: Response to maximal and submaximal exercise testing in patients with congenital complete heart block (abstract). Circulation 56[Suppl. III]:171, 1977.
13. Cueto L, Moller JH: Haemodynamics of exercise in children with isolated aortic valvular disease. Br Heart J 35:93–98, 1973.
14. Cumming GR: Exercise studies in clinical pediatric cardiology. In: Lavallée H, Shephard RJ (eds.) Frontiers of Activity and Child Health. Pélican, Quebec, 1977, pp. 17–45.
15. Cumming GR: Hemodynamics of supine bicycle exercise in normal children. Am Heart J 93:617–622, 1977.
16. Cumming GR: Exercise studies in children after corrective surgery for tetralogy of Fallot. In: Lavallée H, Shephard RJ (eds.) Frontiers of Activity and Child Health. Pélican, Quebec, 1977, pp. 371–384.
17. Cumming GR: Maximal exercise capacity of children with heart defects. Am J Cardiol 42:613–619, 1978.
18. Cumming GR: Maximal supine exercise haemodynamics after open heart surgery for Fallot's tetralogy. Br Heart J 41:683–691, 1979.
19. Cumming GR, Mir GH: Effects of propranolol on the resting and exercise hemodynamics of pulmonary stenosis. Can J Physiol Pharmacol 47:137–142, 1969.
20. Cumming GR, Mir GH: Exercise haemodynamics of coarctation of the aorta. Acute effects of propranolol. Br Heart J 32:365–369, 1970.
21. Delisle G, Olley PM: Épreuve d'effort sous-maximal chez les enfants atteints de tétralogie de Fallot: avant et après correction chirurgicale. Union Med Can 103:886–889, 1974.
22. Dlin RA, Hanne N, Silverberg DS, Bar-Or O: Follow-up of normotensive men with exaggerated blood pressure response to exercise. Am Heart J, in press.
23. Doyle EF, Arumugham P, Lara E, Rutkowski MR, Kiely B: Sudden death in young patients with congenital aortic stenosis. Pediatrics 53:481–489, 1974.
24. Duffie ER, Adams FH: The use of the working capacity test in the evaluation of children with congenital heart disease. Pediatrics 32:757–768, 1963.
25. Epstein SE, Beiser GD, Goldstein RE, et al: Hemodynamic abnormalities in response to mild and intense exercise following operative correction of an atrial septal defect or tetralogy of Fallot. Circulation 45:1065–1075, 1973.
26. Falkner B, Lowenthal DT: Dynamic exercise response in hypertensive adolescents. Int J Pediatr Nephrol 1:161–165, 1980.
27. Finnegan P, Ihenacho HN, Singh SP, Abrams LD: Haemodynamic studies at rest and during exercise in pulmonary stenosis after surgery. Br Heart J 36:913–918, 1974.
28. Fisch C: Complete heart block. A study of two cases in veterans of World War II. N Engl J Med 238:589–592, 1948.

29. Fixler DE, Laird WP, Brown R, et al: Response of hypertensive adolesents to dynamic and isometric exercise stress. Pediatrics 64:579–583, 1979.
30. Fixler DE, Laird WP, Fitzgerald V, et al: Effect of isometric and dynamic exercise stress on hypertensive adolescents (abstract). Pediatr Res 12:364, 1978.
31. Flamm MD, Braniff BA, Kimball R, Hancock EW: Mechanism of effort syncope in aortic stenosis (abstract). Circulation 35, 36[Suppl. II]:109, 1967.
32. Garson A Jr, Gillette PC, Gutgesell HP, McNamara DG: Stress-induced ventricular arrhythmia after repair of tetralogy of Fallot. Am J Cardiol 46:1006–1012, 1980.
33. Garson A, Nihill MR, McNamara DG, Cooley DA: Status of the adult and adolescent after repair of tetralogy of Fallot. Circulation 59:1232–1240, 1979.
34. Gilliam TB, Burke MB: Effects of exercise on serum lipids and lipoproteins in girls, ages 8 to 10 years. Artery 4:203–213, 1978.
35. Gilliam TB, MacConnie SE, Geenen DL, et al: Exercise programs for children: a way to prevent heart disease? Physician Sportsmed 10:96–108, 1982.
36. Glueck CJ: Detection of risk factors for coronary artery disease in children: Semmelweis revisited? Pediatrics 66:834–837, 1980.
37. Godfrey S: Exercise Testing in Children. Applications in Health and Disease. W.B. Saunders, Philadelphia, 1974.
38. Gold WM, Mattioli LF, Price AC: Response to exercise in patients with tetralogy of Fallot with systemic-pulmonary anestomoses. Pediatrics 43:781–793, 1969.
39. Goldberg SJ, Adams FH, Hurwitz RH: Effect of cardiac surgery on exercise performance. J Pediatr 71:192–197, 1967.
40. Goldberg SJ, Mendes F, Hurwitz R: Maximal exercise capability of children as a function of specific cardiac defects. Am J Cardiol 23:349–353, 1969.
41. Goldberg SJ, Weiss R, Kaplan E, Adams FH: Comparison of work required by normal children and those with congenital heart disease to participate in childhood activities. J Pediatr 69:56–60, 1966.
42. Goldring D, Hernandez A, Choi S, et al: Blood pressure in a high school population: II. Clinical profile of the juvenile hypertensive. J Pediatr 95:298–304, 1979.
43. Hagberg JM, Ehsani AA, Heath GW, et al: Beneficial effects of endurance exercise training in adolescent hypertension (abstract). Presented at 29th Annual Meeting of the American College of Cardiology, 1980.
44. Halloran KH: The telemetered exercise electrocardiogram in congenital aortic stenosis. Pediatrics 47:31–39, 1971.
45. Hanne N, Drory Y, Kellermann JJ: Complete heart block and physical performance. Submitted for publication.
46. Hirschfeld S, Tuboku-Metzger AJ, Borkat G, et al: Comparison of exercise and catheterization results following total surgical correction of tetralogy of Fallot. J Thorac Cardiovasc Surg 75:446–451, 1978.
47. Hohn AR, Riopel DA, Keil JE, et al: Blood pressure and humoral factors in children of hypertensive parents (abstract). Pediatr Res 12:383, 1978.

48. Holmgren A: Vasoregulatory asthenia. Can Med Assoc J 96:904–905, 1967.
49. Holmgren A, Jonsson B, Levander M, et al: Effect of physical training in vasoregulatory asthenia and in neurosis without heart symptoms. Acta Physiol 165:891–902, 1952.
50. Holmgren A, Karlberg P, Pernow B: Circulatory adaptation at rest and during muscular work in patients with complete congenital heart block. Acta Med Scand 164:119–130, 1959.
51. Hossack KF, Neilson GH: Exercise testing in congenital aortic stenosis. Aust NZ J Med 9:169–173, 1979.
52. Howitt G: Hemodynamic effects of exercise in pulmonary stenosis. Br Heart J 28:152–160, 1966.
53. Hugenholtz PG, Lees MM, Nadas AS: The scalar electrocardiogram, vectorcardiogram and exercise electrocardiogram in the assessment of congenital aortic stenosis. Circulation 26:79–91, 1962.
54. Hugenholtz PG, Nadas AS: Exercise studies in patients with congenital heart disease. Pediatrics 32:769–775, 1963.
55. Ikkos D, Hanson JS: Response to exercise in congenital complete atrioventricular block. Circulation 22:583–590, 1960.
56. James FW: Effects of physical stress on adolescents with normal or abnormal cardiovascular function. Postgrad Med 56:53–59, 1974.
57. James FW: Exercise testing in children and young adults: an overview. Cardiovasc Clin 9:187–203, 1978.
58. James FW, Kaplan S: Systolic hypertension during submaximal exercise after correction of coarctation of aorta. Circulation 49, 50[Suppl. II]:27–34, 1974.
59. James FW, Kaplan S: Exercise testing in children. Primary Cardiol 3:34–40, 1977.
60. James FW, Kaplan S, Chou T-C: Unexpected cardiac arrest in patients after surgical correction of tetralogy of Fallot. Circulation 52:691–695, 1975.
61. James FW, Kaplan S, Schwartz DC: Ischemic ST segments during exercise in children after coarctectomy (abstract). Am J Cardiol 37:145, 1976.
62. Jandová R, Widimsky J, Ressl J: Hemodynamics in juvenile hypertension at rest and during exercise. Cor Vasa 22:22–32, 1980.
63. Johnson AM: Impaired exercise response and other residue of pulmonary stenosis after valvotomy. Br Heart J 24:375–388, 1962.
64. Jokl E: Sudden death during exercise due to congenital anomaly of aortic valve. In: Jokl E (ed.) Medicine and Sport, Vol. 5. Karger, Basel, 1971, pp. 148–149.
65. Jonsson B, Lee SJK: Haemodynamic effects of exercise in isolated pulmonary stenosis before and after surgery. Br Heart J 30:60–66, 1968.
66. Joransen JA, Lucas RV Jr, Moller JH: Postoperative haemodynamics in tetralogy of Fallot. A study of 132 children. Br Heart J 41:33–39, 1979.
67. Kellerman JJ, Winter I, Kariv I: Effect of physical training on neurocirculatory asthenia. Israel J Med Sci 5:947–949, 1969.
68. Laird WP, Fixler DE, Swanbom CD: Cardiovascular effects of weight training in hypertensive adolescents (abstract). Med Sci Sports 11:78, 1979.

69. Lambert J, Ferguson RJ, Gervais A, Gilbert G: Exercise capacity, residual abnormalities and activity habits following total correction for tetralogy of Fallot. Cardiology 66:120–131, 1980.
70. Lewis AB, Heymann MA, Stanger P, et al: Evaluation of subendocardial ischemia in valvar aortic stenosis in children. Circulation 49:978–984, 1974.
71. Lind AR: Cardiovascular responses to static exercise. Circulation 41:173–176, 1970.
72. Lopez-Cuellar MR, Diaz Arauzo AE, Gaxiola Romero A, Perez Neria J: Effort test in children with aortic stenosis (abstract). Excerpta Med 518:164, 1973.
73. Mark AL, Kioschos JM, Abboud FM, et al: Abnormal vascular responses to exercise in patients with aortic stenosis. J Clin Invest 52:1138–1146, 1973.
74. Maron BJ, Redwood DR, Hirshfeld JW, et al: Postoperative assessment of patients with ventricular septal defect and pulmonary hypertension. Response to intense upright exercise. Circulation 48:864–874, 1973.
75. Michaelson M, Engle MA: Congenital complete heart block: an international study. The natural history. Cardiovasc Clin 4:85–101, 1972.
76. Miller WW, Young DS, Blomquist CG, et al: Physical training in children with congenital heart disease. In: Lavallée H, Shephard RJ (eds.) Frontiers of Activity and Child Health. Pélican, Quebec, 1977, pp. 363–369.
77. Mocellin R, Bastanier C: Zur Frage der Zuverlässigkeit der $\dot{W}_{170}$ als Mass der körplichen Leistungsfähigkeit bei der Beurteilung von Kindern mit Herzkrankheiten. Eur J Pediatr 122:223–239, 1976.
78. Mocellin R, Bastanier C: Funktionelle Untersuchungen bei Kindern und Jugendlichen mit angeborenem AV-Block. Z Kardiol 66:298–302, 1977.
79. Mocellin R, Bastanier C, Hofacker W, Bühlmeyer K: Exercise performance in children and adolescents after surgical repair of tetralogy of Fallot. Eur J Cardiol 4:367–374, 1976.
80. Mocellin R, Friedman J, Sebering W, Buhlmeyer K. Funktionelle Untersuchungen in Ruhe und während Belastung bei Kindern und Jugendlichen mit Ventrikelseptumdefekt und pulmonärer Hypertonie. Z Kardiol 64:1036–1052, 1975.
81. Mocellin R, Rutenfranz J, Bühlmeyer K: Untersuchungen über die körpliche Leistungsfähigkeit gesunder und kranker Heranwachsender. IV. Die Leistungsfähigkeit von Kindern und Jugendlichen mit angeborenen und erworbenen Herzfehlern. Z Kinderheilkd 108:265–287, 1970.
82. Moffat RJ, Gilliam TB: Serum lipids and lipoproteins as affected by exercise: a review. Artery 6:1–19, 1979.
83. Moller JH, Rao S, Lucas RV: Exercise hemodynamics of pulmonary valvular stenosis (study of 64 children). Circulation 46:1018–1026, 1972.
84. Monties JR, Mouly A, Goudard A, et al: The fate of children who have undergone correction of Fallot's tetralogy. Report of fifty long-term assessments (in French) (abstract). Sem Hop Paris 55:453–457, 1979.
85. Moss AJ, Quivers WW: Use of isoproterenol in the evaluation of aortic and pulmonic stenosis. Am J Cardiol 11:734–737, 1963.
86. Nadas AS, Fyler DC: Pediatric Cardiology, 3rd ed. W.B. Saunders, Philadelphia, 1972.

87. Neal WA, Lucas RV, Rao S, Moller JH: Comparison of the hemodynamic effects of exercise and isoproterenol infusion in patients with pulmonary valve stenosis. Circulation 49:948–951, 1974.
88. Nora JJ: Identifying the child at risk for coronary disease as an adult: a strategy for prevention. J Pediatr 97:701–714, 1980.
89. Nudel DB, Gootman N, Brunson S, et al: Exercise performance of adolescents with essential hypertension (abstract). Pediatr Res 12:366, 1978.
90. Nudel DB, Gootman N, Brunson SC, et al: Exercise performance of hypertensive adolescents. Pediatrics 65:1073–1078, 1980.
91. Orsmond GS, Bessinger FB, Moller JH: Rest and exercise hemodynamics in children before and after aortic valvotomy. Am Heart J 99:76–86, 1980.
92. Pedersen EB: Abnormal renal haemodynamics during exercise in young patients with mild essential hypertension without treatment and during long-term propranolol therapy. Scand J Clin Lab Invest 38:567–571, 1978.
93. Pernot C, Worms AM, Dambrine P, et al: L'épreuve d'effort dans les stenoses aortiques congénitales. Arch Mal Coeur 71:517–525, 1979.
94. President's Council on Physical Fitness and Sports: Update: exercise and some coronary risk factors. Phys Fitness Res Dig 9:1979.
95. Riopel DA, Hohn AR: Age effect on treadmill blood pressure responses in aortic stenosis (abstract). Pediatr Res 11:163, 1977.
96. Rose KD: Soccer and the student with asymptomatic idiopathic hypertrophic subaortic stenosis (reply to question). JAMA 225:1000, 1973.
97. Rosenthal A, Freed MD, Keane JF: Isometric exercise in adolescents with congenital aortic stenosis (abstract). Circulation 53, 54[Suppl. II]:48, 1976.
98. Saris WHM: Aerobic power and daily physical activity in children with special reference to methods and cardiovascular risk indicators. Doctoral dissertation, Catholic University, Nijmegen, Krips Repro Meppel 1982.
99. Schieken RM, Geller DF: The cardiovascular effect of isometric exercise in children (abstract). Clin Res 26:741A, 1978.
100. Shoenfeld Y, Shapiro Y, Drory Y, et al: Rehabilitation of patients with NCA (neurocirculatory asthenia) through a short term training program. Am J Phys Med 57:1–8, 1978.
101. Stone FM, Bessinger FB, Lucas RV, Moller JH: Pre- and post-operative rest and exercise hemodynamics in children with pulmonary stenosis. Circulation 49:1102–1106, 1974.
102. Strieder DJ, Aziz K, Zaver AG, Fellows KE: Exercise tolerance after repair of tetralogy of Fallot. Ann Thorac Surg 19:397–405, 1975.
103. Strong WB: Hypertension and sports. Pediatrics 64:693–695, 1979.
104. Strong WB, Spencer D, Miller MD, Salehbhai M: The physical working capacity of healthy black children. Am J Dis Child 132:244–248, 1978.
105. Sutton J, Seldon WA, Gunning JF: The hyperkinetic heart syndrome. Med J Aust 1:1039–1041, 1972.
106. Task Force on Blood Pressure Control in Children: Treatment. Pediatrics 59:808–810, 1977.
107. Taylor A, Halushka P, Privitera P, et al: Effects of exercise on urinary prostaglandin E like material, kallikrein and blood pressure in coarctation of the aorta and normal children (abstract). Circulation 53, 54:48, 1976.

108. Taylor MRH: The response to exercise of children with congenital heart disease. PhD Thesis, University of London, 1972.
109. Taylor MRH: The ventilatory response to hypoxia during exercise in cyanotic congenital heart disease. Clin Sci Molec Med 45:99–105, 1973.
110. Taylor MRH, Godfrey S: Exercise studies in congenital heart block. Br Heart J 34:930–935, 1972.
111. Taylor SH, Donald KW: Circulatory studies at rest and during exercise in coarctation of the aorta before and after operation. Br Heart J 22:117–139, 1960.
112. Thorén C, Herin P, Vávra J: Studies of submaximal and maximal exercise in congenital complete heart block. Acta Paediatr Belg 28[Suppl.):132–143, 1974.
113. Thorland WG, Gilliam TB: Comparison of serum lipids between habitually high and low active pre-adolescent males. Med Sci Sports 13:316–321, 1981.
114. Truccone NJ, Steeg CN, Dell R, Gersony WM: Comparison of the cardiocirculatory effects of exercise and isoproterenol in children with pulmonary or aortic valve stenosis. Circulation 56:79–82, 1977.
115. Tuboku-Metzger A, Hirschfeld S, Borkat G, Liebman J: Hemodynamic correlates of exercise testing in children with aortic stenosis (abstract). Circulation 53, 54[Suppl. II]:48, 1976.
116. Turner MC, Ruley EJ, Buckley KM, Strife CF: Blood pressure elevation in children with orthopedic immobilization. J Pediatr 95:989–992, 1979.
117. Välimäki I, Hursti ML, Pihlakoski L, Viikari J: Exercise performance and serum lipids in relation to physical activity in school children. Int J Sport Med 1:132–136, 1980.
118. Vartia A, Välimäki I: The effect of some chronic cardiac arrhythmias on the physical working capacity of children. Acta Paediatr Scand 58:555–556, 1969.
119. Voller RD, Strong WB: Pediatric aspects of atherosclerosis. Am Heart J 101:815–836, 1981.
120. Wagner HR, Weidman WH, Ellison RC, Miettinen DS: Indirect assessment of severity in aortic stenosis. Circulation 56[Suppl. 1]:20–23, 1977.
121. Watson G, Freed D, Strieder J: Cardiac output during exercise in children with idiopathic complete heart block. In: Lavallée H, Shephard RJ (eds.) Frontiers of Activity and Child Health. Pélican, Quebec, 1977, pp. 393–400.
122. Weiner BH, Starobin OE, Mills RM, Waksmowski CA: Exercise testing in congenital aortic valve disease (abstract). Chest 76:370, 1979.
123. Wessel HU, Paul MH: Exercise performance in tetralogy of Fallot 1 to 12 years after intracardiac repair (abstract). Circulation 49, 50[Suppl. III]:210, 1974.
124. Whitmer JT, James FW, Kaplan S, et al: Exercise testing in children before and after surgical treatment of aortic stenosis. Circulation 63:254–263, 1981.
125. Winkler RB, Freed MD, Nadas AS: Exercise-induced ventricular ectopy in children and young adults with complete heart block. Am Heart J 99:87–92, 1980.

5

Endocrine Diseases

Diabetes Mellitus

Metabolic Response to Acute Exercise

The biochemical response to acute exercise of the patient with diabetes depends on his or her metabolic state prior to exercise; food intake; type, time, and site of insulin injection; the intensity and duration of exertion; and the type of diabetes. In this section we shall concentrate on insulin-dependent diabetes mellitus (IDDM), which is the type commonly found in the juvenile patient.

Blood Glucose Lowering Effect of Exercise. In a reasonably controlled patient with IDDM, prolonged exercise will induce a gradual decline in blood glucose level.[3,63,83,89] A case in point is that of a 13-year-old boy with a 3-year history of IDDM, whose blood glucose response to prolonged exercise is shown in Fig. 5.1. This boy came to our clinic with a vague story of attacks of weakness and hunger during long-distance cycling—his favorite sport. He denied the possibility that these might reflect hypoglycemia. A simulation of a 90-min road ride was performed on a cycle ergometer, resistance and pedaling rate being chosen by the patient. Exercise intensity was moderate, raising his heart rate to 140–150 beats/min. Ninety minutes prior to the ride he ate breakfast and injected into the thigh his dose of medium-acting insulin.

As seen in Fig. 5.1, blood glucose markedly declined, reaching 58 mg% (3.2 mmol/liter) by the end of the ride. Seventy minutes after the start, he began to complain of heat, followed by hunger, fatigue, and sore thighs. Five minutes before the conclusion of the ride, he developed a fine tremor in various muscle groups. These complaints disappeared within 10–15 min post-exercise.

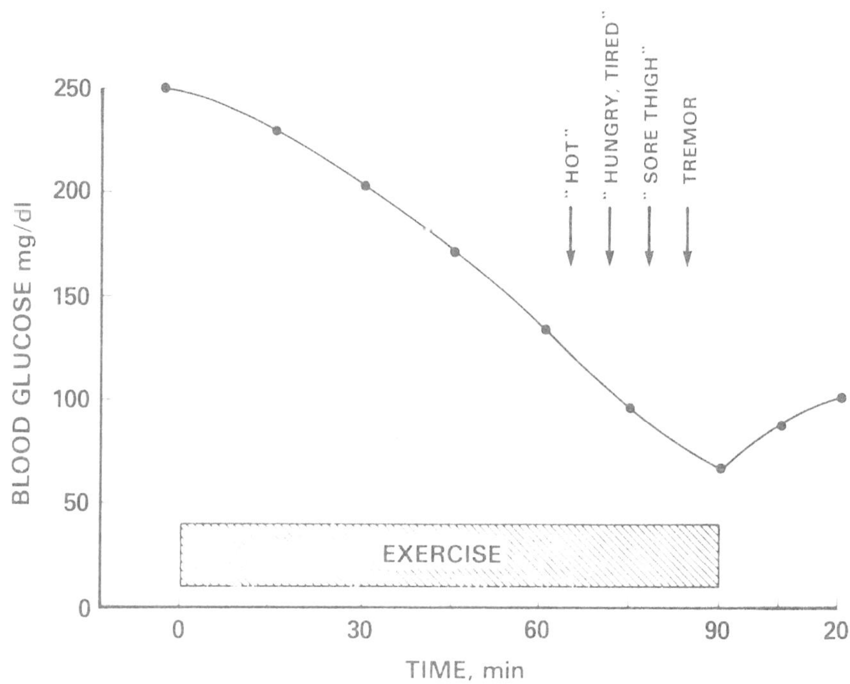

Figure 5.1. Hypoglycemia during prolonged moderate exercise in a juvenile diabetic. Blood samples were taken through an indwelling antecubital venous catheter. For details see text. Data from MacMillan and Bar-Or, unpublished.

Intravenous glucose tolerance during[15] and after[16] exercise is increased in insulin-treated IDDM adolescents and adults but not in insulin-deprived adolescents.[17] Fig. 5.2 demonstrates the exercise-induced increase in glucose clearance from the blood in the adolescent with IDDM.

The declining blood sugar level reflects an imbalance between increased glucose utilization by the exercising muscle on the one hand[69] and an insufficient increase in its production by the liver on the other.[89] The degree of such an imbalance in the IDDM patient depends on the availability of exogenous insulin and on the sensitivity of insulin receptors in the exercising muscle. When circulating insulin level is high, there is a greater suppression of hepatic glucose production, a greater uptake by the muscle, and a faster decline in blood glucose level. Factors that can modify the level of circulating insulin during exercise include: the elapsed time between insulin administration and exercise; the type of insulin; the presence of insulin antibodies; the site of injection; and route of administration. Among these, the route and site of insulin injection seem to be of major importance.

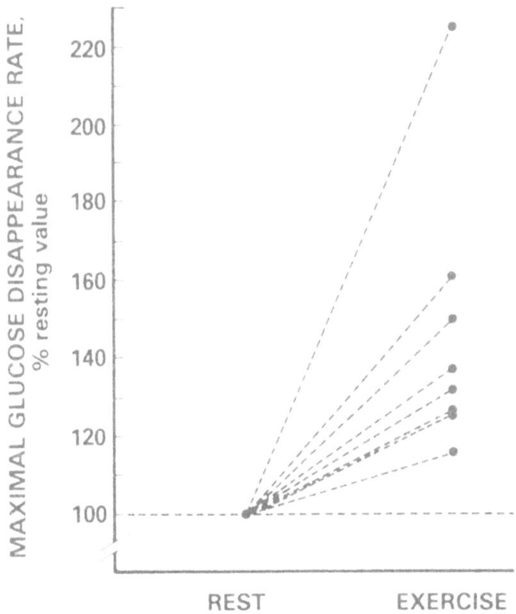

Figure 5.2. Exercise and intravenous glucose tolerance. Maximal disappearance rate of glucose from the blood at rest, and during 35 min cycling at 50% of maximal O_2 uptake. Subjects were eight 12- to 18-year-old diabetic nonobese males who were given IV infusion of insulin and 50% glucose. Individual values are expressed as a percentage of maximal glucose disappearance at rest. Data from Dorchy et al.[15]

Route and Site of Insulin Injection. In Table 5.1 are summarized the effects of these on blood glucose level and on related variables. When insulin is injected subcutaneously (SC) to the exercising limb, there is a faster release of insulin from the depot during exercise than at rest.[4] Likewise, release is faster than from a nonexercising site.[43,89] This results in increased plasma immunoreactive insulin levels, reduced glucose pro-

Table 5.1. Effects of Route and Site of Insulin Injection on Blood Glucose Level and Related Variables of an Exercising Child with Insulin-Dependent Diabetes Mellitus

Route of Injection	*Site of Injection*	*Plasma IRI**	*Glucose Production by Liver*	*Glucose Utilization by Muscle*	*Blood Glucose Level*
IV	Any	nc†	↑↑	↑↑	nc
SC	Nonexercising limb	nc	↑↑	↑↑	nc‡
SC	Exercising limb	↑	↑	↑↑	↓

* = Immune-reactive insulin
† nc = No change
‡ = A mild reduction in blood glucose level may take place

duction by the liver, and a drop in blood glucose level. In contrast, SC injection into a nonexercising site, or a continuous IV infusion, does not impede glucose production by the liver and does not result in reduction of blood glucose level (some reduction may take place but to a lesser extent than following SC injection to the exercising limb). The faster release of insulin from a depot that overlies an exercising muscle could be due to a local increase in temperature, hyperemia, or enhanced lymph flow. The exact mechanism is not known.

Insulin Binding to Receptors. A bout of exercise causes an increase in insulin binding to receptors, located on monocytes or erythrocytes, in sedentary healthy individuals[73] and in patients with IDDM.[59,60] Under some conditions a correlation exists between *in vivo* insulin sensitivity and insulin binding to monocytes. One may therefore speculate that an exercise-induced rise in insulin binding to monocytes reflects a similar increased binding to receptors in exercising muscle cells. This could be another mechanism for the glucose-lowering effect of exercise and for the improved glucose tolerance after exercise.

Exercise-induced Hyperglycemia. Exercise does not always cause hypoglycemia. Diabetics who are hyperglycemic, ketotic, or deprived of insulin often respond to exercise with an *elevation* of glucose level,[2,3,24,55,64,69,283] even with ketoacidosis.[3,83] Fig. 5.3 is an example of this phenomenon in adolescent diabetics who omitted their morning insulin dose and then exercised. The cause of exercise-induced hyperglycemia in the insulin-deprived patient with IDDM is an increased hepatic production of glucose in excess of its uptake by the muscle.[69] Such production is normally inhibited by the presence of insulin. The patient who is ketotic at rest has a greater than normal production of ketone bodies during exercise.[24] Even though uptake of ketone bodies in the muscle is high in such a patient,[83] there is a net increase in blood level of these substances, especially during strenuous exercise.[3,83] Another possible cause for exercise ketosis in the insulin-deprived patient with IDDM is a rise in plasma glucagon and cortisol.[3]

Metabolic Response to Chronic Exercise

A great deal has been written about the beneficial effects to the diabetic of regular physical exercise. In the pre-insulin era, exertion was considered a major mode of treatment. The following statements, for example, appeared in a medical journal in 1915:

"[Diabetic] dogs which for months had regularly shown glycosuria whenever they were given 100 grams of bread, on exercise became able to take 200 grams of bread as a regular daily ration without glycosuria." And, as to humans: "In a patient free from glycosuria with persistent

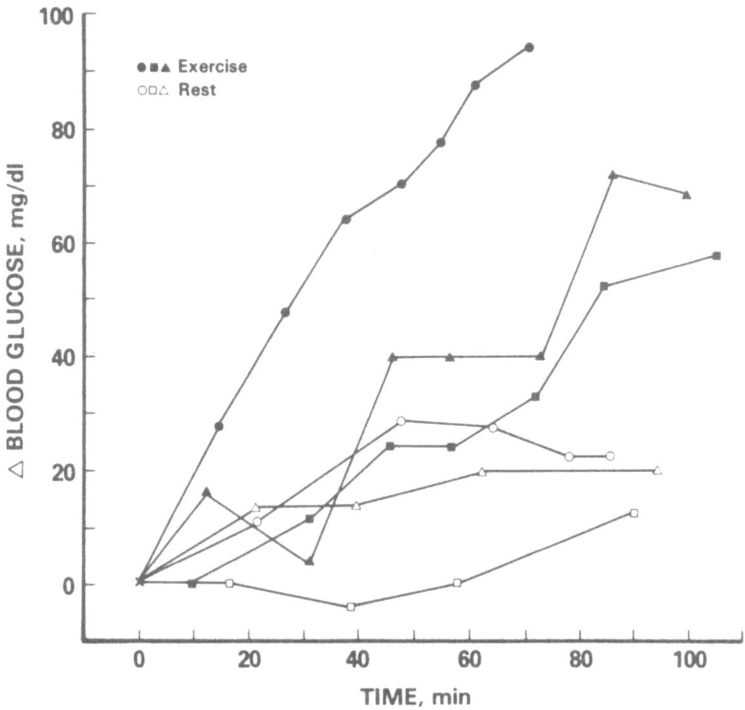

Figure 5.3. Insulin deprivation and exercise-induced elevation of blood glucose concentration. Three 13.1- to 13.8-year-old boys were tested on two mornings following omission of their insulin dose since the preceding evening. Each line represents individual changes in concentration (Δ blood glucose) during the experiment. The empty symbols are values at rest. Values represented by solid symbols were recorded on another morning when the child performed two or three bouts of a 5-min stair-climbing exercise. Adapted from Marble and Smith.[55]

hyperglycemia, one fast day with exercise may reduce the blood sugar as much as several fast days without exercise."[2]

Exercise and Insulin—The Synergistic Effect. The synergistic effects on blood glucose level of insulin and exercise were first reported in 1926: exercise practiced for some weeks helped diabetics double their daily carbohydrate consumption without any concomitant increase in their insulin dose. Some patients could even *reduce* their insulin dosage despite a greater carbohydrate intake.[48] Such observations have since been reproduced in many studies.[1,5,21,38,40,46,47]

The reduction in insulin requirement is sometimes quite dramatic: we have been seeing young diabetics who require 40–60 units daily during fall and winter but only 5–10 units during spring and summer (see also Ref. 38). A similar trend is seen in Fig. 5.4 for a child who almost halved his insulin intake during three consecutive summer seasons.[36]

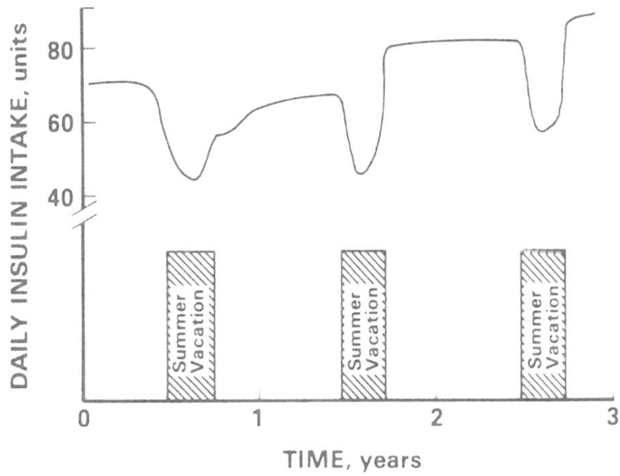

Figure 5.4. Insulin intake and habitual activity. Record of a boy with insulin-dependent diabetes mellitus. Adapted from Jackson and Kelly.[36]

The mechanism that enables the active child with IDDM to use less insulin is not clear. There is strong evidence that *in vivo* sensitivity to insulin and *in vitro* binding to monocytes are higher among trained individuals and can increase in sedentary people who undergo a conditioning program.[44,51,72] The validity of these data for the diabetic child has yet to be shown. Moreover, the greater sensitivity to insulin in the trained individual is short-lived and may disappear within 1–2 days following the last training session.[11] Thus, it may reflect increased glucose tolerance following a single exercise bout, rather than a trait of the trained person. If this assumption is correct, a diabetic can expect to enjoy the insulin-sparing effect of exercise only as long as he is *regularly* active.

Exercise and Control of Diabetes. A relationship has been suggested between control at a young age and the future risk of diabetic complications, such as neuropathy, nephropathy, and retinopathy. The assumption is that strict daily control is important for the reduction of such a risk. It is therefore pertinent to ask whether physical activity can affect the control of diabetes.

Taking the frequency of glycosuria as an indicator, regular physical activity seems to increase control.[5,21,40,47,48] An example often quoted is that of special camps for diabetics, in which children who become more active retain or even improve their metabolic control in spite of a lower insulin dosage and higher calorie intake.[1,47] Fig. 5.5 is a summary of observations from such a training camp: daily calorie intake increased by 50–100% and yet the rate of glycosuria decreased from some 65% to less

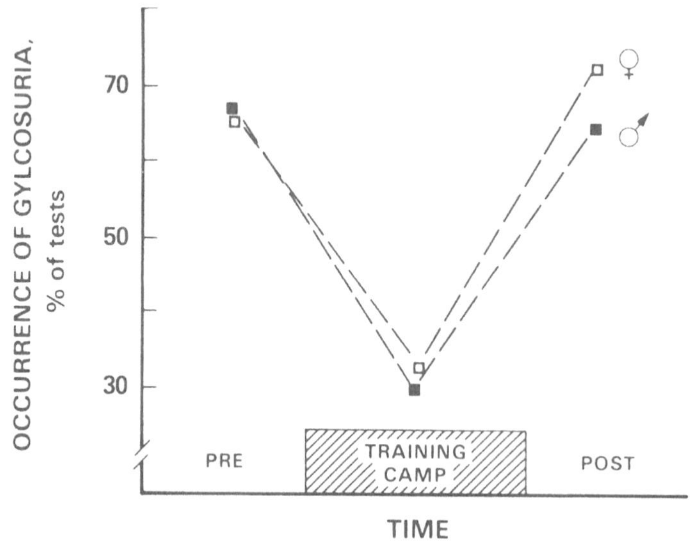

Figure 5.5. Physical activity and daily control of diabetes. Occurrence of glycosuria in diabetic girls and boys before, during, and following a training camp. Data from Sterky as quoted by Larsson et al.[47]

than 30% of tests. A statistical correlation has been reported between the *extent* of activity and an "index of control" based on two to four daily urinalyses.[52] The common variance, however, between the two variables was low, suggesting that even if the role of regular activity is important, other factors strongly affect control of diabetes.

Another index of control is the concentration of hemoglobin A_{1c} (HbA_1) in the blood. HbA_1 dropped in children who took part in exercise-intervention programs.[5,13,29] In one study, HbA_1 in patients who underwent a 5-month conditioning program dropped from 15.1 to 13.8%.[13] In another, it changed from 7.5 to about 6.0%.[29] Although statistically significant, such a reduction is small biologically and its actual benefit hard to assess.

In conclusion, *regular* physical activity is probably of benefit in the daily control of diabetes. Such benefit might, however, be indirect because children who become physically active may be changing other daily habits which themselves affect metabolic control. When an increased activity is *irregular,* control may, in fact, be *disrupted* due to variations in day-to-day metabolism.

Habitual Activity and Diabetes Mellitus

Perhaps more than in any other disease, in the control of juvenile diabetes physical activity is of the utmost importance. In addition to its direct effect on control of the disease, a change in physical activity may affect the other two components of diabetes management—insulin dosage and calorie intake. A physician should therefore become thoroughly familiar with the activity pattern and intensity of each of his diabetic patients: participation in physical education classes and extracurricular activities at school; out-of-school activities during weekdays and weekends; and seasonal variations such as the summer holidays or other extended vacations, when activity is bound to intensify.[36] Most diabetic children and prepubertal adolescents pursue normal activity habits. In contrast, pubertal and postpubertal diabetics are less active than their healthy peers. Fig. 5.6 summarizes the frequency of participation in physical education classes of diabetic boys and girls and of controls.[75] While the 7- to 14-year-old diabetics were quite compliant, the older ones had an extremely high level of absenteeism. As for leisure time sports, the 15- to 20-year-old diabetic boys were considerably less active than their healthy counterparts. Among the girls, both postpubertal groups, healthy and diabetic, reported a rather low level of voluntary activity.

Why would the older diabetics be reluctant to exert? One possibility is their fear of hypoglycemic crisis.[36] They have come to realize that, by keeping their activity level consistently low, they remove one uncertainty regarding diabetic control. Unfortunately some physicians, to whom exercise is a "black box," concur with and encourage a policy of physical

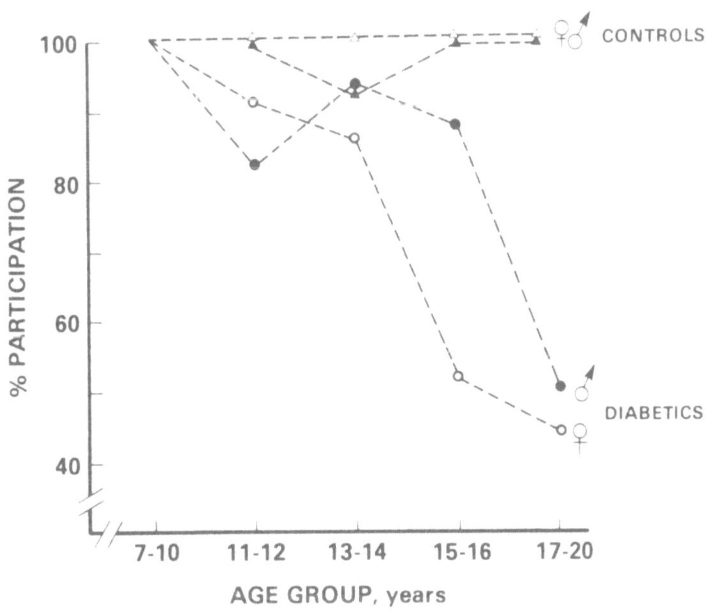

Figure 5.6. Physical activity in diabetes. Frequency of participation in physical education classes of 136 diabetic school children and of 123 age-matched controls. Data from Sterky.[75]

passivity for the young diabetic. Adolescent patients, especially girls, often gain weight[20,75] in excess of height gain. If obesity develops in such individuals, it could be another cause—and effect—of hypoactivity. In a survey from Sweden,[52] patients down-played the fear that too much exercise may disrupt their diabetic control. While almost all recognized the importance and fun of exercise and sports, 30–50% admitted that they did not push themselves to be active on a regular basis.

The above findings are culture-specific and cannot be deduced for all diabetic populations. Even so, this survey points out the gap between *knowing* about the importance of exercise on the one hand and *being involved* in regular activities on the other. This is true for many adolescents with diabetes.

Physical Working Capacity

A number of fitness components have been found deficient among diabetic children and, especially, adolescents. These include maximal aerobic power[20,46,47,67,75] and a battery of performance items designed by the American Association of Health, Physical Education and Recreation such as strength, speed, and muscle endurance.[23] No relationship seems to exist between the age of onset, severity, control, or duration of diabetes and the working capacity of the affected child.[20,67] However, the *combined* effects of early onset and long duration of the disease do seem to contribute to the low maximal aerobic power of diabetics.[75]

To establish a sound strategy toward exercise prescription, one should know whether lower fitness is an inherent characteristic of the disease. Apparently, this is not the case here. The deficient motor performance of some diabetics can be explained, in part, by their lower body stature, as compared with age-matched children or adolescents.[18,75] Indeed, when corrected for body size, differences in maximal aerobic power between diabetics and healthy controls diminish[75] or disappear altogether.[13,31,67]

Another indication that low fitness is not due to the disease *per se* is that only some age-groups are unfit, notably the pubertal and postpubertal patients. In contrast, the younger adolescents and children are as fit as their healthy peers.[20,47,75] The lower exercise performance of pubertal or postpubertal diabetics could be due to the lower level of their habitual activity,[75] as described above. Indeed, when such patients are exposed to a conditioning program, their maximal aerobic power and other functions improve and can equal those of healthy youths.[12,47] Large numbers of elite athletes in various sports are diabetic,[6] which further indicates that low physical fitness is not inherent to the disease.

There are several differences in exercise-related physiologic functions between diabetic and healthy youths, but these do not seem to induce any appreciable functional handicap. Some[34] but not all[62] adolescents with IDDM respond to exercise with a greater degree of albuminuria than do healthy controls. In young adults with IDDM, muscle tissue capillarization is lower than normal, as is the activity of their muscle oxidative enzymes. Muscle glycogen concentration and glycogen utilization during exercise are normal,[68] and so is their ability to resynthesize muscle glycogen following depletion by prolonged exercise.[54] The implication of such normal resynthesis is that patients with good control can train hard and recover well following prolonged intense bouts of exercise.

Exercise in Daily Management

The Triad—Food, Insulin, Exercise. In no other disease are food, exercise, and medication so closely linked as in DM. A change in any one automatically calls for adjustment in the other two. Many diabetics who exercise habitually can tailor their insulin and food intake to their *anticipated* exercise calorie expenditure on any given day.[6] An example is the postman quoted by Joslin.[40] This man used to ride 25 km daily on his bicycle. "Before he starts out in the morning he sticks his head out of the window and if the wind is blowing against him then he lowers the insulin 20 units." Others are known to anticipate the amount of walking, swimming, or physical work done on a given day and modify their food intake accordingly.

This practice is suitable not only for the adult or adolescent patient who, through the years, has gained experience by trial and error. It can also be applied for the young, less experienced patient, who can be taught to adjust insulin and food according to anticipated activities. Emphasis must be given to days when especially strenuous or prolonged activity is anticipated (e.g., a hike, a field day at school, or an athletic event). Experience has shown that on such days one should increase food intake, rather than decrease the insulin dose, and do so *before* the anticipated exercise session[14,30,36,47,66,74] Food items to be added can be calculated as fruit exchanges and starch exchanges. For examples see Table 5.2.

On a day of competition or major practice session, meals should be eaten not less than three hours before warm-up time[14] and should include a higher than normal carbohydrate content.[66] For a regular practice session, a 60–80 min interval is sufficient. During prolonged activities, sugar-containing fluids (sucrose or fructose) must be taken at least once per hour (and supplemented by fluids to maintain water balance as per Chapter 9, section entitled Water and Electrolyte Replenishment).

Another useful precaution against exercise-induced hypoglycemia is a change in the site of insulin injection. For instance, when an activity is to be performed primarily by the lower limbs (e.g., in running), the patient should, on that day, inject insulin into his arm. When both upper and

Table 5.2. Fruit and Starch "Exchanges" to Be Taken by a Child with IDDM Prior to an Extra Physical Activity*

		20 kg				30 kg				40 kg			
Activity	*Duration (min)*	*Energy Expended (kcal)*	*Carbo-hydrate (g)*	*"Exchange" Fruit*	*"Exchange" Starch*	*Energy Expended (kcal)*	*Carbo-hydrate (g)*	*"Exchange" Fruit*	*"Exchange" Starch*	*Energy Expended (kcal)*	*Carbo-hydrate (g)*	*"Exchange" Fruit*	*"Exchange" Starch*
Basketball (game)	20	100	15		1	200	30		2	300	45		3
	40	200	30	1	1½	400	60	1	3	600	90	2	4
Cross-country ski	30	70	10		⅔	140	20		1½	210	30		2
(leisure pace)	60	140	20	½	1	280	40	½	2	420	60	1	3
Cycling—10 km/hr	30	45	7		½	75	10		⅔	120	20		1½
—15 km/hr	30	65	10		⅔	120	18		1	180	30		2
Figure skating	20	80	12		¾	160	25		1⅔	240	35		2
(practice)	40	160	25	½	1	320	50	1	2	480	70	½	4
Ice hockey (on-ice)	10	50	10		⅔	100	20		1½	150	30		2
	20	100	20		1⅓	200	40		2½	300	60		4
Running— 8 km/hr	30	110	17		1	200	30		2	270	40		2½
—12 km/hr	30	—	—	—	—	270	50		3	370	70		4½

Snow shoeing	30	100	15		1	200	30		2	300	45		3
	60	200	30	½	1½	400	60	1	3	600	90	1½	4½
Soccer (game)	30	110	17		1	215	32		2	320	50		3½
	60	220	35	½	1½	430	65	1	3	640	95	2	4½
Swimming (breast stroke—	20	60	10		⅔	120	18		1	230	35		2
30 m/min)	40	120	20	½	1	240	36	½	2	460	70	1	3½
Tennis	30	75	12		¾	130	20		1½	190	30		2
	60	150	24	½	1½	260	40	1	2	380	55	1	3
Walking—4 km/hr	40	75	12		¾	105	16		1	135	20		1½
—6 km/hr	40	105	16		1	135	20	½	1	170	26	⅓	1½

* Table constructed with the help of Mrs. Karen Chelswick.

Purpose: To instruct children with diabetes mellitus—and their parents—about the compensatory increase in calorie intake for physical activities which are *above and beyond* those practiced daily. Examples are given for children of 20, 30, and 40 kg body weight.

Assumptions:
1. These added activities have not been taken into consideration in the *regular* daily food intake.
2. Carbohydrate equivalent of activities is 60% of energy expenditure, apart from ice hockey and running at 12 km/hr, where the carbohydrate equivalent is 75% of total energy expenditure.
3. The additional food is taken *before* the activity (for activities that last more than 1 hr, periodic food intake should be practiced *during* the activity).

lower limbs are expected to exercise (e.g., cross-country skiing), the abdomen will be the site of choice. An alternative approach may be the use of continuous subcutaneous infusion by an "insulin pump," which prevents uncontrolled release from the site of injection.[35] The feasibility of this approach for the exercising diabetic child has not yet been ascertained.

Danger of exercise-induced *hyperglycemia* and ketoacidosis is limited to those insulin-dependent patients who have omitted to take their insulin, or to the metabolically uncontrolled patient. It will not occur in a patient who is in good control and who has received at least some insulin.[17,64] At special risk are those children who suffer from "brittle IDDM" and become, unpredictably, out of control. The safest approach for preventing hyperglycemia and ketoacidosis in these patients is to *avoid exercise under such conditions* and to take appropriate amounts of insulin.

Exercise Exchange Menu. To be able to adjust calorie and insulin intake according to changes in activity, one must know the calorie equivalent of various activities. This value depends on the type, intensity, and duration of the activity, as well as on the body weight of the child. In Table III.1 (Appendix III) are listed calorie equivalents of common activities, each performed for 10 min. The table is subdivided into body weight groups to show how many calories a given activity is worth for a particular child.

Based on our experience, the concept of the *exercise exchange menu* can be used in analogy to the "food exchange" menu. Each exercise exchange is equivalent to an expenditure of 100 kcal (420 kJ). Taking the carbohydrate contribution as 60% of total energy fuel, one exchange is equivalent to 60 kcal or 15 g of carbohydrate. Thus, in terms of carbohydrate balance, one exercise exchange equals one food exchange.

A list of exercise exchanges is given in Table 5.3 for children of different body weights. Such a list can be used for educational purposes with parents and patients. The child can create his own "exercise-exchange menu," combining games and other activities of his choice. Within the setting of an outpatient clinic or a diabetic day-care center, it is the nurse or the dietitian who will be thoroughly familiar with the exercise exchange concept. However, the physician must also understand this concept and be able to prescribe activities accordingly.

The above may be the optimal approach, but it may not be effective for all children, especially the younger age-groups. By nature, a young child is spontaneous and unpredictable and may not abide by overly regimented instructions. Pushing such a child to adhere to the exercise exchange concept may be counterproductive. It is for the therapist and the parents to sense how far, and how fast, they should introduce this approach with any given child. Success will be achieved once the child

Table 5.3. "Exercise Exchanges" of 100 kcal (420 kJ) in Children of Various Body Weights.*

Activity	*Body Weight*		
	20 kg	*40 kg*	*60 kg*
Basketball (game)	30	15	10
Calisthenics	75	40	25
Cross-country ski (leisure)	40	20	15
Cycling—10 km/hr	65	40	25
—15 km/hr	45	25	15
Field hockey	35	20	15
Figure skating	25	15	10
Horseback riding			
canter	110	60	40
trot	45	25	15
gallop	35	20	10
Ice hockey (ice time)	20	10	5
Judo	25	15	10
Running			
8 km/hr	25	15	10
10 km/hr	20	15	10
12 km/hr		10	10
14 km/hr			5
Sitting—complete rest	125	100	85
—quiet play	90	65	55
Snow shoeing	30	15	10
Soccer	30	15	10
Squash		10	10
Swimming—30 m/min			
breast stroke	55	25	15
front crawl	40	20	15
back stroke	60	30	20
Table tennis	70	35	25
Tennis	45	25	15
Volleyball (game)	50	25	15
Walking—4 km/hr	60	40	30
—6 km/hr	40	30	25

* Values are the number of minutes that a certain activity should be sustained.

exercises daily within the framework of exercise exchange and still enjoys the fun of sports.

Growth Hormone (GH) Deficiency

Using Exercise in the Diagnosis

Rationale for Provocation Testing. The definitive diagnosis of GH deficiency depends on the serum level of the hormone. During most hours of the day this level is low in both the healthy and the GH-deficient child, and a random blood test will not reveal a difference between the two. To overcome this, a number of provocation procedures have been developed, by which the pituitary acidophils are stimulated to produce and release GH in increased amounts.[25,41] A child whose hypothalamic-pituitary function is adequate will usually respond with an elevation of serum GH, while one with hypopituitarism will have little or no response. Post-stimulus concentrations of 5–8 ng/ml or more have been considered indicative of normality.[25,41,56]

Provocation Tests at Rest. The provocation tests in use are: insulin-induced hypoglycemia, intravenous administration of arginine, intramuscular glucagon, oral *l*-dopa or propranolol, a combination of *l*-dopa and propranolol, sleep, and exercise. Whereas the last two are considered physiologic stimuli, the others involve pharmacologic intervention in doses that exceed physiologic levels. It is therefore unclear whether the response of a child to any of the above pharmacologic stimuli indeed reflects physiologic processes within the hypothalamic-pituitary axis. Furthermore, insulin provocation, which is considered the most reliable and potent stimulus,[25,41,50,78] causes discomfort and is not without risk due to the hazards of hypoglycemia and of its treatment. Especially at risk are those children with pituitary insufficiency, who may respond with blood glucose levels of 20 mg% (1.1 mmol/liter) or less. Insulin provocation cannot readily be performed in the physician's office. The other pharmacologic stimuli, while better tolerated by the child, are less reliable and potent than insulin and, as a result, 10–30% of healthy children respond to them without an increase in GH.[25] A properly conducted sleep test requires simultaneous EEG monitoring to synchronize blood sampling with specific stages of sleep. It is therefore logistically demanding and is not widely used for clinical purposes.

Exercise as a Provocation Test. It was first shown by Roth et al[65] that increased physical activity induces a rise in serum GH. Possible mechanisms for such a rise have been reviewed elsewhere.[71] Since the early 1970s, various pediatric laboratories have adopted exercise as a screening test of GH deficiency.[7,9,10,19,27,37,42,53,56,61,82,86,87,88] Figure 5.7 depicts

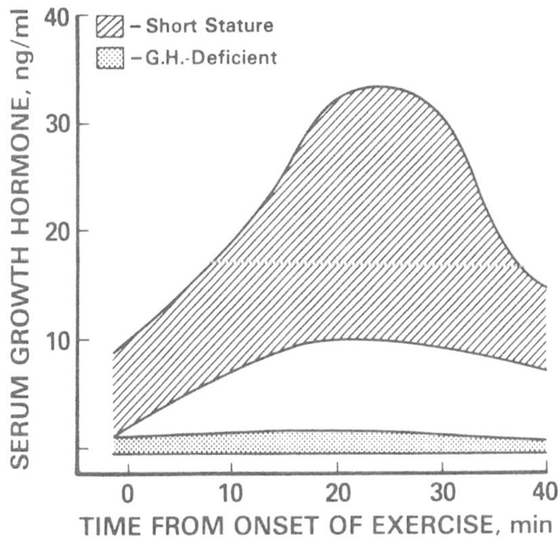

Figure 5.7. Changes in serum growth hormone following exercise provocation in growth-hormone-deficient (dot pattern) and non-growth-hormone-deficient (diagonal line pattern) children. A composite graph of the range of response. Based on data from Eisenstein et al,[19] Garlaschi et al,[27] Keenan et al,[42] Lacey et al,[45] Pombo et al,[61] Shanis and Moshang,[70] and Winter.[86]

the range of GH rise, as a function of the time elapsed since the start of exercise. In some protocols exercise is used in combination with a pharmacologic stimulus.[49,53,70] Figure 5.8 shows an example in which children were given propranolol prior to the exercise test.

The purpose of a screening test is to filter out those children who are not GH deficient, thus obtaining a smaller number of children who need a more definitive evaluation. A good screening test should be highly

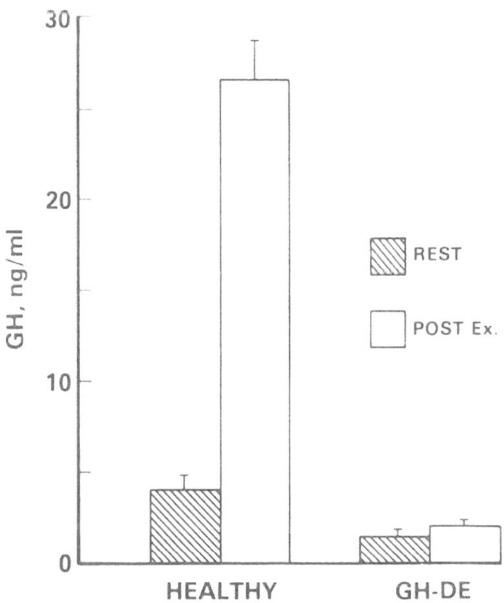

Figure 5.8. Exercise in diagnosis of growth-hormone deficiency. Thirty-two healthy, short-statured 3- to 15-year-old children and five children with growth-hormone deficiency (GH-DE) performed a 20-min exercise, 2 hr after receiving orally 24–40 mg of propranolol. Blood was sampled immediately after exercise. Vertical lines denote 1 S.E.M. Data from Shanis and Moshang.[70]

sensitive (few false negative results) but also specific (few false positive results). In Table 5.4 are summarized the protocols and findings of 12 studies in which short-statured children were exercise-tested. Altogether there were 544 children and adolescents with no GH deficiency and 57 who had proven hypopituitarism based on history, physical characteristics, hormonal evaluation (mostly by the use of an insulin test), and therapeutic trials. Occurrence of false negative results (i.e., post-exercise increase in GH in spite of proven GH deficiency) was zero in all but one study, which indicates that exercise is a highly sensitive provocation. Out of 544 patients who had no hypopituitarism, exercise induced a normal rise of GH in 84.4% with only 15.6% false positive responses. Such specificity compares favorably with any of the other pharmacologic or physiologic tests, excluding insulin hypoglycemia, which has higher specificity.

The major reasons for a false positive result are: 1) the exercise stress is too short, 2) its intensity is too low or too high, 3) blood is sampled too early or too late, and 4) the test is done at a time when the pituitary is refractory to further stimuli. In addition, it may be that the pathways by which exercise triggers release of GH are different from those following nonphysiologic stimuli. If so, then some deficiencies may be manifested by one response to a physiologic stimulus and a different response to a pharmacologic one. A case in point is a boy whose serum GH concentration rose normally (19 ng/ml) following a combined arginine-insulin test, but remained low (5–6 ng/ml) following exercise. Nevertheless, GH therapy was initiated and growth rate accelerated to 7.5 cm/yr, compared with 2.5 cm/yr before therapy. The implication was that the child had suffered from partial GH deficiency, which could be shown by a physiologic but not by a pharmacologic test.[87] A similar response has been described by others.[56,78,81]

Optimizing the Exercise Protocol. A major reason for variation in the response to testing is the lack of standardization of exercise intensity. Loads of 75–90% of maximal aerobic power yield a greater GH rise than do milder or all-out loads.[7,33,76,77,78] It is not clear whether the low release of GH during all-out exercise is due to factors inherent to intense muscular activity, or whether the duration of such tasks is too short to trigger GH release. As shown in Table 5.4, some laboratories have been using stairway-running, skipping, or "walking with a parent."[10,19,42,61,70] Such protocols are not recommended because exercise intensity is hard to control.

GH response is also dependent on the fitness level of the patient. When an identical task is given to individuals who differ in fitness, the less fit will show a greater GH rise than the fit.[7,79] The *relative* exercise intensity should therefore be equated among individuals. This can be achieved by exercising all children at a heart rate of 170–185 beats/min

Table 5.4. Studies in Which Exercise Has Been Used as a Provocation for GH Release in Children with Short Stature

Patients			*Exercise Protocol*		*PEAK GH (ng/ml)*					
Age	*"Short"*	*GH Deficient*	*Duration min*	*Type of Exercise*	*"Short"*	*GH Deficient*	*Diagnostic GH Value (ng/ml)*	*False Positive (%)*	*False Negative (%)*	*Reference*
3–15	25	2	25	Walk & Stairway	13.7	<1.0	6	6	0	42
6–25	23	9	~20	Stairway	?	?	7.6	63	0	56
<10–20	54	5	5–10	Cycle or skip, run	?	?	5	18	0	9
11–12	95	2	10	Cycle all-out	~14.2	1.5	10	32	0	45
13–18	15	4	15–30	Cycle all-out	~35.0	1.0	?	0	0	86
8–11	49	—	10	Cycle—submax	15.6	—	5	2	—	27
5–15	15	3	20	Stairway	>24.6	<3.0	7	0	0	53*
11–20	18	5	?	Stairway	9.3	1.9	7	50	0	87
?	24	—	10–20	Cycle—all-out	?	?	7	0	0	88
3–15	32	8	20	Walk, jog, stairway	26.6	2.0	7	0	0	70*
?	176	18	25	Walk and stairway	29.6	5.1	10	7	6	19
5–15	18	1	20	Stepladder	10.9	0.5	7	6	0	61*
6–14	25	3	10	Cycle—all-out	?	?	10	?	?	82
2–19	58	13	20	Walk and stairway	16.0	1.7	5	13	0	37
2–16	53	4	20	Jump, run "vigorous"	13.9	—	6	8	0	39
3–16	152	9	30	Stairway	—	<3.0	5	29	0	58

* A combination of a pharmacologic and an exercise provocation.

(the better conditioned children will require higher exercise intensity to reach such heart rate).

Obese children respond to exercise (and to pharmacologic stimuli) with less elevation of GH than do nonobese children.[28,85] Juvenile diabetics, on the other hand, have a normal, or even enhanced GH response to exercise.[28,32,80] Neither the mechanisms nor the implications of such responses are clear.

The concept of the "refractory period" has been introduced to denote the time during which a normal pituitary gland will not respond sufficiently to a stimulus for GH release. Such refractoriness can occur following a previous exercise stimulus.[22] The duration of the post-exertional refractory period has not been determined. It would be prudent, however, to ascertain that no strenuous activity has been performed by the child 4–6 hr prior to testing and that he be rested physically and emotionally for 45–60 min before such testing starts.

It has been stated that one post-exercise blood sample is sufficient for screening, with no need for a pre-exercise determination.[39,84] Such a practice is not advisable because some children have a high pre-exercise GH level and a paradoxic *lowering* of the serum level following exercise. If only the latter is determined, a false positive result may be obtained.

High ambient temperature is in itself a stimulus for a rise in serum GH.[26,57] A summative effect on GH of high ambient temperature and of exercise has been observed experimentally,[26] and lowering of environmental temperature caused attenuation of GH release.[8] One should therefore attempt to control climatic conditions when standardizing a clinical test.

An additional point in standardization is the timing of blood sampling. Ideally, one should obtain a sample when serum GH concentration is at its peak—apparently, about 25–35 min after the start of the exercise, irrespective of its duration[7] (unpublished data also from Bar-Or et al). Thus, when the task is very brief, a peak may be reached after its cessation. On the other hand, when the exercise lasts more than 40 min, the peak may be reached while the individual is still exercising.[7,86]

Based on the above, the following protocol is recommended:

1. Have the child rest and avoid emotional stress for at least 45–60 min before exercise. Draw blood for a baseline value at the end of this period.
2. Use an ergometer (cycle, treadmill, step) for which the load can accurately be determined.
3. Apply a "single stage with adjustments" protocol. (See Fig. II.1(f) and section entitled Examples of Exercise Protocols in Appendix II.) Aim for an intensity of 80–90% of the individual maximal aerobic power (HR = 170–185 beats/min).

4. Duration of exercise should be 15 min, or shorter if the child is exhausted, but not less than 10 min.
5. Room air temperature should be 23–25°C.
6. A post-exercise blood sample should be drawn 25–30 min from the *start* of the test.

Whenever feasible, insert a venous catheter and take a number of post-exercise blood samples, to account for individual differences in peaking time.

References

1. Akerblom HK, Koivukangas R, Ilkka J: Experiences from a winter camp for teenage diabetics. Acta Paediatr Scand Suppl 283:50–56, 1980.
2. Allen FM: Note concerning exercise in the treatment of severe diabetes. Boston Med Surg J 173:743–744, 1915.
3. Berger M, Berchtold P, Cuppers HJ, et al: Metabolic and hormonal effects of muscular exercise in juvenile type diabetics. Diabetologia 13:355–365, 1977.
4. Berger M, Berchtold P, Gries FA, Zimmermann H: Die Bedeutung von Muskelarbeit und training für die Therapie des Diabetes Mellitus. Dtsch Med Wochenschr 103:439–443, 1978.
5. Bergstad I, Dahl-Jorgensen K, Stanghelle JK, et al: The effect of intensive physical training in young insulin dependent diabetic patients (abstract). Diabetologica 19:257, 1980.
6. Bierman J, Toohey B: The Diabetic's Sports and Exercise Book. Jove/HBJ, New York, 1977.
7. Buckler JMH: Exercise as a screening test for growth hormone release. Acta Endocrinol 69:219–229, 1972.
8. Buckler JM: The relationship between changes in plasma growth hormone levels and body temperature occurring with exercise in man. Biomedicine 19:193–197, 1973.
9. Buckler JMH: Plasma growth hormone response to exercise as a diagnostic aid. Arch Dis Child 48:565–567, 1973.
10. Buckler JMH: Exercise as a physiological stimulus to growth hormone release (letter). Arch Dis Child 50:830, 1975.
11. Burstein R: Changes in insulin resistance in trained athletes upon cessation of training. MSc Thesis, McMaster University, 1982.
12. Costill DL, Cleary P, Fink WJ, et al: Training adaptations in skeletal muscle of juvenile diabetics. Diabetes 28:818–822, 1979.
13. Dahl-Jorgensen K, Meen HD, Hanssen KF, Aagenaes O: The effect of exercise on diabetic control and hemoglobin A_1 (HbA_1) in children. Acta Paediatr Scand Suppl 283:53–56, 1980.
14. De Mondenard J-P: Principes alimentaires d'un sportif diabétique—À propos du cyclisme. Vie Med Can Francais 8:643–648, 1979.

15. Dorchy H, Ego F, Baran D, Loeb H: Effect of exercise on glucose uptake in diabetic adolescents. Acta Paediatr Belg 29:83–85, 1976.
16. Dorchy H, Haumont D, Loeb H, et al: Decline of the blood glucose concentration after muscular effort in diabetic children. Acta Paediatr Belg 33:105–109, 1980.
17. Dorchy H, Niset G, Ooms H, et al: Study of the coefficient of glucose assimilation during muscular exercise in diabetic adolescents deprived of insulin. Diabete Metab 3:31–34, 1977.
18. Draminsky-Petersen H, Korsgaard B, Deckert T, Nielsen E: Growth, body weight and insulin requirement in diabetic children. Acta Paediatr Scand 67:453–457, 1978.
19. Eisenstein E, Platnick LP, Lee PA, et al: Evaluation of the growth hormone (GH) exercise test (abstract). Pediatr Res 10:338, 1976.
20. Elo O, Hirvonen B, Peltonen T, Välimäki I: Physical working capacity of normal and diabetic children. Ann Paediatr Fenn 11:25–31, 1965.
21. Engerbretson DL: The effect of physical conditioning upon the regulation of diabetes mellitus. PhD Dissertation. Pennsylvania State University, 1970.
22. Eriksson BO, Persson B, Thorell JI: The effects of repeated prolonged exercise on plasma growth hormone, insulin, glucose, free fatty acids, glycerol, lactate and -hydroxybutyric acid in 13-year-old boys and in adults. Acta Paediatr Scand Suppl 217:142–146, 1971.
23. Etkind EL, Cunningham L: Physical abilities in diabetic boys. Israel J Med Sci 8:848–849, 1972.
24. Felig P, Wahren J: Fuel homeostasis in exercise. N Engl J Med 293:1078–1084, 1975.
25. Frasier SD: A review of growth hormone stimulation tests in children. Pediatrics 53:929–937, 1974.
26. Frewin DB, Frantz AG, Downey JA: The effect of ambient temperature on the growth hormone and prolactin response to exercise. Aust J Exp Biol Med Sci 54:97–101, 1976.
27. Garlaschi C, del Guercio MJ, di Natale B, et al: Muscular exertion: a test of pituitary function in children. Acta Paediatr Scand 64:752–754, 1975.
28. Garlaschi C, di Natale B, del Guercio MJ, et al: Effect of physical exercise on secretion of growth hormone, glucagon and cortisol in obese and diabetic children. Diabetes 24:758–761, 1975.
29. Ginsberg-Fellner F, Witt ME: The effects of exercise on hemoglobin A_1C and blood lipids in diabetic children (abstract). Diabetes 27:436, 1978.
30. Guthrie DW: Exercise, diets and insulin for children with diabetes. Nursing 7:48–54, 1977.
31. Hagan RD, Marks JF, Warren PA: Physiologic responses of juvenile-onset diabetic boys to muscular work. Diabetes 28:1114–1119, 1979.
32. Hansen AP: Abnormal serum growth hormone response to exercise in juvenile diabetics. J Clin Invest 49:1467–1478, 1970.
33. Hartley LH, Mason JW, Hogan RP, et al: Multiple hormonal responses to graded exercise in relation to physical training. J Appl Physiol 33:602–606, 1972.
34. Hermansson G, Ludvigsson J: Renal function and blood pressure reaction during exercise in diabetic and non-diabetic children and adolescents. A pilot study. Acta Paediatr Scand Suppl 283:86–94, 1980.

35. Huttunen NP, Koivisto VA, Nikkila EA: Exercise-induced hypoglycaemia and albuminuria at rest and during exercise are decreased by continuous subcutaneous insulin infusion therapy (CSI) in juvenile onset diabetics (JOD) (abstract). Diabetologia 19:284, 1980.
36. Jackson RL, Kelly HG: A study of physical activity in juvenile diabetic patients. J Pediatr 33:155–166, 1948.
37. Johanson AJ, Morris GL: A single growth hormone determination to rule out growth hormone deficiency. Pediatrics 59:467–468, 1977.
38. Johansson C: The diabetic's own view on physical exercise as a part of life. Acta Paediatr Scand Suppl 283:117–119, 1980.
39. Johnsonbaugh RE, Bybee DE, Georges LP: Exercise tolerance test. Single-sample screening technique to rule out growth hormone deficiency. JAMA 240:664–666, 1978.
40. Joslin EP: The treatment of diabetes mellitus. In: Joslin EP, Root, White, Marble A (eds.) The Treatment of Diabetes Mellitus, 10th ed. Lea and Fehiger, Philadelphia, 1959, pp. 243–300.
41. Kaplan SL, Abrams CAL, Bell JJ, et al: Growth and growth hormone. I. Changes in serum level of growth hormone following hypoglycemia in 134 children with growth retardation. Pediatr Res 2:43–63, 1968.
42. Keenan BS, Killmer LB, Sode J: Growth hormone response to exercise. A test of pituitary function in children. Pediatrics 50:760–764, 1972.
43. Koivisto VA, Felig P: Effects of leg exercise on insulin absorption in diabetic patients. N Engl J Med 298:79–83, 1978.
44. Koivisto VA, Soman VR, Defronzo R, Felig P: Effects of acute exercise and training on insulin binding to monocytes and insulin sensitivity in vivo. Acta Paediatr Scand Suppl 283:70–78, 1980.
45. Lacey KA, Hewison A, Parkin JM: Exercise as a screening test for growth hormone deficiency in children. Arch Dis Child 48:508–512, 1973.
46. Larsson Y, Persson B, Sterky G, Thorén C: Functional adaptation to vigorous training and exercise in diabetic and non-diabetic adolescents. J Appl Physiol 19:629–635, 1964.
47. Larsson YAA, Sterky GCG, Ekengren KEK, Moller TGHO: Physical fitness and the influence of training in diabetic adolescent girls. Diabetes 11:109–117, 1962.
48. Lawrence RD: The effect of exercise on insulin action in diabetes. Br Med J 1:648–650, 1926.
49. Liberman B, Cesar FP, Wajchenberg BL: Human growth hormone (hGH) stimulation tests: the sequential exercise and *l*-Dopa procedure. Clin Endocrinol 10:649–654, 1979.
50. Lin T, Tucci JR: Provocation tests of growth-hormone release. A comparison of results with seven stimuli. Ann Intern Med 80:464–469, 1974.
51. Lohmann D, Liebold F, Heilmann W, et al: Diminished insulin response in highly trained athletes. Metabolism 27:521–524, 1978.
52. Ludvigsson J: Physical exercise in relation to degree of metabolic control in juvenile diabetes. Acta Paediatr Scand Suppl 283:45–49, 1980.
53. Maclaren NK, Taylor GE, Raiti S: Propranolol-augmented, exercise-induced human growth hormone release. Pediatrics 56:804–807, 1975.
54. Maehlum S, Hostmark AT, Hermansen L: Synthesis of muscle glycogen

during recovery after prolonged severe exercise in diabetic and nondiabetic subjects. Scand J Clin Lab Invest 33:309–316, 1977.
55. Marble A, Smith RM: Exercise in diabetes mellitus. Arch Intern Med 58:577–588, 1936.
56. Okada T, Hikita T, Ishitobi K, et al: Human growth hormone secretion after exercise and oral glucose administration in patients with short stature. J Clin Endocrinol Metab 34:1055–1058, 1972.
57. Okada Y, Matsuoka T, Kumahara Y: Human growth hormone secretion during exposure to hot air in normal adult male subjects. J Clin Endocrinol Metab 34:759–763, 1972.
58. Okada Y, Watanabe K, Takeuchi T, et al: Re-evaluation of exercise as a screening test for ruling out human growth hormone deficiency. Endocrinol Jpn 25:437–442, 1978.
59. Pedersen O, Beck-Nielsen H, Heding L: Increased insulin receptors after exercise in patients with insulin-dependent diabetes mellitus. N Engl J Med 302:886–892, 1980.
60. Pedersen O, Beck-Nielsen H, Schwartz, Sorensen N, Heding L: Effects of exercise on insulin receptors on erythrocytes and monocytes from insulin dependent diabetics. Acta Paediatr Scand Suppl 283:81–85, 1980.
61. Pombo M, Martinson JM, Tato F, Pena J: Propranolol and exercise test for growth hormone assays (letter). Pediatrics 60:778, 1977.
62. Poortmans J, Dewancker A, Dorchy H: Urinary excretion of total protein, albumin and β_2-microglobulin during exercise in adolescent diabetics. Biomedicine 25:273–274, 1976.
63. Pruett EDR, Maehlum S: Muscular exercise and metabolism in male juvenile diabetics. I. Energy metabolism during exercise. Scand J Clin Lab Invest 32:139–147, 1973.
64. Richardson R, Case AL: Factors determining the effect of exercise on blood sugar in the diabetic. J Clin Invest 13:949–961, 1934.
65. Roth J, Glick SM, Yalow RS, Berson SA: Secretion of human growth hormone: physiologic and experimental modification. Metabolism 12:577–579, 1963.
66. Round Table: Diabetes and Exercise. Physician and Sportsmed 7:47–64, 1979.
67. Rutenfranz J, Mocellin R, Bauer J, Herzig W: Untersuchungen über die korpliche Leistungsfähigkeit gesunder und kranker Heranwachsender. II. Die Leistungsfähigkeit von Kindern und Jugendlichen mit Diabetes Mellitus. Z Kinderheilkd 103:133–156, 1968.
68. Saltin B, Houston M, Nygaard E, et al: Muscle fiber characteristics in healthy men and patients with juvenile diabetes. Diabetes 28:93–99, 1979.
69. Sanders CA, Levinson GE, Abelman WH, Freinkel N: Effect of exercise on peripheral utilization of glucose in men. N Engl J Med 271:220–225, 1964.
70. Shanis BS, Moshang T: Propranolol and exercise as a screening test for growth hormone deficiency. Pediatrics 57:712–714, 1976.
71. Shephard RJ, Sidney KH: Effects of physical exercise on plasma growth hormone and cortisol levels in human subjects. In: Wilmore JH, Keogh JF (eds.) Exercise and Sport Sciences Reviews, Vol. 3. Academic Press, New York, 1975, pp. 1–30.
72. Soman VR, Koivisto VA, Deibert D, et al: Increased insulin sensitivity and

insulin binding to monocytes after physical training. N Engl J Med 301:1200–1204, 1979.
73. Soman VR, Koivisto VA, Grantham P, Felig P: Increased insulin binding to monocytes after acute exercise in normal man. J Clin Endocrinol Metab 47:216–219, 1978.
74. Sprague RG: Physical activity and the control of diabetes mellitus. In: Hamwi GJ, Danowski TS (eds.) Diabetes Mellitus: Diagnosis and Treatment, Vol. II. American Diabetes Association, Inc., New York, 1967, pp. 117–119.
75. Sterky G: Physical work capacity in diabetic schoolchildren. Acta Paediatr (Scand) 52:1–10, 1963.
76. Sutton JR, Jones NL, Toews CJ: Growth hormone secretion in acid-base alterations at rest and during exercise. Clin Sci Mol Med 50:241–247, 1976.
77. Sutton JR, Lazarus L: Effect of adrenergic blocking agents on growth hormone response to physical exercise. Horm Metab Res 6:428–429, 1974.
78. Sutton JR, Lazarus L: Growth hormone in exercise: comparison of physiological and pharmacological stimuli. J Appl Physiol 41:523–527, 1976.
79. Sutton JR, Young JD, Lazarus L, et al: The hormonal response to physical exercise. Aust Ann Med 18:84–90, 1969.
80. Tamborlane WV, Sherwin RS, Koivisto V, et al: Normalization of the growth hormone and catecholamine response to exercise in juvenile-onset diabetic subjects treated with a portable insulin infusion pump. Diabetes 28:785–788, 1979.
81. Tanner JM, Whitehouse RH, Hughes PCR, Vince FP: Effect of human growth hormone treatment for 1 to 7 years on growth of 100 children with growth hormone deficiency, low birthweight, inherited smallness, Turner's syndrome and other complaints. Arch Dis Child 46:745–781, 1971.
82. Turnheim E, Ogris E, Swoboda W: Erfahrungen mit dem sogenannten Ergometertest als Screeningverfahren bei Kleinwuchs im Kindesalter. Wien Klin Wochenschr 87:608–611, 1975.
83. Wahren J, Hagenfeldt L, Felig P: Splanchnic and leg exchange of glucose, amino acids, and free fatty acids during exercise in diabetes mellitus. J Clin Invest 55:1303–1314, 1975.
84. Wilson TA, Solomon IL, Schoen EJ: Exercise screening of short children for growth hormone deficiency in a family practice setting. J Fam Pract 11:547–549, 1980.
85. Wilkinson PW, Parkin JM: Growth-hormone response to exercise in obese children (letter). Lancet 2:55, 1974.
86. Winter JSD: The metabolic response to exercise and exhaustion in normal and growth-hormone-deficient children. Can J Physiol Pharmacol 52:575–582, 1974.
87. Wise PH, Burnet RB, Geary TD, Berriman H: Selective impairment of growth hormone response to physiological stimuli. Arch Dis Child 50:210–214, 1975.
88. Wise PH, Burnet RB, Geary RD, Berriman H: Exercise as a physiological stimulus to growth hormone release (reply). Arch Dis Child 50:830, 1975.
89. Zinman B, Murray FT, Vranic M, et al: Glucoregulation during moderate exercise in insulin treated diabetics. J Clin Endocrinol Metab 45:641–652, 1977.

6
Nutritional Diseases

Anorexia Nervosa (AN)

Physical Working Capacity

The self-inflicted undernutrition of AN is accompanied by physiologic and performance characteristics similar to those found during imposed undernutrition, even though patients with AN are often physically hyperactive and only in advanced stages may have iron-deficiency anemia and become hypoactive.

A summary of physiologic characteristics at rest and during exercise is given in Table 6.1 to show that it is not only body adiposity that is reduced. Other dimensions such as lean body mass, heart volume, and blood volume are also reduced,[36] as are the resting metabolic rate, respiratory rate, heart rate, blood pressure, and cardiac output.[42]

Although reduction in dimensions has a direct bearing on performance, it cannot fully explain the impairment in physical working capacity in AN. Peak power output and maximal O_2 output are low, even when expressed per body weight.[36,45,46] For example, a maximal O_2 uptake of 35.0 and 34.3 ml/kg × min was found in boys and girls with AN, respectively. These values are about 2 S.D. below the expected mean for the boys and 1 S.D. below mean for the girls. The deficit of maximal O_2 uptake in AN is more extreme than expected from the small heart volume[43,45]: on the average, a healthy adolescent with a heart volume of 500 ml will have a maximal O_2 uptake of nearly 2 liters/min, compared with only 1.3 liters/min in a patient with the same heart volume.[45] There is no valid explanation why maximal heart rate of these patients seldom exceeds 180 beats/min (compared with 195–200 among healthy adolescents).[106] It may be due to the smaller than normal muscle mass that takes part in exercise, or due to a high vagal activity. Whatever the cause,

Table 6.1. Some Physiologic Characteristics of the Adolescent with Anorexia Nervosa

At Rest	
Percent fatness	Low
Lean body mass	Low
Heart volume	Low
Blood volume	Low
Core temperature	Low
O_2 uptake	Low
Respiratory rate	Low
Heart rate	Low
Blood pressure	Low
Cardiac output	Low
ECG voltage	Low
Blood lactate	High
During exercise	
Peak power output	Low
O_2 uptake—submax and max	Low
Heart rate—submax and max	Low
Blood pressure—submax and max	Low
Cardiac output—max	Low
Cardiac output per O_2 uptake	Normal
Blood (muscle?) lactate—max	Normal

the end result is a subnormal maximal cardiac output and a reduced O_2 transporting capacity. The thermoregulatory capability of the exercising patient with AN is deficient, as discussed in Chapter 9.

Longitudinal data are unavailable to compare maximal aerobic power during the premorbid phase with that during undernutrition. However, girls and boys with AN were followed up during a period of nutritional repletion[43] and, as shown in Fig. 6.1, markedly improved their peak power output and maximal aerobic power. These changes accompanied the return to normal of body weight, lean body mass, and other dimensions. One can therefore assume that the low working capacity of the patient with AN is a direct result of undernutrition and that it was normal at the premorbid phase.

Exercise Perception

Patients with AN often engage in strenuous and prolonged physical activities which, to a bystander, may seem incompatible with their malnutrition and low working capacity. It has been suggested[22] that, in analogy to their distorted body image and denial of feeling tired, these patients also deny being overly active. A possibility has therefore been

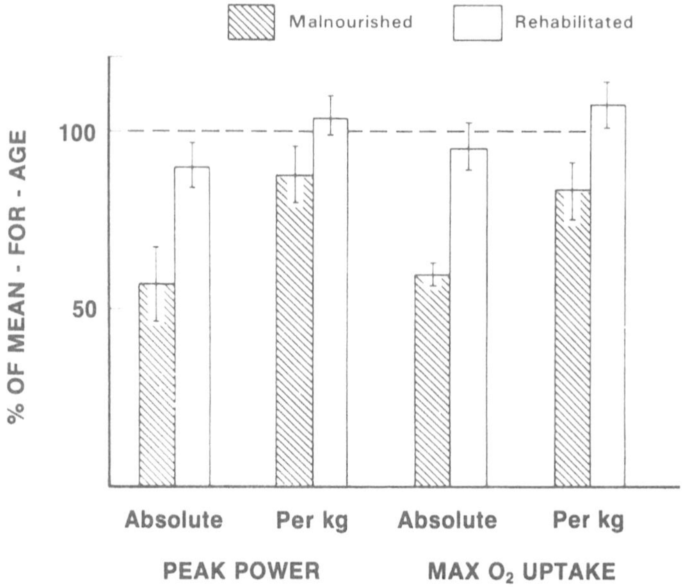

Figure 6.1. Effect of nutritional rehabilitation after anorexia nervosa on performance. Eight adolescent patients (5 boys, 3 girls) were tested on a cycle ergometer before and after nutritional repletion. Values are presented as percent of expected mean (taken from Figs. I.1–I.4 in Appendix I). Based on data by Fohlin.[44]

raised that these patients may lose their perceptual "acuity" regarding exercise intensity. This has been examined experimentally[34] and found to be unjustified. The adolescent with AN does have the ability to discriminate well among various exertional intensities.

Habitual Activity

Unlike the involuntarily malnourished adolescent, the patient with AN is often *more* active than the healthy adolescent. Only in extreme inanition will the patient be debilitated and rendered inactive.

Excessive engagement in exercise and sports is part and parcel of an overall calorie-expending "strategy" of many patients with AN. Concomitantly with their food-rejecting obsession they may develop subtle, and often clandestine, techniques for energy expenditure. While some run long distances, others may volunteer for house chores such as working in the garden or walking the dog. Some practice in their bedrooms, whereas others prefer to work out away from family and friends, so that they will not be accountable for such "forbidden" behavior. The devotion to and dependence on exercise and sports can be used by the therapist as the basis for behavior-modification therapy, in which exercise is allowed as a reward whenever weight gains have been made.

Malnutrition

Growth and Exercise Performance

Malnutrition, especially the protein-calorie type, is a very prevalent childhood disorder in various geographic areas and by far the most common cause of growth failure during infancy and childhood. The relationship between malnutrition in early life and eventual growth failure is so distinct that anthropometric indices have been used as the sole criteria for malnutrition.[89] In addition to stunted growth, the severely malnourished child may display flabby muscles and postural abnormalities.

Will subpar growth and muscle development affect the physical working capacity of a malnourished child? Based on simple physical principles, a small body size should be a handicap for those fitness components that depend on height or absolute muscle mass: when a small and a large child are to push an equal resistance (e.g., a wheelbarrow), to throw an object (e.g., pitch a baseball or shot-put), or to reach heights (e.g., high-jump, volleyball), the small child will be at a disadvantage. The same goes for more prolonged activities, where a given task (e.g., rowing) will be sustained longer by a larger individual.

As might be expected, data on this important topic are derived from populations of underdeveloped countries. Such studies indeed indicate that, merely because of their stunted growth, malnourished children (or older individuals who were malnourished earlier in life) have a deficient capacity for some, but not all, physical tasks.[2,32,89,97] Among Ethiopian boys who varied in nutritional state, $\dot{W}_{170}$ was related to body size or to the nutritional state.[2] Among other East African malnourished teenagers, the maximal aerobic power was subnormal, but when calculated per body mass, or per lean leg volume, it was within the East African norm.[32] Only in one study[97] was maximal O_2 uptake, relative to body weight, lower in malnourished children (6-year-olds) than in well-nourished ones. The same investigators[7] have shown that, in severely malnourished adults (serum albumin <2.5 g/dl, creatinine excretion <450 mg/liter/hr/square meter skin area), deficiency in maximal aerobic power is greater than can be explained merely by body mass or muscle-cell mass.

One may conclude, in general, that the low exercise performance of the malnourished child is directly related to the reduced body mass and, specifically, the lean mass.[32] In especially severe levels of malnutrition, low maximal aerobic power may be related also to a deficiency in the O_2 transport system. Iron-deficiency anemia is another cause of deficient exercise performance in the malnourished child. This aspect is discussed in Chapter 8.

There are limited data from developed countries on the effects of undernutrition on working capacity. A study performed in the author's laboratory compared fitness of male Israeli high school graduates who differed in adiposity. When the 50 leanest boys (3–7% fat), out of a population of 2000 graduates, were compared to those 50 with average adiposity (10–14% fat), the former had as good a performance in 100 and 2000 m runs, and an even better performance in pull-ups and "dips." As expected, the absolute maximal O_2 uptake of the thin boys was lower, but not when expressed per kilogram of body weight. A similar trend was observed in a study comparing the physical performance of Italian preschool children from the south with the better nourished ones from the center of the country.[41]

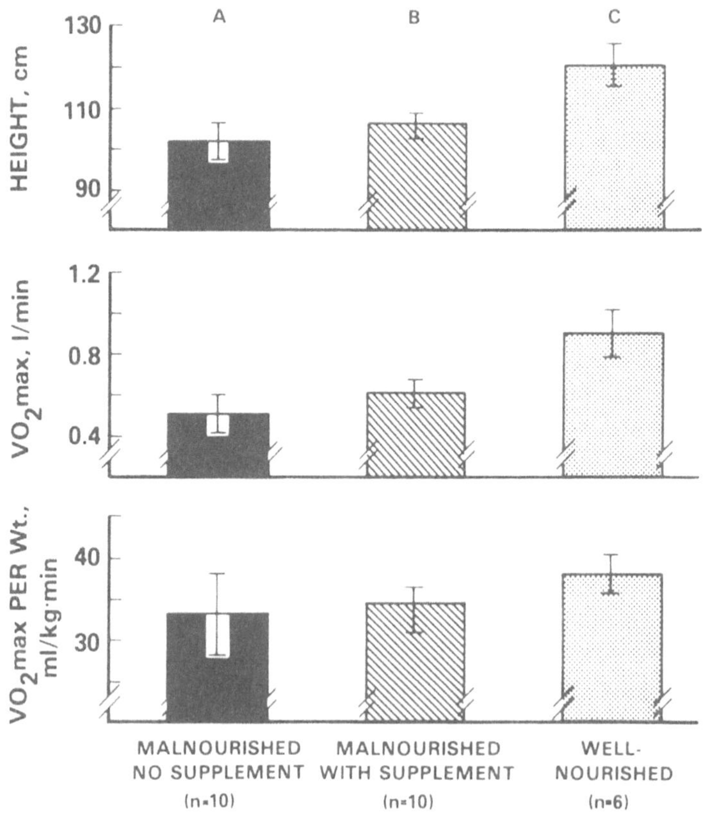

Figure 6.2. Nutrition and working capacity. Body height and maximal aerobic power capacity ($\dot{V}O_2$ max) of 6-year-old Colombian children. Group A had been found malnourished at age 3 and had not been given nutritional supplement since then. Group B had been found malnourished at age 3 and had been given a nutritional supplement, health care, and psycho-educational stimulation in a special school. Group C had been well-nourished throughout their childhood. Data from Spurr et al.[97]

These data suggest that low body weight and marked leanness among children and adolescents who are generally well nourished is *per se* not detrimental to working capacity. Furthermore, adolescents with average adiposity within a well-nourished population are probably *overnourished,* when physical working capacity is the criterion.

The question of whether nutritional rehabilitation in later years can improve the physical working capacity of the previously malnourished child is of clinical and public health importance. Fig. 6.2 summarizes data on 6-year-old Colombian children. Group A had been malnourished at age 3 and ever since. Group B had been malnourished at age 3 and since then were nutritionally rehabilitated. Group C were well nourished throughout their life. In spite of nutritional rehabilitation, Group B still had stunted growth and a reduced maximal aerobic power ($\dot{W}_{170}$). Other studies[89] also indicate that the eventual maximal aerobic power of the previously malnourished child depends on whether his body size can reach normal standards. Studies on rats[83,84] show that in spite of nutritional rehabilitation, the animal that was malnourished in the first weeks of life will have a deficient muscle endurance and glycolytic activity (see Fig. 6.3), as well as a reduced endurance time for swimming.

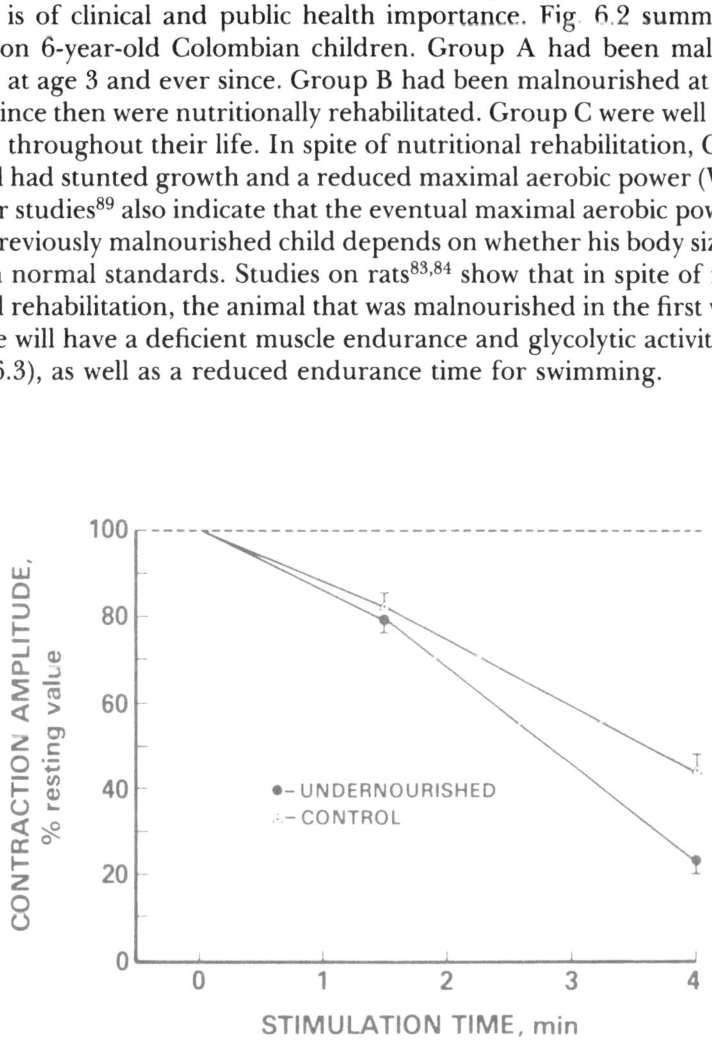

Figure 6.3. Muscle fatigability and nutritional history. Reduction in the amplitude of contraction of the *in vitro* gastrocnemius of mature rats. One group ("undernourished") had been given a reduced diet for the first 13 weeks of life, and then was nutritionally rehabilitated for 15–20 weeks. The other group ("control") had been given a normal diet. Values are expressed as percent of the initial contraction amplitude. Mean ± 1 S.E.M. Data from Raju.[83]

Which malnutrition is most detrimental to physical working capacity? Due to their retrospective nature, studies on humans cannot easily be designed where specific entities of malnutrition are compared. It seems, though, that the major type of malnutrition that induces stunted growth and reduced working capacity is a combined protein-calorie malnutrition. In a study on mature rats,[53] endurance time of swimming was found *higher* in those given a low protein diet for 10 weeks than in those given a normal diet. This seemingly paradoxical result can be explained by the greater starch consumption in the protein-deficient animals, which may have increased the glycogen content of their muscle, thus improving endurance time. A similar phenomenon has been suggested for humans.[7]

Habitual Activity

Based on starvation-related data for adults,[60] one would assume that a child debilitated by a marked protein-calorie deficiency will not be as active as his well-nourished peers. Not enough data are available to confirm such a notion.

In a pilot study[55] the cumulative number of heartbeats of malnourished Colombian children was lower than that of well-nourished controls during a comparable period, suggesting less activity among the former. In a longitudinal study two groups of Mexican infants were followed from birth till 2 years. In one group both mother and child were getting a supplement to their food, while the other had the regular diet consumed by low socioeconomic families in a typically poor rural area. The supplemented babies increased their activity level with age, while the less nourished ones had a consistently lower activity level. The difference between the groups grew with time so that, by the age of 24 months, the well-nourished children were six times as active as the malnourished ones.[27] Figure 6.4 represents the relationship between infant activity and the state of nutrition. Both obese and malnourished children are less active than the well nourished but lean ones.

An hypothesis has been put forth[55] that the subpar intellectual performance of malnourished, or previously malnourished, children is induced by restricted activity and, as a result, limited stimuli and learning opportunities. Because of the immense numbers of malnourished children worldwide, such an hypothesis deserves a thorough test and should be a challenge to anthropologists, nutritionists, and exercise physiologists.

Effect of Conditioning

To what extent can physical working capacity of nutritionally deficient children be improved by physical conditioning? No intervention studies

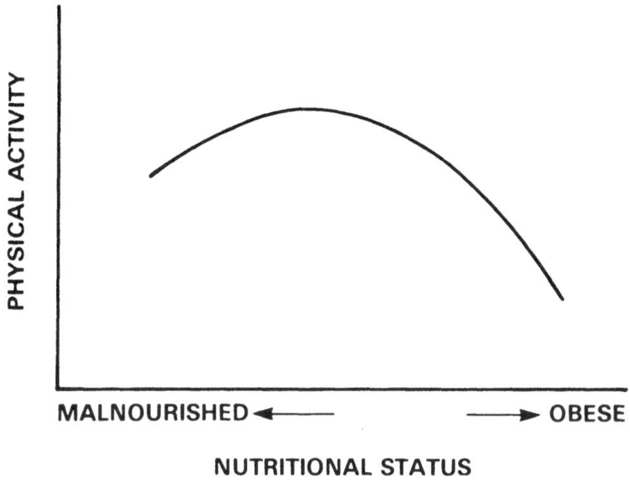

Figure 6.4. Habitual activity in relationship to the level of nutrition in children and infants. A schematic representation.

are available in which malnourished children underwent a conditioning regimen. There is, however, evidence that malnourished boys who are physically active have a better exercise performance than their less active malnourished peers.[2,89] Malnourished rats, when given a swimming conditioning program, had a distinct improvement in muscle endurance and in swimming-endurance time.[84] No concomitant increase was observed in their body weight.

From the practical point of view, the above data suggest that exercise performance of previously malnourished children can be improved by conditioning, even if body dimensions remain small.

Obesity

Response to Acute Exercise

An obese child can seldom perform physical activities at par with his leaner peers. One reason for his deficient performance is a greater metabolic cost of exercise. In physiologic terms, obese children require a higher O_2 uptake to perform a given task. Their *maximal* O_2 uptake, in contrast, is often lower than that of leaner children.[35,80] Thus, at a given level of exercise the obese operate at a higher percentage of their maximal aerobic power and their "metabolic reserve" is lower. For elaboration of the metabolic reserve concept see Chapter 1, section entitled

Mechanical Efficiency and Economy of Movement. The high metabolic cost of exercise to the obese individual is not limited to weight-bearing tasks such as walking, running, or climbing. It also appears during cycling, where the body is in large part supported by the seat.[3,33]

A high submaximal O_2 uptake might reflect either reduced mechanical efficiency at the muscle subcellular level or mechanical "wastefulness" due to the carrying of excessive weight and to their clumsy execution of movement. Based on current knowledge, the latter seems to be true.[40]

Maximal O_2 uptake is closely related to the mass of the muscle that performs the activity, as has been shown for healthy children during cycling.[33] Maximal O_2 uptake of obese children, however, is *lower* than expected from their muscle mass.[35] It is not clear whether such a deficiency represents a specific trait of body-fat excess or is merely an indication of detraining.

In line with the high metabolic cost of their activities, obese children also exercise at a higher percentage of their maximal heart rate,[11,70,86] suggesting a low cardiac reserve. It has been estimated that, among adults, each 10% increase in body fat is accompanied by a 10 beat/min rise in heart rate.[9] Whether the same is true for children has not yet been established. Nor are there data on the effects of childhood obesity on exercise cardiac output or stroke volume. Systolic arterial blood pressure is excessively elevated during exercise in the obese child, more so than at rest.[5,8]

Respiratory function in the exercising obese child seems adequate, even though some have a low ventilatory response to CO_2 at rest.[26] Obese adults have a somewhat reduced tidal volume, excessive ventilation and respiratory rate, as well as increased alveolar-arterial O_2 differences during submaximal exercise.[37]

The combined stresses of exercise and heat induce a high physiologic strain on the obese child. For a full discussion see Chapter 9, section entitled Obesity.

Habitual Activity

The relationship between childhood obesity and physical hypoactivity is of unique importance: unlike the case with other diseases, hypoactivity may play a role in the *etiology*, as well as be an *outcome*, of childhood obesity. In this section we shall first present descriptive reports on the habitual activity of children who differ in adiposity. This will be followed by an analysis of the possible *causal* relationships between hypoactivity and obesity.

Adiposity, Activity, and Calorie Intake. The idea that physical hypoactivity is typical of obese people and may play a role in the etiology of obesity was first suggested in 1907.[108] Various authors have since discussed this relationship,[51] but the first who evaluated the activity patterns of overweight children was Bruch in 1940[21]: in a study of 140 overweight girls and boys, 2–14 years old, she assessed their activities through case histories, reports of parents, and direct observations in the clinic. Sixty-eight percent of the girls and 76% of the boys were classified as inactive. These were subject to greater parental protectiveness and had considerably fewer social contacts than had the more active ones. Most studies since 1940 have confirmed the low activity level among obese children and adolescents, compared with their leaner peers, be it at school,[30,58] out of school,[66] in summer camp,[24] or at home.[104,111]

Even when obese youths do participate in sports, the *intensity* of their activity is low.[24] This is especially so in nonregimented play,[30,66] when the individual child, rather than the teacher or instructor, determines his or her own level of involvement. Hypoactivity is found in both sexes of obese youngsters but is more prevalent among girls.[38,109]

Contrary to common belief, obese children often *eat less* and have a lower daily calorie turnover than do lean children.[19,30,58,88,100] An example of such low turnover is presented in Fig. 6.5 for adolescent girls. Sixty-eight percent of the overweight group expended and ate less than 2000 kcal/day (8400 kJ/day), compared with 11% among the lean. In contrast, 53% of the lean and only 11% of the overweight spent more than 2500 kcal/day (10,500 kJ/day).

Although the great majority of published reports confirm a low calorie turnover in the obese child, there are some that find no relationship

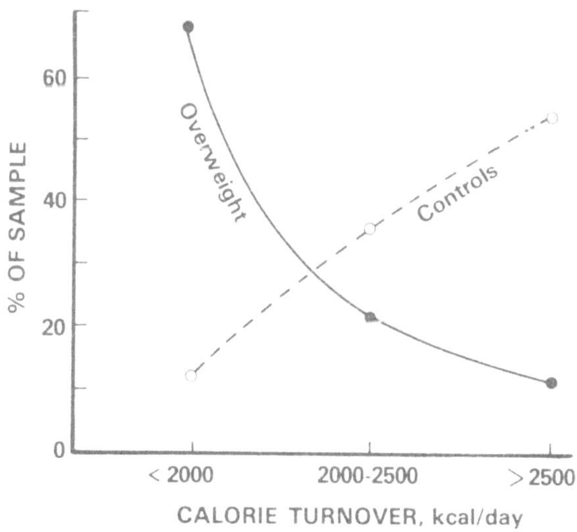

Figure 6.5. Frequency distribution of 29 overweight and 28 control adolescent girls according to their daily calorie turnover. Based on interviewer-administered recall questionnaires. Schematic adaptation from Johnson et al.[58]

between activity level and the degree of adiposity.[100,103,110,114] Such a discrepancy may result from the variety of methods used to determine activity and dietary habits. In addition, activity patterns differ when a child is exposed to different stimuli. For example, obese children were found to be less active than their leaner siblings at home, but not at school.[111]

Is Hypoactivity a Cause of Childhood Obesity? To assess a possibly *causal* relationship between obesity, eating habits and hypoactivity, ideally one should gather data during the earliest stages of life, preferably before an individual becomes obese. When eating patterns of 288 Canadian babies were observed during their first year of life, neither calorie intake nor the type of feeding, frequency of meals, or age at which solids were introduced was a predictor of skinfold thickness at age one.[47] The authors concluded that monitoring of physical activity during the first year of life is indispensable in any future study on the etiology of infantile obesity.

With new developments in measuring techniques, information is gradually becoming available on the activity pattern of infants, newborns, and even fetuses. Among infants, as in schoolchildren and adolescents, there is an inverse relationship between adiposity and activity. When 4- to 6-month-old babies were subdivided according to adiposity, the most obese had the fewest limb motions, as assessed by actometers (Fig. 6.6). They expended less than 20% of their total energy on physical activity, compared with 35–40% among the leanest infants. Total calorie intake, as well as "extra" intake (i.e., calories consumed above maintenance requirements) were also inversely related to the level of adiposity.[87] In another study, five newborns of obese mothers were observed during the first 8 weeks of their life. Weight gain in these formula-fed babies was inversely related to their activity, as assessed by actometers. Gain in length, on the other hand, was positively correlated with activity.[65]

The above studies indicate that hypoactivity in obese children starts very early in life. One still needs the perspective of time to determine which comes first—the large body weight and adiposity or the low level activity. Of relevance is a carefully designed study by Griffiths and Payne,[52] who chose their 4- to 5-year-old subjects according to their parents' adiposity rather than their own. The offspring of obese parents were distinctly less active and ate less than those of leaner parents (Fig. 6.7). These findings are important because, at the time of testing, the children of both parental groups had similar anthropometric and body composition characteristics. Bearing in mind that the offspring of obese parents have a high probability of becoming obese later in life,[48] we may speculate that—in this particular sample—hypoactivity preceded obesity.

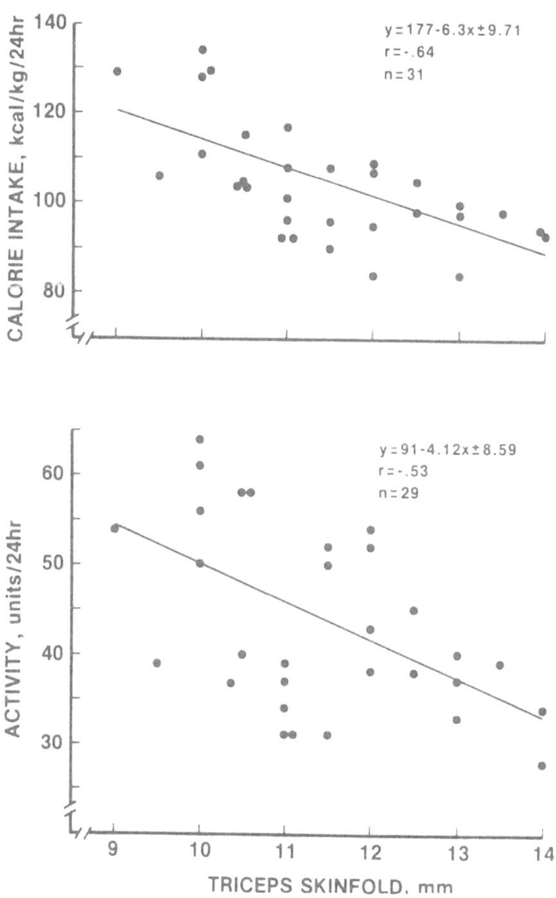

Figure 6.6. Physical activity and calorie intake in relationship to adiposity (triceps skinfold) in 4- to 6-month-old infants. Activity is represented by actometer units. Actometers were attached to wrist and ankle. Data from Rose and Mayer.[87]

Causes of Hypoactivity. Why are obese youngsters insufficiently active? While there may be still unexplained hereditary or other physiologic mechanisms that determine level of activity early in life, there are also well-documented psychosocial and fitness-related factors that may influence the willingness of obese children to pursue an active life. Parental attitude toward physical activity is paramount: parents of obese children and adolescents were found to discourage social contacts and any physical activities that connoted "risk." Furthermore, a relationship was found between such parental overprotection and the lack of activity among the children. Families of obese adolescent girls pursued fewer common rec-

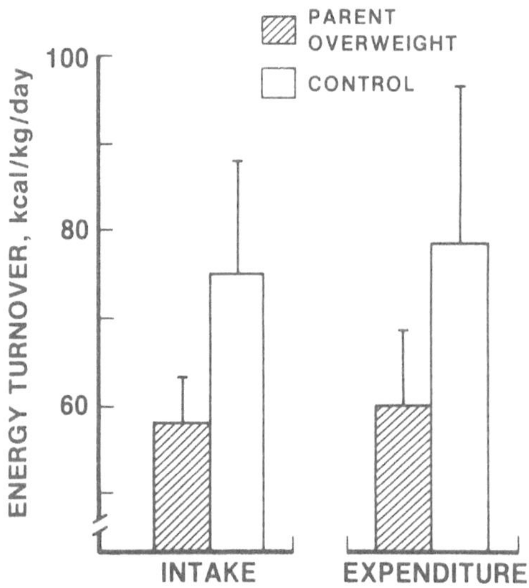

Figure 6.7. Energy balance of 4- to 5-year-old children, grouped according to weight excess of parents. Adiposity of children was determined from skinfold thickness; activity, by recording of a heart rate profile during the day; and calorie intake, by duplicating consumed food. Vertical lines denote 1 S.D. Data from Griffiths and Payne.[52]

reational activities than those of lean girls. The latter seemed also to have more intersibling ties and friendships and to report a more unified family.[21]

Projective and other tests indicated that obese girls were preoccupied with the thought that they were overweight; they showed "passivity" and "withdrawal," and reacted in a way typical to minority ethnic groups, perceiving themselves as "victims of prejudice."[71] Obese girls were aware of their low levels of activity but had no perception of their *degree* of inactivity.[23] Direct observations and questionnaires revealed that obese youths and adolescents perceived that their non-obese peers considered them "too awkward" and "physically limited."[1] And, indeed, obese children and adolescents are not appreciated by their peers,[11] who will not invite them to take part in games and will often ridicule them.

Obese schoolchildren perceive a given exercise level as harder than do their lean counterparts (unpublished data from the author's laboratory). This greater perceived strain renders the obese individual reluctant to pursue strenuous or prolonged activities.

In conclusion, hypoactivity is prevalent among most groups of obese infants, children, and adolescents of both sexes,[67] as shown also for female adults. Animal studies, as well as information on infants and on pre-obese children, suggest that inactivity may be a major *etiologic* factor in the genesis of childhood obesity. We still lack definitive confirmation of such a causal relationship and no clear-cut explanation is available as to *why* babies, or fetuses, vary in their activity. Is it maternal behavior, the availability of food, a subtle interaction between baby and mother, or perhaps genetic disposition? In later childhood, and in adolescence, hypoactivity could be a cause—but also a result—of obesity. Whatever the initial stage, hypoactivity and obesity create an unhappy symbiosis through which the afflicted child may enter a vicious circle of inactivity–positive calorie balance–obesity–reduced fitness–further inactivity.

Physical Working Capacity

The exercise performance of children, whether measured in the laboratory or in the field, is strongly related to their level of adiposity. Maximal aerobic power, determined in the laboratory[15,70,72,85] or as performance in long-distance running,[50] is inversely related to percentage of body fatness. The top half of Fig. 6.8 is an example of such a relationship. From Fig. 6.9a it is apparent that maximal aerobic power ($\dot{W}_{170}$) among grossly overweight children and adolescents is below the expected levels. This subnormal maximal aerobic power is evident even when data are normalized for lean body mass (Fig. 6.9b)[85] or for heart volume.[15]

Fitness components other than maximal aerobic power are also deficient in obese children. Even though taller and, in absolute terms, stronger than their peers, they are weaker per kilogram body weight. As a result, when performing a task that requires lifting of their own body, they are at a disadvantage. An example is given in Fig. 6.10, which summarizes performance of three groups of 50 adolescents each who were tested in the author's laboratory. One group comprised very lean individuals (mean fatness = 5.0%), another, those of average adiposity (11.5%), and the third, mildly obese subjects (22.8%). Whereas the lean and the average group had similar scores in strength-related events, the performance of the obese was dramatically inferior. Some 70% of the latter could not pull themselves up on the horizontal bar even once and could not complete a single "dip" on the parallel bars ("dip" = the change of position from straight-arm support to bent-elbow support and back). Among obese college students (30.9% fat), the combined strength of four muscle groups, expressed per kilogram lean body mass, was only 85% that of an average adiposity group (13.5% fat).[61] Since muscle mass is closely related to lean body mass, it is not clear why one kilogram of muscle of an obese person will generate less force than a kilogram of muscle of a lean person. No information is available on the infrastructure of muscles, or the recruitment of motor units, of children who vary in adiposity. We therefore cannot state whether the above difference in strength results from morphologic, physiologic, or motivational differences.

Motor skills of obese schoolchildren differ little from controls. However, their improvement in skills following a training program seems to

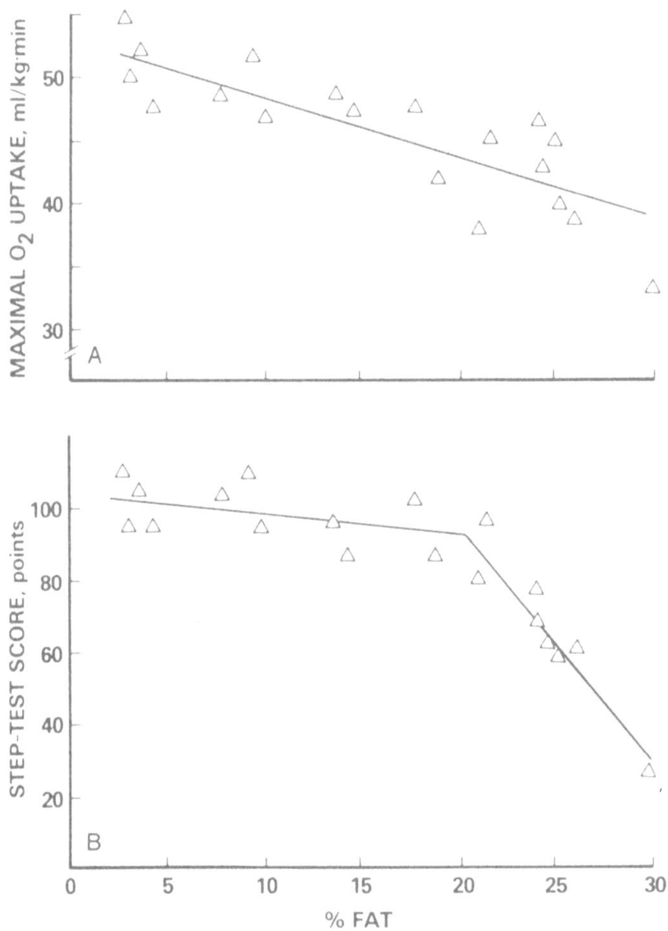

Figure 6.8. Maximal O_2 uptake measured during a treadmill test (**A**), and the score in a modified Harvard Step Test (**B**), in relationship to body fatness. Individual performance of 19 nonathletic adolescent boys who ranged in adiposity from very thin to moderately obese. (Bar-Or and Zwiren, unpublished report).

be lower.[11] Obese schoolchildren tend to achieve low grades in physical education classes,[66,85] as shown in Fig. 6.11. Among the leanest of 518 schoolchildren, more than 80% were high achievers. The pattern for the most overweight group was practically reversed: the low achievers outnumbered the high achievers by more than three to one.

Is there any *linear* relationship between the level of adiposity and the impaired performance? As seen from Fig. 6.8, maximal O_2 uptake declines linearly with the increase in percentage of fat. Other fitness components, however, do not have a linear relationship with percent fatness. The lower half of Fig. 6.8 is a case in point: the "fitness index," as

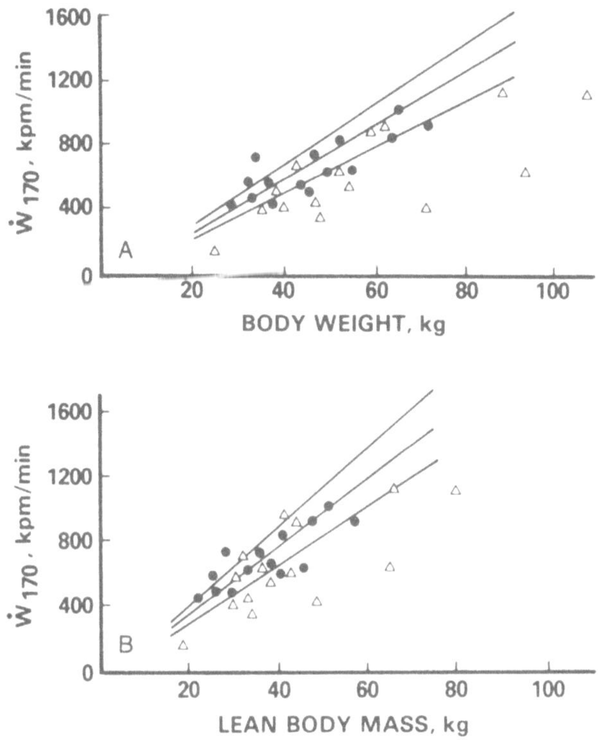

Figure 6.9. Relationship in overweight 6- to 18-year-old boys between power output in kilopound meters per min (kpm/min) at heart rate of 170 beats/min ($\dot{W}_{170}$) and body weight **(A)**, and lean body mass **(B)**. The boys were divided into those with 1–2 S.D. of overweight (●) and those more than 2 S.D. overweight (△) as compared with a standard population. Adapted by permission from Rehs et al.[85]

calculated from the step test (this index is positively related to the number of steps that the individual manages to complete and is inversely related to the post-exercise heart rate), was a little lower in mildly obese adolescents than in lean ones. For those boys who had more than 23% fat, however, the score became markedly low. Thus the more obese adolescents were "penalized" twice: their maximal aerobic power was low and they also had to lift a heavier body, which caused a clumsy up-and-down motion and early fatigue in their legs. In other fitness items where body weight has to be supported, we also find that the decline in performance is more dramatic from mild to moderate obesity than it is from normal adiposity to mild obesity.

It has been suggested[107] that mild obesity is not necessarily accompanied by decreased performance because the extra weight carried by the child serves as a training stimulus. Such a rationale may hold only if the child is active enough. If, on the other hand, he is hypoactive (as shown

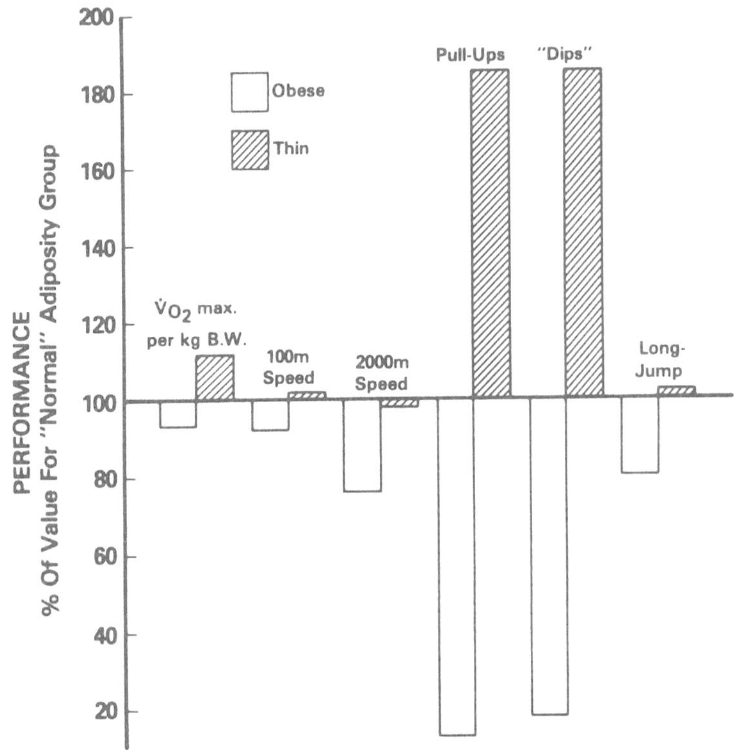

Figure 6.10. Maximal aerobic power ($\dot{V}O_2$ max), speed, muscle strength, and power of obese (22.8% fat) and very thin (5.0% fat) male adolescents. Values expressed as percentage performance of adolescents with medium adiposity (11.5% fat). (Bar-Or and Zwiren, unpublished report).

for most obese children) then his extra weight is only detrimental to performance.

Beneficial Effects of Conditioning

One of the most tangible effects of long-term physical conditioning is upon body composition. Conditioning induces in most individuals an increase in lean body mass and a concomitant decrease in fat. Overall body mass may be reduced whenever the decrease in adipose mass exceeds the increase in fat-free mass.

In this section emphasis will be given to the following questions: How effective is conditioning as a reducing regimen? How does the conditioning effect compare with restriction of calorie intake? Does an increase in energy expenditure affect appetite? What is the long-term effect of fat reduction by conditioning? What are the recommended physical activities for the obese child?

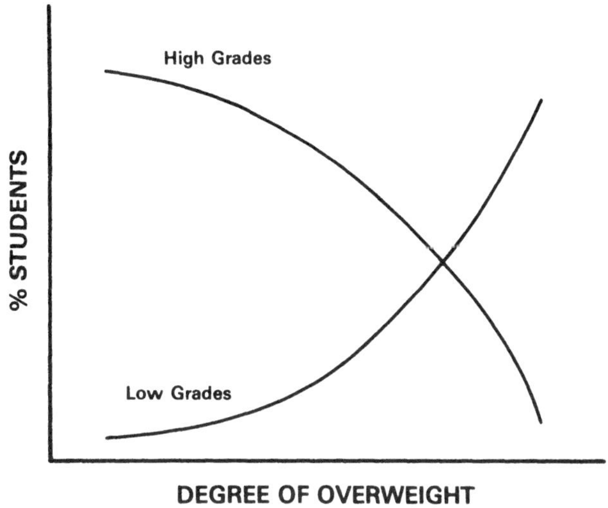

Figure 6.11. Relationship between grades obtained in physical education classes and the level of overweight among 518 six- to 18-year-old girls and boys. Schematic adaptation from Rehs et al.[85]

Effectiveness of Conditioning as a Reducing Regimen. Exercise is by far the single most flexible component of energy expenditure in the energy balance equation. With a mild activity such as walking or jogging for 45–60 min, one can raise the daily calorie expenditure by as much as 20–25% (see also Appendix IV for calorie equivalent of various activities). More intense or prolonged exertion, as practiced by athletes, can raise daily expenditure by 100% and more.

Are such increases sufficient to affect body weight and fatness? One often hears parents and, occasionally, health practitioners rationalize that a reducing regimen by exercise "isn't worth it" because of the low calorie equivalent of various activities. Indeed, to lose 1 kg of adipose tissue (or 7000 kcal), a 40-kg child may have to run 140 km or play tennis for 26 hours. If these tasks were to be completed in one week, such an objection might be valid. However, calorie expenditure has a cumulative effect. By walking or running, for example, 4 km a day (equivalent to 200 kcal in a 40-kg child), the child can lose 1 kg within 35 days, or 5 kg in half a year, even without a reducing diet. In Table 6.2 are summarized studies in which obese children and adolescents participated in a conditioning program of 3–29 weeks' duration. Although these studies differ in subject selection and in conditioning dosage, they show some 5–10% decrease in body weight and, more importantly, a 15–30% decrease in adiposity. The *rate* of fat reduction markedly varies among studies, ranging between 1 and 5% of the initial fat mass per week. This rate depends neither on the intensity nor on the type of exercise. It is the

Table 6.2. Effects of Physical Conditioning on Body Adiposity and Weight of Obese Children and Adolescents

Subjects			*Conditioning*		*Dietary Treatment*	*Effects*		*Reference*
Age (yr)	*No. & Sex*	*Adiposity**	*Activity*	*Frequency & Duration*		*Body Wt.*	*Body Fat*	
12.7	15 F 18 M	35.3% (F) 29.6% (M)	Intensive P.E.† summer camp	Daily—7 weeks	1700 kcal/day	12.8% ↓ (F) 11.1% ↓ (M)	27% ↓ (F) 26% ↓ (M)	80
8–10	22 M	>2 S.D. wt. for ht.	Extra P.E. at school	1–2 weeks, 17 weeks	—	No change	Skinfold ↓	15
10–12	7 M	30%	Intensive P.E. summer camp	Daily, 7 weeks	1700 kcal/day	11.5% ↓	17% ↓	96
13–14	49 M	41% overweight	Various (home) + basketball	Daily, 18 months	Nutrition education	5% less increase than control	27% ↓ (of overweight)	28
8–11 8–12 11–14	40 F 48 M 101 F	23.3 mm‡ 22.6 mm 27.4 mm	Extra P.E. at school	5 weeks, 5–6 months	Nutrition education	0–7% ↓ (vs. control)	0–18% ↓ (vs. control)	91
11–12	20 M	43% overweight	Extra P.E. at school	2 weeks, 9 months	—	25% ↓ (of overweight)	?	105
15–17	28 F	39%	Walk & jog at school	4 weeks, 15–29 weeks	—	Minimal	21–24% ↓	73
11–13	7 F 7 M	28% overweight	Intensive training	Daily, 3 weeks	1000 kcal/day	4.7 kg ↓	?	113
13–16	4 F 3 M	overweight	Various, summer camp	Daily, 6 weeks	1000 kcal/day	9% ↓	?	112

* Adiposity is determined in percent of body weight, unless stated otherwise.
† P.E. = physical education classes.
‡ Triceps skinfold.

overall calorie expenditure that really matters. Following a strenuous, prolonged exercise bout, there may be a rise in basal metabolic rate for as much as 15 hours.[39] Such a rise is probably due to stimulation of the rate of carbohydrate substrate cycles.[74] Therefore, the calorie equivalent of an exercise bout will be an *underestimation* of the actual additional energy expenditure.

For reasons unknown, boys seem to benefit more than girls from weight-reducing conditioning programs.[79,91,94] Such a sex-related difference has also been shown for rats.[56] Nevertheless, weight and fat reduction are feasible also for adolescent girls who undergo short-term conditioning regimens.[73,79,80,91,113]

In adults, obesity of the hypercellular type (hyperplastic obesity) is more retractable to exercise treatment than the hypertrophic type, in which the size rather than the number of cells is increased.[12,62] There is no evidence for such a trend among children.

The effectiveness of an exercise reducing regimen may be enhanced by exposure to a cold climate. This has been suggested for young adults who had been exposed to a 10-day mission in the Arctic,[75] and confirmed by a study in a climatic chamber cooled to −40°C.[92] The long-term effects of this phenomenon are not known, nor has its potential as a reducing regimen been studied.

Dietary Restriction vs. Exercise Therapy. Low calorie intake and increased activity can, separately or in tandem, induce loss of weight and fat. The use of diet *per se* as a reducing regimen will not be discussed here. This topic has been amply covered elsewhere. Our purpose is to compare the merits and disadvantages of diet and exercise as modes of treatment of childhood obesity. These are summarized in Table 6.3.

A major concern in any reducing regimen is that it might interfere with normal growth. Ideally, one would like to achieve a reduction in

Table 6.3. A Comparison Between Dietary Restriction and Physical Exercise as Reducing Regimens in Childhood Obesity

Effect	*Diet*	*Exercise*
Weight loss	Yes	Yes
Fat loss	Yes	Yes
Fat-free mass	Loss	Gain
Growth retardation	Possibly (extreme diets)	No
Increased fitness	None or some	Yes
Reduction of adipocyte size	Yes	More than in diet(?)
Feeling of hunger	Yes	No
Rate of weight loss	Fast or slow	Slow
Carry-over effect	No	No

body fat without a concomitant loss of protein, water, minerals, or vitamins. A very low calorie regimen (e.g., 1000 kcal/day in an 11-year-old child), although effective for short-term weight reduction, may impede growth velocity.[20,79] A low-calorie diet, even when not extreme, has catabolic effects manifested, under some circumstances, by a negative nitrogen balance and by the loss of fat-free mass. These effects can be reduced, or even reversed, when exercise is incorporated.[90,95,117] Such an anabolic effect of exercise is especially of relevance to the growing child and adolescent.

As discussed under Conditioning Effects Other than Fat Reduction, below, weight loss through exercise increases fitness level. Weight loss by diet may also improve fitness, but an extremely low calorie intake, such as in prolonged semistarvation,[60] may induce weakness and a reduction in exercise performance. Obese children subjected to marked calorie restriction as the sole method of treatment may complain of lethargy and show a distinct reduction in their spontaneous habitual activity.

Either a low-calorie diet or conditioning may cause a decrease in adipocyte size. Exercise treatment may induce a greater reduction in adipocyte size than does diet alone.[12] The practical implications of this difference are not clear.

Neither exercise nor a dietary regimen is effective if the child does not pursue them on a long-range basis. To some obese children and adults the hunger that accompanies dietary restriction is unbearable, so that even a well-motivated patient may not be able to comply with the treatment: in this respect exercise has the advantage of not being accompanied by hunger. On the other hand, obese children may be reluctant to take part in conditioning programs where they are exposed to the watchful eye and sharp tongue of their peers.

Perhaps the most important advantage of diet is that some effect can be shown within a few days. The overall rate of weight loss is faster with diet than it is with exercise therapy alone. To avoid disappointment, the slow effect of exercise must be explained to the parents and child prior to the start of a program.

The moral of the above is that a low-calorie diet and exercise should be *combined* in a reducing regimen. Such a combination will maximize the dividends and counteract the deficiencies of either treatment.

Exercise Therapy and Changes in Appetite. A question commonly asked by parents is whether the effects of increased activity will not be nullified by a concomitant increase in appetite. It is true that appetite often increases with a rise in energy expenditure[57]: athletic children who train extensively consume more calories than nonathletic children.[4,10] However, the positive relationship between the level of activity and appetite does not hold throughout the activity spectrum. Studies with rats,[31,69,102] a dog,[81] and monkeys[6] have indicated that when a highly sedentary animal increases its level of physical activity, appetite may *decrease* rather than increase. In adult humans, cross-sectional[69] and longitudinal[63] data confirm this pattern.

The same phenomenon has also been described for children. Overweight 8- to 10-year-old boys attended a 4-month program of one or two extra physical education periods weekly. By the end of the program their daily calorie intake had decreased by 12% (from 2129 to 1874 kcal/day). Those who had two extra periods weekly had a greater decrease in calorie intake than did those having only one extra period.[15]

The mechanism for such an apparent paradox is not clear. A threshold activity level may exist (see Fig. 6.12), below which appetite is not well regulated and the individual eats more than is justified by his calorie expenditure. Only when this threshold is surpassed does appetite compensate for changes in activity. It has been speculated[68] that from the perspective of evolution, a sedentary lifestyle in humans and in some animals is below the physiologic range for adequate appetite regulation.

Whatever the mechanism, it is evident that changes in appetite do not compensate for increased activity of the sedentary obese. Even for individuals whose food intake does increase with exercise therapy, the end result is a negative calorie balance.[12,25]

Conditioning Effects Other than Fat Reduction. In addition to its effects on adiposity, a conditioning regimen induces other changes, as listed in Table 6.4. While some are specific to the obese individual, others apply to anyone who undergoes conditioning.

Biochemical changes are evident in carbohydrate, lipid, and protein metabolism alike. Plasma insulin concentration is reduced with conditioning: it can be as low as half that found in the inactive individual. This drop, coupled with better glucose tolerance, reflects a greater cell sensitivity to insulin in various tissues.[12,13,93,101] The improved sensitivity to

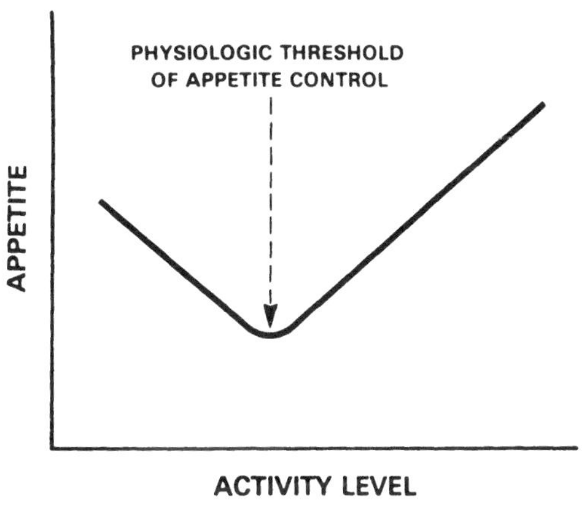

Figure 6.12. Appetite and physical activity. A schematic presentation of a concept. See text for details.

Table 6.4. Conditioning Effects in the Obese Child, Other than on Body Composition

Function	*Effect**
Biochemical	
Plasma insulin	−
Cell sensitivity to insulin	+
Glucose tolerance	+
FFA mobilization	+
Lipolysis in tissues	+
Serum triglyceride	−
Total serum cholesterol	−/o
Low-density lipoproteins	−
High-density lipoproteins	o
Nitrogen sparing	+
Anabolic steroids	+
Cardiorespiratory	
Heart rate—rest and submax	−
Systolic blood pressure—rest	−
Ventilatory equivalent—submax	−
O_2 uptake—submax	−
O_2 uptake—max	+
Psychologic	
Self-image; body image	+
Rating of perceived exertion	−
Self-confidence	+
Sociability	+

* − = decrease; o = no change; + = increase.

insulin can result from weight loss *per se.* But it is also a result of the actual increase in fitness, irrespective of body weight changes,[14] and is highly correlated to the improvement in maximal aerobic power.[93]

Changes in lipid metabolism include an enhanced lipolysis, free fatty acid mobilization from adipose tissue, and a decrease in plasma low-density lipoprotein cholesterol. An increase in high-density lipoprotein cholesterol and a decrease in total cholesterol and in triglycerides have been found in some studies, but not in others.[49,113,115] No information is available as to the long-range carry-over effects of conditioning on the lipid profile of obese children. In the light of the relationship between lipid profile and the risk for coronary heart disease, this question is of paramount importance.

The nitrogen-sparing effect of exercise in juvenile obesity has been discussed above. From biochemical analyses and assessment of lean body mass, it seems obvious that conditioning is accompanied by anabolic processes—specifically, protein synthesis.[16,59,62,90,94,95,117]

Cardiorespiratory changes that occur with conditioning and weight loss are compatible with a decreased physiologic strain during exercise. Heart rate at rest and at submaximal exercise drops,[80,101,112,115] as does the resting systolic blood pressure.[112] No data are available on changes in stroke volume and cardiac output. Ventilation at submaximal exercise drops,[80,112] especially at power loads beyond the anaerobic threshold.[112] The work of breathing decreases with reduction in thoracic and abdominal wall mass, which further contributes to the lower O_2 cost of exercise.

Psychosocial changes that occur with weight reduction have been studied mostly in conjunction with dietary restriction. Conditioning-related changes include improvement of self-image and body image,[99] self-confidence, and ability to adjust to peer society.[82] Hitherto withdrawn children develop social awareness and an ability to integrate in, and enjoy, group activities. We have seen obese adolescent females who, prior to weight loss, limited their social activities to playing with children much younger than themselves—the adolescent assuming a motherly role. After successfully losing weight, they actively sought company of their own age. Such individuals often take up sports and recreational activities in which they were previously reluctant to participate. They also seem to take better care of their appearance, motivated by their new (smaller size!) clothing and perception of their improved physical appearance.

In an experiment performed in our laboratory in Israel, 10- to 12-year-old obese girls and boys underwent a 6-week program in which two physical education periods were added to the regular two physical education periods at school. Rating of perceived exertion, using the Borg scale,[17] decreased by some 2 scale units, at power loads of 50 and 100 watts. The inference was that a given task seemed easier to the now conditioned subjects.

A rewarding turn of events is when a child's success in losing weight draws other family members, notably overweight parents, to similar activity programs. Such a development indicates a good prognosis for further, more prolonged weight loss by the child.

Functional changes that result from conditioning of the obese child are similar to those found in any child who improves his or her fitness level. Observations made specifically on obese youngsters show an improvement in their maximal aerobic power.[15,78,80,96,105,115] O_2 uptake for a given task drops[80] in proportion to weight loss.[112] With such an improved economy of work, endurance time in a prolonged activity (30 min or more) increases. The ability of the child to lift his own weight, as in pull-ups,[28] improves due to reduction of body weight and increase of muscle strength.

Conditioning and Adipose Tissue Cellularity. The total mass of adipose tissue is a product of the number of adipocytes (fat cells) and their average mass. Understanding the growth and replication of adipocytes may shed light on the genesis of childhood obesity and aid in its prevention.

There are no data on the effect of physical conditioning on the adipocytes of obese children. There is, however, an elegant study on the long-term effect of conditioning on the adipocytes of rats: starting on their fifth day of life, groups of rats were exposed to daily swimming or a diet regimen, until week 28 of their life. All animals then remained sedentary until week 62. Measurements taken at that stage conclusively show, as summarized in Fig. 6.13, that those who had swum during their growth retained a lower body weight, fat weight, and number of adipocytes than control animals. The adipocyte count was also lower than that of the rats who had dieted.[76,77] This study demonstrates that exercise during growth is effective in reducing the rate of replication of adipocytes in the mature rat and the effect is long-lasting. While such results cannot be extrapolated for humans, they pose a definite challenge for a similar intervention study in children.

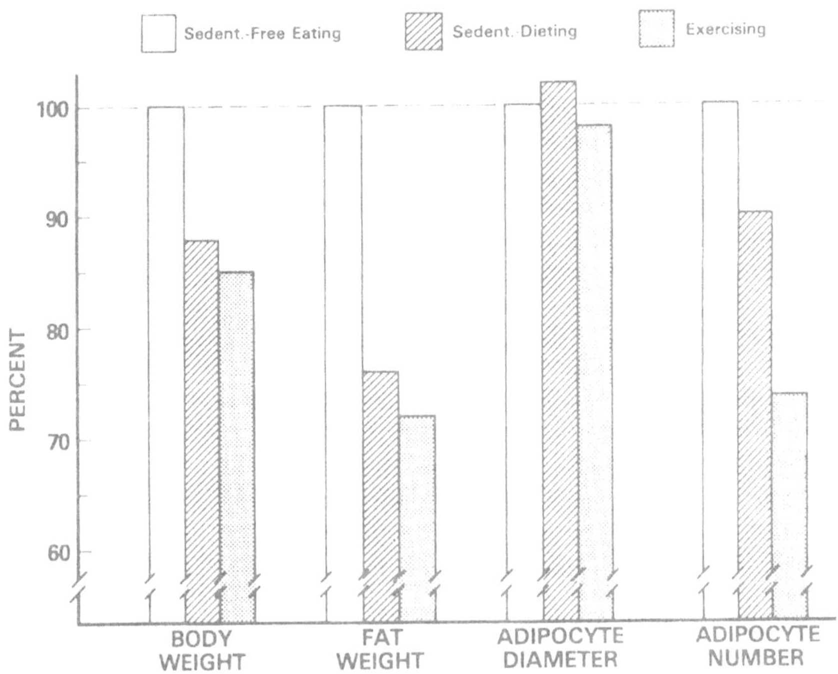

Figure 6.13. Long-term effect of exercise or low-calorie diet during growth on body weight, body fatness, adipocyte size, and adipocyte number in the mature rat. Values were obtained at week 62 of life and are expressed as percentage of values obtained in rats that neither exercised nor dieted during growth. Data from Oscai et al.[76]

Lasting Effects of Conditioning. As with diet, an increase in physical activity requires changes in habits. Many obese children can be motivated to join, and adhere to, activity programs as long as they are brief and closely supervised. Such conditions prevail, for example, in summer camps or in hospital. It is unrealistic, however, to expect compliance once the child is back at home, without the guidance and support of an instructor. It is therefore pertinent to question whether conditioning programs have a carry-over effect on the obese child beyond their duration.

Unfortunately, available information is not encouraging. Investigators from Czechoslovakia, for example, have reported fat loss and increased work performance in obese children during a 7-week diet and exercise summer camp. The benefits virtually disappeared when the children returned to their home routine.[79] A similar school-based program was found effective for weight control of obese boys. However, during a break in the program (a 3-month vacation), the experimental group gained more weight than obese controls who had not taken part in the program.[28] Another successful reducing regimen was carried out in a public school for 4 years.[91] In a follow-up survey, conducted 3 years later, the beneficial effects of the program had all but disappeared.[67]

We may conclude that, in spite of some few exceptions,[79,101] the effects of exercise and diet do not carry over beyond the duration of the programs.[29] There are some indications that, in adults, behavior modification techniques can prolong exercise therapy effects for up to a year beyond the duration of the treatment.[54,98] Whether this approach will succeed with children has not yet been established. This is a highly promising area for future research. Finally, we must emphasize that any long-term effect of a reducing program is contingent upon the full cooperation, including change of habits, of the parents.

Recommended Activities. Three principles should guide the selection of activities for the obese child. They should be *effective* for fat reduction, *feasible* of performance, and *enjoyable*.

Effectiveness. Activities for fat reduction should emphasize calorie expenditure, which is a function of *work* rather than power. As large a body mass as possible, preferably the whole body, should be transported from one place to another as in walking, running, dancing, cross-country skiing, or swimming.

Activities that are less suitable for weight control are, for example, strength training, gymnastics, or downhill skiing. These involve a high power output but the overall calorie expenditure is not high because the actual period of exercise is short.

In walking or running, it is distance rather than speed that determines overall calorie expenditure. Thus, one should emphasize to the child and parents that walking 1 km is as effective as running 1 km, the only difference being that walking takes more time. As with diet, activities can be exchanged for one another, based on their calorie equivalents. Table 6.5 shows eight recommended activities that require 200 kcal when performed by a 40-kg child. The heavier the child, the greater is the calorie expenditure in a given task. This is especially true for tasks that involve transporting the whole body. Appendix III presents a more detailed list of activities and their calorie equivalent for children whose body weight ranges between 20 and 60 kg. Such a list can be used by the practitioner for exercise prescription (see example in Chapter 2, section entitled Exercise Prescription).

It has been claimed that strength training of a muscle group can induce a *local* reduction of adipose tissue (skinfold overlying the muscle). Careful analysis indicates that this is not the case.[62]

To maximize weight reduction by exercise one would like to achieve the highest possible energy expenditure for any given task. An interesting question is whether a particular exercise yields the same calorie expenditure at different times of the day. Circadian rhythms of energy expenditure have been reported for adult obese women, who expended some 20% more energy in afternoons and evenings than in mornings, while performing an identical cycling task. These differences were above and beyond those found in their resting metabolic rate.[116] If similar circadian rhythms exist in children and adolescents, one may be able to plan reducing programs accordingly.

Table 6.5. Equicaloric Activities, Recommended for Obese Children*

Activity	*Duration (min)*
Cross-country skiing (leisure)	50
Cycling—10 km/hr	80
Ice hockey (on ice)	20
Running— 8 km/hr	30
—10 km/hr	25
Soccer (game)	30
Swimming (breast stroke)	
30 m/min	50
Tennis	45
Walking—4 km/hr	80
Walking—6 km/hr	60

* Each activity consumes 200 kcal (840 kJ) when performed for stated duration by a 40-kg child.

Feasibility. Not all sports are equally feasible for the obese child. Some activities will "penalize" him because of his physique, and *a priori* put him at a disadvantage. For example, in jumping, rope-climbing, or maneuvers on the parallel bars, excess body weight is a distinct handicap (see also Figs. 6.8 and 6.10). Such activities should not be prescribed.

Recommended activities for the obese child should exploit certain physical *advantages*: his high stature, large lean body mass (thus high absolute strength), buoyancy, and high subcutaneous thermal insulation. These may enhance the child's performance in such activities as shot-put, discus, some positions in football and basketball, and, especially, swimming and water polo.

Many obese adolescents choose to swim not only because of their high buoyancy and thermal comfort, but also because their body is submerged and not exposed to view. We have seen, for example, an obese 13-year-old boy who refused to take part in physical education classes because his "large breasts" could be seen when he wore a T-shirt. This boy happily joined a swimming program. A year later, following weight reduction, he joined in other sports.

Recreational Aspects. To increase motivation and compliance, activities must be enjoyable. Those which are preferred by obese children and adolescents include: sailing, dancing, horseback riding, archery, cross-country skiing, ice skating, cycling, and, particularly, swimming. Such sports share a number of common denominators: 1) they involve the individual rather than a team, thus avoiding peer disapproval; 2) they are recreational rather than competitive: this aspect was found especially important among adolescent females;[82] 3) in each of these sports, intensity and overall amount can be individually tailored according to inclination and ability: this aspect is especially critical at the start of a new program, when the emphasis should be put on gradual progression. Dance-oriented exercise has been found to be highly successful with obese people of all ages. If performed correctly it can induce high levels of energy expenditure, combined with pleasant recreation. It can be done within a group, but also in the privacy of the home.[64]

Within what framework should an activity program for obese children be conducted? Ideally, one would want them to integrate with nonobese children at school or at the sports club. This approach, unfortunately, has not been found successful. An obese child, like other children with an "abnormal" appearance, is more often than not rejected by his healthy peers. More success has been shown in programs where all the participants were obese. It seems that, as with adults, obese children or adolescents accept each other quite readily and are less inhibited than when in the company of the nonobese.[82,91] Programs exclusively for obese children were tried in special summer camps;[80,82,96,112,113] in hospital;[98] and at school.[15,18,28,73,91,101] Such programs often combine exercise with nutrition education[18,28] or with a low-calorie diet.[80,96,99,112,113]

School-based intervention programs have a number of advantages over other frameworks: they can cater to large numbers of patients; can operate over most of the year; the large number of participants makes recruitment of specialized personnel, such as psychologists, nutritionists, or adapted physical educators, economically feasible; and there is less reliance on the initiative of parents.[28,29,91] Summer camp- or hospital-based programs have the advantage of affording *combined intensive* regimens of diet and exercise which are indicated for the extremely obese youngster. Home-based programs offer the least direct professional control, and their success is greatly dependent on cooperation and acceptance by parents. Their advantage is that the whole family can be induced to change its habits, with long-standing dividends. Home-based programs are also suitable for behavior modification techniques.[98]

In our newly developed Children's Therapeutic Exercise and Health Centre, we have been trying to spur obese patients to perform a prescribed activity at home or to join a two or three periods per week dance-exercise program, in combination with dietary changes. Each child is seen by a nurse-practitioner, a nutritionist, and a physician every 2–3 weeks. We encourage parents to accompany the child. Measurement of weight, height, skinfold thickness, and aerobic and anaerobic performance are taken periodically for follow-up, and to motivate the child and the parents. In cases where parental cooperation is inadequate, a social worker joins the team. The skill, enthusiasm, and ability of the paramedical staff to motivate and reinforce success is paramount to the success of the program.

References

1. Allon N: Self-perceptions of the stigma of overweight in relationship to weight-losing patterns. Am J Clin Nutr 32:470–480, 1979.
2. Areskog NH, Selinus R, Vahlquist B: Physical work capacity and nutritional status in Ethiopian male children and young adults. Am J Clin Nutr 22:471–479, 1969.
3. Åstrand I, Åstrand PO, Stunkard A: Oxygen intake of obese individuals during work on a bicycle ergometer. Acta Physiol Scand 50:294–299, 1960.
4. Åstrand PO, Engström L, Eriksson B, et al: Girl swimmers. Acta Paediatr (Scand) Suppl 147, 1963.
5. Backman L, Freyschuss U, Hallberg D, Melcher A: Cardiovascular function in extreme obesity. Acta Med Scand 193:437–446, 1973.
6. Baile CA, Zinn W, McLaughlin C: Exercise, blood lactate and food intake. Experientia 26:1227–1235, 1970.

7. Barac-Nieto M, Spurr GB, Maksud MG, Lotero H: Aerobic work capacity in chronically undernourished adult males. J Appl Physiol 44:209–215, 1978.
8. Barta L, Szoke L, Vandor-Szobotka V: Working capacity of obese children. Acta Paediatr Acad Sci Hung 9:17–21, 1968.
9. Bassey EJ, Bryant JC, Clark E, et al: Factors affecting cardiac frequency during self paced walking: body composition, age, sex and habitual activity. J Physiol (Lond) 291:46P, 1979.
10. Berg K: Body composition and nutrition of adolescent boys training for bicycle racing. Nutr Metab 14:172–180, 1972.
11. Berndt I, Rehs H-J, Rutenfranz J: Sportpädagogische Gesichtspunkte zur Prophylaxe der Adipositas im Kindesalter. Öff Gesundh-Wesen 37:1–9, 1975.
12. Björntrop P: Exercise in the treatment of obesity. Clin Endocrinol Metab 5:431–453, 1976.
13. Björntorp P: Obesity and physical exercise in relation to glucose tolerance and plasma lipids. In: Carlson LA, Pernow B (eds.) Metabolic Risk Factors in Ischemic Cardiovasular Disease. Raven Press, New York, 1982.
14. Björntorp P, de Jounge K, Sjoström L, Sullivan L: The effect of physical training on insulin production in obesity. Metabolism 19:632–638, 1970.
15. Blomquist B, Borjeson M, Larsson V, et al: The effect of physical activity on the body measurements and work capacity of overweight boys. Acta Paediatr Scand 54:566–572, 1965.
16. Boileau RA, Buskirk ER, Horstman DH, et al: Body composition changes in obese and lean men during physical conditioning. Med Sci Sports 3:183–189, 1971.
17. Borg G: Physical Performance and Perceived Exertion. Gleerup, Lund, 1962.
18. Botvin GJ, Cantlon A, Carter BJ, Williams CL: Reducing adolescent obesity through a school health program. J Pediatr 15:1060–1062, 1979.
19. Bradfield RB, Paulos J, Grossman L: Energy expenditure and heart rate of obese high school girls. Am J Clin Nutr 24:1482–1488, 1971.
20. Brook CGD, Lloyd JK, Wolf OH: Rapid weight loss in children. Br Med J 3:44–45, 1974.
21. Bruch H: Obesity in childhood. IV. Energy expenditure of obese children. Am J Dis Child 60:1082–1109, 1940.
22. Bruch H: Eating disorders: Obesity, anorexia nervosa and the person within. Basic Books, New York, 1973.
23. Bullen BA, Monello LF, Cohen H, Mayer J: Attitudes towards physical activity, food and family in obese and nonobese adolescent girls. Am J Clin Nutr 12:1–11, 1963.
24. Bullen BA, Reed RB, Mayer J: Physical activity of obese and non-obese adolescent girls appraised by motion picture sampling. Am J Clin Nutr 14:211–223, 1964.
25. Buskirk ER: Increasing energy expenditure: The role of exercise. In: Wilson NL (ed.) Obesity. F.A. Davis, Philadelphia, 1969, pp. 163–176.
26. Chaussain M, Gamain B, LaTorre AM, et al: La fonction respiratoire au repos chez l'enfant obèse. Bull Eur Physiopathol Respir 13:599–609, 1977.
27. Chavez A, Martinez C, Bourges H: Nutrition and development of infants

from poor rural areas. 2. Nutritional level and physical activity. Nutr Rep Int 5:134–144, 1972.

28. Christakis G, Sajecki S, Hillman RW, et al: Effect of a combined nutrition, education and physical fitness program on the weight status of obese high school boys. Fed Proc 25:15–19, 1966.
29. Coates TJ, Thoresen CE: Treating obesity in children and adolescents: a review. Am J Public Health 68:143–151, 1978.
30. Corbin CB, Pletcher P: Diet and physical activity patterns of obese and nonobese elementary school children. Q Assoc Health Phys Educ 39:922–928, 1968.
31. Crew EI III, Fuge KW, Oscai LB, et al: Weight, food intake and body composition: effect of exercise and of protein deficiency. Am J Physiol 216:359–363, 1969.
32. Davies CTM: Physiological responses to exercise in East African Children. II. The effects of schistosomias, anaemia and malnutrition. Environ Child Health 19:115–119, 1973.
33. Davies CTM, Barnes C, Godfrey S: Body composition and maximal exercise performance in children. Hum Biol 44:195–214, 1972.
34. Davies CTM, Fohlin L, Thoren C: Perception of exertion in anorexia nervosa patients. In: Berg K, Eriksson BO (eds.) Children and Exercise IX. University Park Press, Baltimore, 1980, pp. 327–332.
35. Davies CTM, Godfrey S, Light M, et al: Cardiopulmonary responses to exercise in obese girls and young women. J Appl Physiol 38:373–376, 1975.
36. Davies CTM, Von Dobeln W, Fohlin L, et al: Total body potassium, fat free weight and maximal aerobic power in children with anorexia nervosa. Acta Paediatr Scand 67:229–234, 1978.
37. Dempsey JA, Reddan W, Balke B, Rankin J: Work capacity determinants and physiologic cost of weight-supported work in obesity. J Appl Physiol 21:1815–1820, 1966.
38. Durnin JVGA: Physical activity by adolescents. Acta Paediatr Scand Suppl 217:133–135, 1971.
39. Edwards HT, Thorndike A Jr, Dill DB: The energy requirement in strenuous muscular exercise. N Engl J Med 213:532–535, 1935.
40. Farebrother MJB: Respiratory function and cardiorespiratory response to exercise in obesity—A review article. Br J Dis Chest 73:211–229, 1979.
41. Ferro-Luzzi A, D'Amicis A, Ferrini AM, Maiale G: Nutrition, environment and physical performance of preschool children in Italy. Bibl Nutr Dieta 27:85–106, 1979.
42. Fohlin L: Body composition, cardiovascular and renal function in adolescent patients with anorexia nervosa. Acta Paediatr Scand Suppl 268:5–20, 1977.
43. Fohlin L: Exercise performance and body dimensions in anorexia nervosa before and after rehabilitation. Acta Med Scand 204:61–65, 1978.
44. Fohlin LPM: The effects of growth, body composition, and circulatory function of anorexia nervosa in adolescent patients. In: Berg K, Eriksson BO (eds.) Children and Exercise IX, University Park Press, Baltimore, 1980, pp. 317–326.
45. Fohlin L, Davies CTM, Freyschuss U, et al: Body dimensions and exercise

performance in anorexia nervosa patients. In: Borms J, Hebbelinck H (eds.) Pediatric Work Physiology, Karger, Basel, 1978, pp. 102–107.
46. Fohlin L, Freyschuss U, Bjake B, et al: Function and dimensions of the circulatory system in anorexia nervosa. Acta Paediatr Scand 67:11–16, 1978.
47. Gagnon G, Brault-Dubuc M, Nadeau M, Demirjian A: Subcutaneous fat and nutrition in the first year of life. Nutr Rep Int 19:541–551, 1979.
48. Garn SM: The origins of obesity. Am J Dis Child 130:465–467, 1976.
49. Gilliam TB, Burke MB: Effects of exercise on serum lipids and lipoproteins in girls, ages 8 to 10 years. Artery 4:203–213, 1978.
50. Gracey M, Hitchcock NE, Wearne KL, et al: The 1977 Busselton children's survey. Med J Aust 2:265–267, 1979.
51. Greene JA: Clinical study of the etiology of obesity. Ann Intern Med 12:1797–1803, 1939.
52. Griffiths M, Payne PR: Energy expenditure in small children of obese and non-obese parents. Nature 260:698–700, 1976.
53. Hansen-Smith FM, Maksud HG, Van Horn DL: Influence of chronic undernutrition on oxygen consumption of rats during exercise. Growth 41:115–121, 1977.
54. Harris MB, Hallbauer ES: Self-directed weight control through eating and exercise. Behav Respir Ther 11:523–529, 1973.
55. Heywood PF, Rur B, Latham MC: Use of the SAMI heart rate integrator in malnourished children. Am J Clin Nutr 24:1446–1450, 1971.
56. Holloszy JO: The effect of endurance exercise on body composition. In: Vague J, Boyer J (eds.) The Regulation of the Adipose Tissue Mass. Excerpta Media International Congress Series No. 315, 1974, pp. 254–258.
57. James WPT, Trayhurn P, Davies H, et al: Interactions of food intake and energy expenditure: an overview. In: Luigi A, et al (eds.) The Body Weight Regulatory System: Normal and Disturbed Mechanisms. Raven Press, New York, 1981, pp. 147–152.
58. Johnson ML, Burke BS, Mayer J: Relative importance of inactivity and overeating in the energy balance of obese high school girls. Am J Clin Nutr 4:37–44, 1956.
59. Jokl E: Physical activity and body composition: fitness and fatness. Ann NY Acad Sci 110:778–794, 1963.
60. Keys A, Brozek J, Henschel A, et al: The Biology of Human Starvation. University of Minnesota Press, Minneapolis, 1950.
61. Kitagawa K, Miyashita M: Muscle strengths in relation to fat storage rate in young men. Eur J Appl Physiol 38:189–196, 1978.
62. Krotkiewski M, Mandroukas K, Sjoström L, et al: Effects of long-term physical training on body fat, metabolism and blood pressure in obesity. Metabolism 28:650–658, 1979.
63. Leon AS, Conrad J, Hunninghake DB, Serfass R: Effects of a vigorous walking program on body composition, and carbohydrate and lipid metabolism of obese young men. Am J Clin Nutr 32:1776–1787, 1979.
64. Linder P: Techniques of management for the inactive obese child. In: Collipp PJ (ed.) Childhood Obesity, 2nd ed. PSG, Littleton, Mass., 1980, pp. 179–205.

65. Mack RW, Kleinhenz ME: Growth, caloric intake and activity levels in early infancy: a preliminary report. Hum Biol 46:345–354, 1974.
66. Markuske H: Adipositas und Schulsport. Sportarzt Sportmed 10:404–406, 1969.
67. Mayer J: Obesity during childhood. In: Winnick M (ed.) Childhood obesity. Wiley, New York, 1975, pp. 73–80.
68. Mayer J, Marshall NB, Vitale JJ, et al: Exercise, food intake and body weight in normal rats and genetically obese mice. Am J Physiol 177:544–548, 1954.
69. Mayer J, Roy P, Mitra KP: Relations between caloric intake, body weight and physical work: studies in industrial male population in West Bengal. Am J Clin Nutr 4:169–175, 1956.
70. Mocellin R, Rutenfranz J: Untersuchungen über die körperliche Leistungsfähigkeit gesunder und kranker Heranwachsender. III. Die Leistungsfähigkeit von Kindern und Jugendlichen mit Adipositas. Z. Kinderheilkd 104:179–196, 1968.
71. Monello LF, Mayer J: Obese adolescent girls: an unrecognized "minority" group? Am J Clin Nutr 13:35–39, 1963.
72. Montoye HJ, Mikkelsen WM, Block WD, Gayle R: Relationship of oxygen uptake capacity, serum uric acid and glucose tolerance in males and females age 10 to 69. Am J Epidemiol 108:274–282, 1978.
73. Moody DL, Wilmore JH, Girandola RN, Royce JP: The effects of a jogging program on the body composition of normal and obese high school girls. Med Sci Sports 4:210–213, 1972.
74. Newsholme EA: A possible metabolic basis for the control of body weight. N Engl J Med 302:400–405, 1980.
75. O'Hara WJ, Allen C, Shephard RJ: Loss of body fat during an arctic winter expedition. Can J Physiol Pharmacol 55:1235–1241, 1977.
76. Oscai LB, Babirak SP, Dubach FB, et al: Exercise or food restriction: effect on adipose tissue cellularity. Am J Physiol 227:901–904, 1974.
77. Oscai LB, Babirak SP, McGarr JA, Spirakis CN: Effect of exercise on adipose tissue cellularity. Fed Proc 33:1956–1958, 1974.
78. Pařizková J, Stanková L, Spřynarová S, Vamberová M: Influence de l'éxercise physique sur certains index metaboliques sanguins chez les garçons obese après l'éffort. Nutr Dieta 7:21–27, 1965.
79. Pařizková J, Vamberová M: Body composition as a criterion of the suitability of reducing regimens in obese children. Dev Med Child Neurol 9:202–211, 1967.
80. Pařizková J, Vanecková M, Vamberová M: A study of changes in some functional indicators following reduction of excessive fat in obese children. Physiol Bohemoslov 11:351–357, 1962.
81. Passmore R: A note on the relation of appetite to exercise. Lancet 1:29, 1958.
82. Peckos PS, Spargo JA, Heald FP: Program and results of a camp for obese adolescent girls. Postgrad Med 27:527–533, 1960.
83. Raju NV: Effect of early malnutrition on muscle function and metabolism in rats. Life Sci 15:949–960, 1974.
84. Raju NV: Effect of exercise during rehabilitation on swimming perfor-

mance metabolism and function of muscle in rats. Br J Nutr 38:157–165, 1977.

85. Rehs HJ, Berndt I, Rutenfranz J: Untersuchungen zur Frage der Leistungsfähigkeit Adipöser unter besonderer Berücksichtigung des Sportunterrichtes. Z Kinderheilkd 115:23–39, 1973.
86. Rehs HJ, Berndt I, Rutenfranz J, Burmeister W: Untersuchungen zur Bestimmung der Hautfaltendicke mit verschiedenen Kalibern. Z Kinderheilkd 120:121–133, 1975.
87. Rose HE, Mayer J: Activity, calorie intake, fat storage and the energy balance of infants. Pediatrics 41:18–29, 1968.
88. Saris WHM: Aerobic power and daily physical activity in children. With special reference to methods and cardiovascular risk indicators. Doctoral dissertation, Catholic University, Nijmegen, Krips Repro Meppel, 1982.
89. Satyanarayana K, Nadamuni Naidu A: Nutrition and menarch in rural Hyderabad. Ann Hum Biol 6:163–165, 1979.
90. Schrub J-C, Wolf L-M, Courtois H, Javet F: Cure de jeune avec exercice musculaire. Évolution du poids et du bilan azoté. Nouv Presse Med 4:875–878, 1975.
91. Seltzer CC, Mayer J: An effective weight control program in a public school system. Am J Public Health 60:679–689, 1970.
92. Shephard RJ, O'Hara WJ, Allen C, Allen G: A controlled study of fat loss in the cold (abstract). Med Sci Sports 11:98, 1979.
93. Soman VR, Koivisto VA, Deibert D, et al: Increased insulin sensitivity and insulin binding to monocytes after physical training. N Engl J Med 301:1200–1204, 1979.
94. Sonka J: Effects of diet or diet and exercise in weight reducing regimens. In: Pářizková J, Rogozkin V (eds.) Nutrition, Physical Fitness and Health. University Park, Baltimore, 1978, pp. 239–247.
95. Sonka J, Gregorova I, Tomosova Z, et al: Plasma androsterone, dehydroepiandrosterone and 11-hydroxycorticoids in obesity. Effects of diet and physical activity. Steroids Lipids Res 3:65–74, 1972.
96. Spřynarová S, Pářizková J: Changes in the aerobic capacity and body composition in obese boys after reduction. J Appl Physiol 20:934–937, 1965.
97. Spurr GB: Childhood undernutrition: implications for adult work capacity and productivity. In: Folinsbee LJ et al (eds.) Environmental Stress. Individual Human Adaptations. Academic Press, New York, 1978.
98. Stalonas PM Jr, Johnson WG, Christ M: Behaviour modification for obesity: the evaluation of exercise, contingency management, and program adherence. J Consult Clin Psychol 46:463–469, 1978.
99. Stanley EJ, Glaser HH, Levin DG, et al: Overcoming obesity in adolescents. A description of a promising endeavor to improve management. Clin Pediatr 9:29–36, 1970.
100. Stefanik PA, Heald FP, Mayer J: Caloric intake in relation to energy output of obese and non-obese adolescent boys. Am J Clin Nutr 7:55–62, 1959.
101. Sterky G: Clinical and metabolic aspects of obesity in childhood. In: Pernow B, Saltin B (eds.) Muscle Metabolism During Exercise. Plenum Press, New York, 1971.

102. Stevenson JAF, Box BM, Feleki V, Beaton JR: Bouts of exercise and food intake in the rat. J Appl Physiol 21:118–122, 1966.
103. Stunkard A, Pestka J: The physical activity of obese girls. Am J Dis Child 103:116–121, 1962.
104. Thomson ME, Cruickshank FM: Survey into the eating and exercise habits of New Zealand pre-adolescents in relation to overweight and obesity. N Z Med J 89:7–9, 1979.
105. Thorén C: Physical training of handicapped school children. Scand J Rehab Med 3:26–30, 1971.
106. Thorén C: Working capacity in anorexia nervosa. In: Borms J, Hebbelinck M (eds.) Pediatric Work Physiology. Karger, Basel, 1978, pp. 89–95.
107. Thorén C, Seliger V, Máček M, et al: The influence of training on physical fitness in healthy children and children with chronic diseases. In: Linneweh F (ed.) Current Aspects of Perinatology and Physiology of Children. Springer-Verlag, Berlin-Heidelberg, 1973.
108. Von Noorden C: Obesity, Metabolism and Practical Medicine, Vol. 3. Kenner, Chicago, 1907, p. 696.
109. Vuille JC, Mellbin T: Obesity in 10-year-olds: an epidemiologic study. Pediatrics 64:564–572, 1979.
110. Watson AW, O'Donovan DJ: Influence of level of habitual activity on physical working capacity and body composition of post-pubertal schoolboys. Q J Exp Physiol 62:325–332, 1977.
111. Waxman M, Stunkard AJ: Caloric intake and expenditure of obese boys. J Pediatr 96:187–193, 1980.
112. Whipp BJ, Ruff WK: The effect of caloric restriction and physical training on the response of obese adolescents to graded exercise. J Sports Med Phys Fitness 11:146–153, 1971.
113. Widhalm K, Maxa E, Zyman H: Effect of diet and exercise upon the cholesterol and triglyceride content of plasma lipoproteins in overweight children. Eur J Pediatr 127:121–126, 1978.
114. Wilkinson PW, Parkin JM, Pearlson G, et al: Energy intake and physical activity in obese children (abstract). Br Med J 1:756, 1977.
115. Ylitalo V: Treatment of obese schoolchildren with special reference to the mode of therapy, cardiorespiratory performance and the carbohydrate and lipid metabolism. Acta Paediatr Scand Suppl 290:1–108, 1981.
116. Zahorska-Markiewicz B: Effects of timing on energy expenditure during rest and exercise in obese women. Nutr Metab 24:238–243, 1980.
117. Zuti WB, Golding LA: Comparing diet and exercise as weight reduction tools. Physician Sportsmed 4:49–53, 1976.

7

Neuromuscular Diseases

Cerebral Palsy (CP)

Physical Working Capacity

Whether spastic, athetotic, or ataxic, the child with CP is limited in his physical abilities. The spastic child may not be able to generate much muscle force or endure effort for a long time but can perform fine movements better than the athetotic child. The latter can better perform activities that require strength and will have a greater walking ability.[33] Children with mixed spastic-athetotic manifestations will be the most limited.

Directly measured maximal aerobic power of children and adolescents with CP is 10–30% lower than in controls.[8,45,46] If maximal aerobic power is indirectly assessed from submaximal heart rate, the performance of children with CP is only some 50% of normal.[42] Apparently, submaximal heart rate is disproportionately high in CP children due to their low mechanical efficiency. The use of submaximal heart rate to predict maximal aerobic power is, therefore, not recommended.

Mechanical Efficiency and Economy of Movement

Mechanical efficiency of patients with CP is low, particularly among the spastic ones, as shown in Fig. 7.1. This is true whether ergometry is done with the legs[11,43,44] or with the arms.[8] Assuming that ATP yield per mole of metabolic fuel is not different between patients and healthy children, the low efficiency must be due to "wasteful" contraction of the spastic or dyskinetic muscles. Whereas the healthy child, in performing a rhythmic task (e.g., walking, running, pedaling, or swimming), alternately relaxes the agonist and antagonist muscles, the spastic child keeps a high muscle

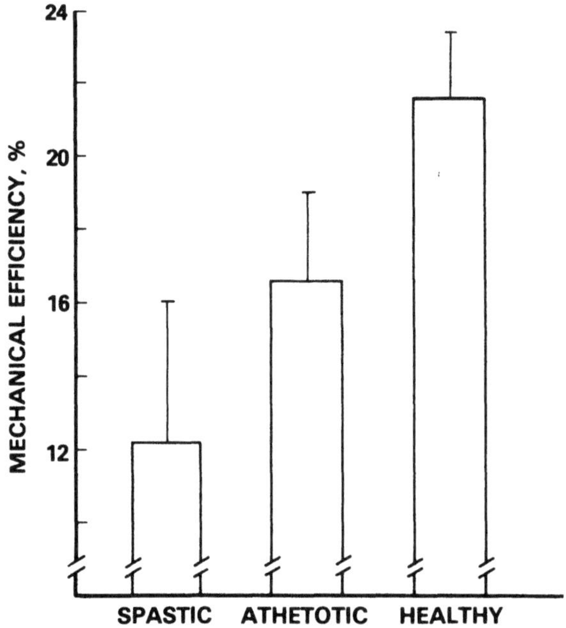

Figure 7.1. Mechanical efficiency in cerebral palsy. Comparison between spastic, athetotic, and healthy adolescents and young adults, all performing a leg cycling task. Data from Lundberg.[44]

tone at all times. We have observed patients who exert force by both arms in two opposite directions, while cranking the pedals, causing a disruption in the continuous flow of cranking. Furthermore, the postural trunk muscles exert excessive force in the child with CP. Such effort does not contribute to overcoming the pedal resistance and will not be included in the calculation of mechanical power output. The O_2 cost may vary considerably from one test to another.[11] Such variability further shows that low efficiency among patients with CP is not due to intrinsic biochemical characteristics but, rather, to less economical movement. An interesting observation was made using light tracing[7] for gait analysis: the patient with CP, in an attempt to compensate for the deficient components of his gait, exaggerates the use of those movements that are easier to perform. The end result is wasteful locomotion.

Habitual Activity

"One of the criteria for success in the rehabilitation of the physically handicapped child is the achievement of verticalization and walking."[64] In the child with CP, a great deal of surgical, physiotherapeutic, and educational effort is invested in the early years to achieve these goals.

The end result is that, with or without the aid of crutches or braces, the child can often be made to walk and perform other useful motor activities. It has therefore been a frustrating experience to note that such success is only short-lived: when these children reach adolescence many of them become progressively reluctant to walk, and they eventually regress to wheelchair status. A deteriorating working capacity seems to be the culprit: body weight and adiposity increase without a concomitant increase in muscle strength; maximal aerobic power decreases due to detraining, and a given task becomes progressively harder. A case in point is a group of adolescents who were followed up for 1–2 years. Their heart rate at a given power load became higher with age, at a rate of 10 beats/min/year.[42] This suggests a progressive detraining and deterioration of cardiovascular adaptation to exercise. When these are superimposed on the high metabolic cost of ambulation,[49] deterioration in mobility soon follows. While such assistive devices as electric wheelchairs, hydraulic beds, and ramps are a convenience, they reduce the patient's reliance on muscle work and induce further detraining. A similar regression in ambulation occurs in adolescents with other motor disabilities, such as post-poliomyelitis.[5]

A quantitative analysis of activity among patients with CP has been made by heart rate telemetry. Calorie expenditure ranged from 950 to 3,500 kcal/24 hr, obese and spastic patients being considerably less active (and eating less) than the leaner and dyskinetic ones.[12] Systematic comparison with healthy controls is not available, but clinical experience shows that the severely affected child with CP is considerably less active than his healthy peers. It is also clear that habitual activities (including regular physical education classes) which are not administered within special supervised programs are seldom intense enough to induce conditioning changes.[9,62]

Exercise Testing

In the Laboratory. Laboratory testing of children with CP requires special protocols and equipment, as well as highly skilled and devoted personnel. Many patients cannot use their legs and must be tested by arm cranking ergometry.[8,11,25,64] Those who can use their legs should have their feet strapped to the pedals. This is especially indicated for the dyskinetic child. Uncoordinated patients find it hard to pedal at a constant pace and, when a mechanically braked ergometer is used, one should continuously count the *actual* number of revolutions. Some young patients have an extremely short concentration span. Others may find it impossible to tolerate a mouthpiece and a nose clip while exercising. As a result, it is sometimes impossible to obtain direct measurements of O_2 uptake, especially at high exercise intensities. A submaximal protocol may have to be resorted to in such cases.

In the Field. The relative complexity of laboratory exercise testing of children with CP warrants the use of fitness tests that can be administered at school or at the sports club. The use of simple and familiar motor tasks will also have a strong motivational effect. In a study of post-poliomyelitic and severely affected patients with CP,[64] some simple field tasks were performed and the results were compared with those obtained in the laboratory. The tasks included:

1. A 25-m walk on the level at a comfortable speed
2. A 25-m walk on the level at maximal speed
3. Walking up a ramp—5 m long, 15 degrees incline
4. Walking down the same ramp

The subjects used their regular assistive devices (crutches or braces), and there was no attempt to make them modify their habitual gait.

Neither the walk at normal speed nor climbing up or down the ramp had any relationship to maximal O_2 uptake. Walking at *maximal* speed did show significant correlation with maximal O_2 uptake, but only 25–30% of the walking speed variance could be explained by maximal aerobic power. Thus success in such walking tasks apparently is not limited by the maximal aerobic power of the severely affected patient. It may be related more to walking skill and perhaps to local muscular strength and endurance.

Even though the physiologic correlates of field tests are not yet fully established, such tests are recommended for their simplicity and motivational impact.

Conditioning

As described above, children and adolescents with CP become progressively immobile and detrained. The rationale for their rehabilitation by conditioning is to slow down and, it is hoped, reverse such deterioration. An increase in working capacity may improve physical well-being, social integration, and occupational potential.

The Physiologic Effect. The hemodynamic, respiratory, and metabolic responses to conditioning of the child with CP are similar to those of healthy youngsters.[8,9,25,46,47,53,64] One conditioning response found among patients with CP, but not in healthy subjects, is an increase in blood flow to the exercising muscles as compared with preconditioning levels.[47] The mechanism for such an increase is not clear. Spastic muscles have subnormal blood flow during exercise,[36] and the postconditioning increase in blood flow could reflect a decrease in spasticity. However, such a decrease has yet to be clinically and objectively confirmed.

One objective measurement assumed to reflect the degree of muscle spasticity is the ratio between the electromyogram (EMG) manifestation of the Hoffman reflex (H-reflex) and the direct motor action potential (M). Both are detected by a skin electrode placed over the belly of the triceps surae muscle, in response to electrical stimulation of the posterior tibial nerve. The maximal amplitude of the H-reflex wave supposedly reflects only those motor neurons that respond to reflex excitation and corresponds to spasticity, whereas maximal M wave amplitude represents all the motor neurons.[6] In a follow-up of seven spastic patients who were taking part in a twice-per-week activity program,[63] the H : M ratio decreased within a 6–8 month period. The corresponding ratio of three sedentary controls increased (Fig. 7.2). This finding could have been interpreted as objective evidence for the beneficial effect of conditioning on spasticity. However, when the group of patients was enlarged (n = 19) and the follow-up extended to 12–30 months,[64] the results became equivocal. Based on the H : M ratio, no definitive conclusion can yet be drawn on the reduction of spasticity by physical conditioning.

Another approach to objective evaluation of spasticity is by the integrated EMG, where a greater average amplitude indicates greater muscle tone. This method was used to evaluate the effects of a one-year rehabilitation program on 15 6- to 8-year-old children with spastic CP. The program included vertical jumps on the trampoline and neuromuscular activation by the Bobath method.[55] Despite an improvement in gait and other skills, there was no corresponding clear-cut reduction in the integrated EMG potential.

In conclusion, the possible beneficial effect of exercise on muscle spasticity in CP has been suggested clinically but not proven objectively.

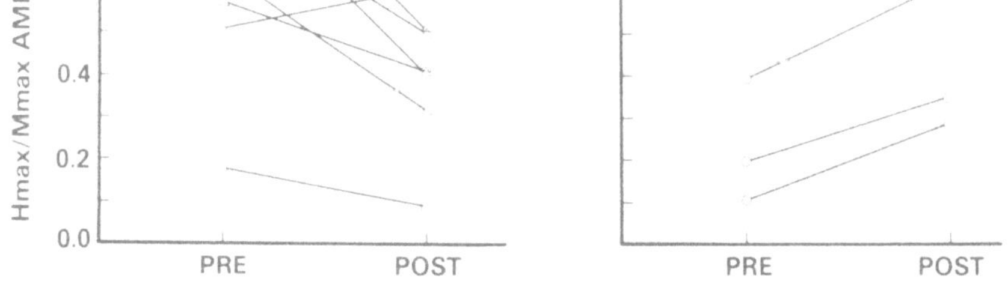

Figure 7.2. The H : M amplitude ratio before (pre) and after (post) a 6–8 month conditioning program of cerebral palsied adolescents. A comparison between individual response of active and sedentary patients. Readings were obtained from the triceps surae muscle. Adapted from Spira.[63]

The Functional Effect. Physical educators or physiotherapists who use sports as a means of rehabilitation do not need scientific tools to discern improvement in the mobility of children with CP. The ability of these youths to train and compete in running, swimming, volleyball, hockey, and many other sports is testimony to their potential for improvement in mobility and other motor tasks.

Functional improvement along with physiologic adaptation to conditioning has been widely documented.[9,55,62,64] Even the very severely disabled individual may enjoy functional improvement. Figure 7.3 summarizes the walking performance of 19 severely disabled spastic and mixed spastic-athetotic adolescents who took part in a 2-year sports rehabilitation program.[64] The improvement in walking speed at a comfortable and at maximal pace was evident. Even more impressive was the 45–55% improvement in their speed of walking up or down a 5-m ramp (15 degree slope). The patients were subdivided into those who at the start of the program could perform short distance walks for daily needs ("functional walkers") and those who could not take more than a few steps, which were insufficient to serve their functional needs ("physiologic walkers"). As seen in Fig. 7.4, the latter group derived the greater functional benefit from the program, with a dramatic improvement in

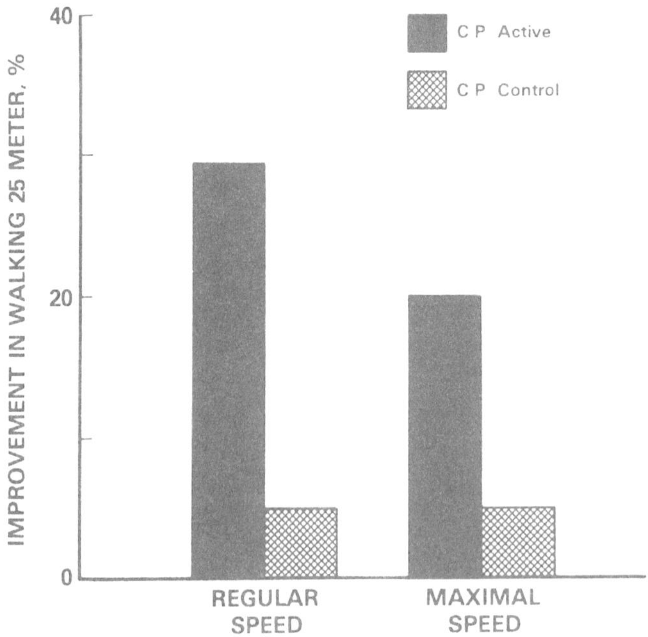

Figure 7.3. Effect of a 2-year conditioning program on the walking speed of cerebral palsied adolescents. A comparison between active participants and sedentary controls. Values are percentage of preconditioning performance. Adapted from Spira and Bar-Or.[64]

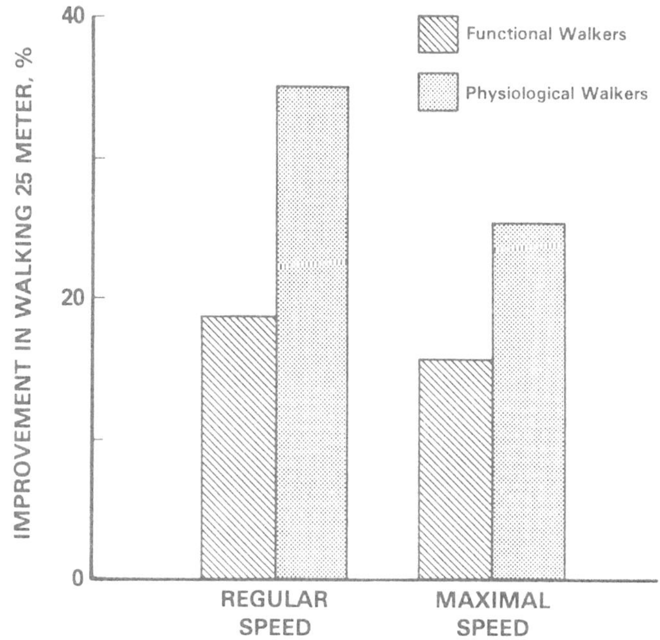

Figure 7.4. Effect of a 2-year conditioning program on the walking speed of cerebral palsied adolescents. A comparison between "functional" and "physiologic" walkers (see text for definition). Values are percentage of preconditioning performance. Adapted from Spira and Bar-Or.[64]

walking ability. When given a questionnaire to evaluate the program, 70% of the participants reported some "relief of their spasm" and of muscle clonus while walking.

Recommended Activities. Neither activities administered at regular or special schools nor habitual activities at home seem sufficient to induce an optimal fitness level in patients with CP. Reinforcement of these activities is needed to induce an increase in physical working capacity. Activities that are found effective in eliciting higher maximal aerobic power among mildly or moderately affected patients include: jogging, running, exercising with medicine balls and pulleys, pedaling or arm cranking a cycle ergometer, riding a tricycle, wheelchair sprinting, wheelchair slalom, and swimming. The more severely affected can be given mat exercises (individualized according to ability), floating, ducking and moving in the water, crawling, and pedaling a cycle ergometer in the supine position—all tailored to the residual ability of the patient.

Unless highly motivated, children with CP will not adhere to prolonged, intense activities. It is important to intersperse the above activi-

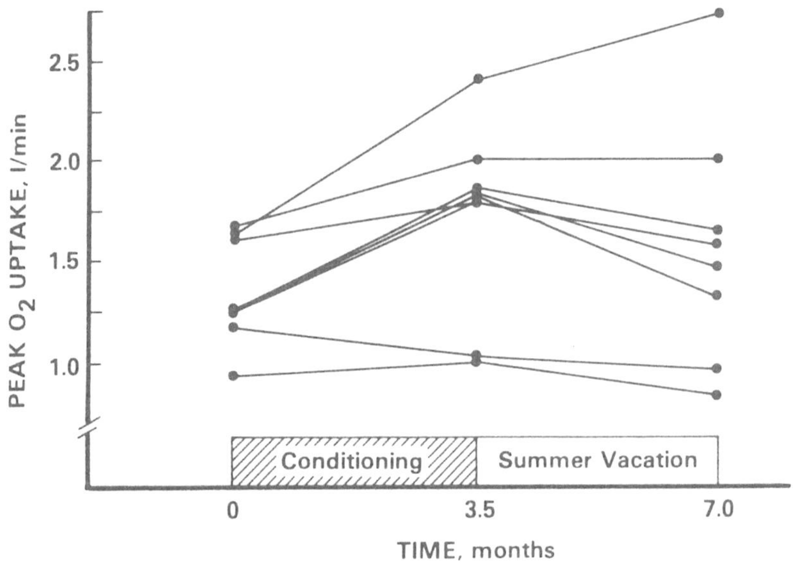

Figure 7.5. Individual changes in maximal aerobic power (peak O_2 uptake) following a 3½-month activity program at school and after a summer vacation. Subjects were 10- to 18-year-old patients with CP. Adapted from Berg.[9]

ties with games that emphasize recreation and fun. Due to their short concentration span, such patients should frequently change from one game to another.

Concentrated intense activities need not last more than 15–20 min. Activities that are less intense and include a recreational element can last up to 90 min. Two sessions per week are sufficient to induce a noticeable conditioning effect within a few weeks.

Figure 7.5 demonstrates changes in maximal aerobic power among school children with CP who had stopped their intervention program for the summer vacation.[9,62] Following the vacation, five out of the eight subjects had a deterioration in peak O_2 uptake that almost nullified the effects of the preceding activities. This further emphasizes the inability of the youngster with CP to retain his fitness unless taking part in special programs throughout the year.

Epilepsy

"Should my epileptic child be allowed to take part in sports?" This is a question often asked by concerned parents. The physician's answer should take into account the following three considerations:

Will fatigue, or hyperventilation that accompanies an intense activity, trigger a seizure?
Will repeated head trauma incurred during collision or contact sports aggravate the state of epilepsy?
Will an epileptic seizure (of any kind) during participation in sports endanger the child or people around him?

There are no definitive epidemiologic studies on the connection between sports and epilepsy, nor are there enough data on the pathophysiologic relationship. The following comments are based on the cumulative experience of clinicians and on very few laboratory observations.

Fatigue

There is no documented evidence that fatigue *per se* will induce a seizure. On the contrary, most seizures occur during sleep or rest, as shown clinically and by continuous electroencephalographic (EEG) monitoring.[30,35,38] Nor is there any reason for an epileptic child to sleep or rest longer hours than his healthy peers. As aptly summarized by Lennox in 1941 (cited in Ref. 38): "Physical and mental activity seems to be an antagonist of seizures. Enemy epilepsy prefers to attack when the patient is off-guard, sleeping, resting or idling."

Hyperventilation and Exercise

In any person, but mostly in children with epilepsy, hyperventilation may cause a decrease in frequency and an increase in amplitude of the EEG activity, lowering the threshold for seizures. Indeed, hyperventilation can trigger an epileptic seizure in a susceptible individual.

Since intense exertion is accompanied by hyperventilation, it is pertinent to ask whether or not it will induce a seizure. Thirty adolescents with epilepsy were followed by use of telemetry EEG.[30] They first hyperventilated at rest, then exercised (20–50 deep knee bends) and again hyperventilated. While the initial hyperventilation induced an increase of EEG abnormalities in all patients, exercise caused *diminution* of slow-frequency waves and of seizure discharge and a decrease in wave amplitude. Such normalization of the EEG pattern lasted 1–2 min after exercise. Furthermore, deliberate hyperventilation 10–15 sec following exercise did not induce as much wave abnormality as that caused by the initial hyperventilation. The above, as yet unconfirmed, study suggests that exercise has a suppressive effect on hyperventilation-induced wave abnormality and seizures. The mechanism for raising the seizure threshold by exercise is not clear. It could be related to the metabolic acidosis of exercise, which counteracts the alkalotic effect of hyperventilation.

Head Trauma

An objection often raised against the participation of epileptics in collision or contact sports is that repeated mechanical blows to the head may aggravate the state of epilepsy. While head trauma, especially penetrating injury, may exacerbate epilepsy, there are no epidemiologic data that evaluate the risk of collision or contact sports. Livingston[38] reported that out of 15,000 young patients under his supervision during 34 years, there was not a single case of seizures due to athletics-related head trauma. The American Academy of Pediatrics, Committee on Children with Handicaps[2] recommended that participation in contact sports should be determined individually for each epileptic child. A trial period should be initiated, during which the response of the child to *specific* sports can be evaluated. The Committee on Medical Aspects of Sports within the American Medical Association has issued a number of statements since 1968, having reversed its stand a number of times. The Committee now does *not* bar youngsters with epilepsy from participation in contact sports.[3,4,41]

Accidents to the Child and Others

A potential hazard exists when a child with epilepsy has a seizure while performing such activities as cycling, horseback riding, rope- or tree-climbing, skin-diving, and swimming. Similarly, if a seizure occurs while the patient is doing sports that involve throwing objects or shooting (e.g., javelin, discus, archery), there is a definite danger to people around him.[41] Precautions should be taken to reduce the likelihood of such accidents. Patients who are not well controlled by drugs should be barred temporarily from engaging in such sports.

A Physician's Dilemma

As discussed above, intense exertion *per se* should not pose a danger to the epileptic child and, in fact, may reduce the risk of seizures. Yet there are young patients who do convulse during exercise and not while at rest.[35] Should the physician then "play it safe" by instructing all children with epilepsy to refrain from intense activities or from collision or contact sports? Before such a simplistic and "safe" conclusion is reached, physicians and parents must realize that, to any child, activity can be a primary avenue for self-expression and social acceptability. Epileptics are not inherently inferior in their physical potential and should therefore be allowed to cultivate this means of physical and psychosocial expression. A striking case is that of a 13-year-old girl with a history of psychomotor seizures who took up long-distance running as a hobby. This girl, to show her ability to overcome her handicap, successfully completed some 2,000-km during a 40-day run and took part in other ultra-marathon races.[20] As pointed out repeatedly,[2,15,16,19,31,35,38,39,40,54] the damage caused by inactivity due to overprotection often outweighs the remote risk of a sport injury or a sport-induced seizure.

Final decisions should, therefore, be made for each individual case, taking into consideration the level of epilepsy control, the ambitions of the child, the availability of other (nonexertional) means of expression, and the likelihood of compliance with an imposed regimen.

Recommendations for Physical Activity

Based on the above considerations, it is impossible to set forth rules that will be binding on all children with epilepsy. The following recommendations should be treated as general guidelines, based on clinical experience and certain physiologic reasoning.

1. A child with medically controlled epilepsy should be encouraged to do as much physical activity as he or she desires.
2. Strenuous activity (long-distance running, a prolonged tennis match) is not contraindicated, even if it causes marked fatigue.
3. Activities such as horseback riding, mountain climbing, swimming, or diving should not be practiced without supervision.
4. Bicycle riding should be limited only if the state of epilepsy is not well controlled by medication.
5. Collision (football, ice hockey, lacrosse, rugby) and contact (baseball, basketball, soccer, wrestling) sports can be practiced in the medically balanced patient. Like any other athlete, he should be coached on the prevention of trauma.
6. Boxing, involving repeated mechanical impact on the head, should be avoided.
7. For unexplained reasons, certain activities may be more epileptogenic than others to a given child. If a certain activity *repeatedly* triggers a seizure of any kind, it should be forbidden.
8. Activities that may cause damage to spectators should not be allowed an uncontrolled epileptic.
9. Give individualized consideration to the child. Final policy should be determined with the cooperation and consent of parents and, preferably, the child.

McArdle's Syndrome

In this rare myopathy, skeletal muscle phosphorylase is absent or deficient. As a result, glycogen is not split down to glucose-1-phosphate and the ability to exercise is dependent upon the blood supply of free fatty acids and glucose. Patients fatigue early during exercise and often suffer muscle cramps, tenderness, and swelling.[48] Their work performance is extremely low. If mild exercise can be sustained for a few minutes, the patient can sometimes increase the intensity without undue fatigue, probably due to hyperemia-induced reinforcement of free fatty acids. This phenomenon has been termed "second wind."[51] Local cooling has also been found effective in delaying fatigue and muscle cramps.[52]

A case has been described of a girl who, since the age of 9, suffered exertion-induced pain and swelling, especially of the deltoid and triceps muscles. These symptoms improved with rest but took a whole day to abate fully. Repeated exercise tests showed a reduction in lactate production, although only during some of the tests. It may be that her phosphorylase activity was *intermittently* deficient.[26]

Progressive Muscular Dystrophy (PMD)

Exercise Limitations

Three fitness components are deficient in the child with PMD: muscle strength, muscle endurance, and maximal aerobic power.

Muscle Strength. This fitness component is crucial to the child's ability to rise, to walk, and to perform other daily functions. Assessment by clinical rating[69] and measurement by objective methods[27,32,61] show a continuous regression in muscle strength. This is exemplified in Fig. 7.6, where the height-related strength of children with Duchenne dystrophy is compared with that of controls. While the latter had a continuous increase in strength, the dystrophic children had only minimal strength changes, so that their absolute values at the age of 16 were similar to or lower than those at 5 years.[27] Such nonprogression in absolute muscle strength during growth is equivalent to a marked drop in function.

The extent of such regression may be appreciated if we compare the strength of children with PMD to that of the 5th percentile of size-matched controls. Such a comparison is shown in Fig. 7.7 for six muscle groups. Almost invariably, patients taller than 120 cm score below the 5th percentile. Thus, strength norms established for healthy children are of little use for the evaluation of a child with PMD.

Muscle Endurance. Muscle strength is the maximal force produced for a short period of time (less than 5 sec). Muscle endurance reflects the ability to *sustain* static or rhythmic contraction for longer periods.

Clinical observations show that children with PMD have a low muscle endurance. This can be judged from their easy fatigability when walking or climbing stairs. An attempt has been made to assess muscle endurance

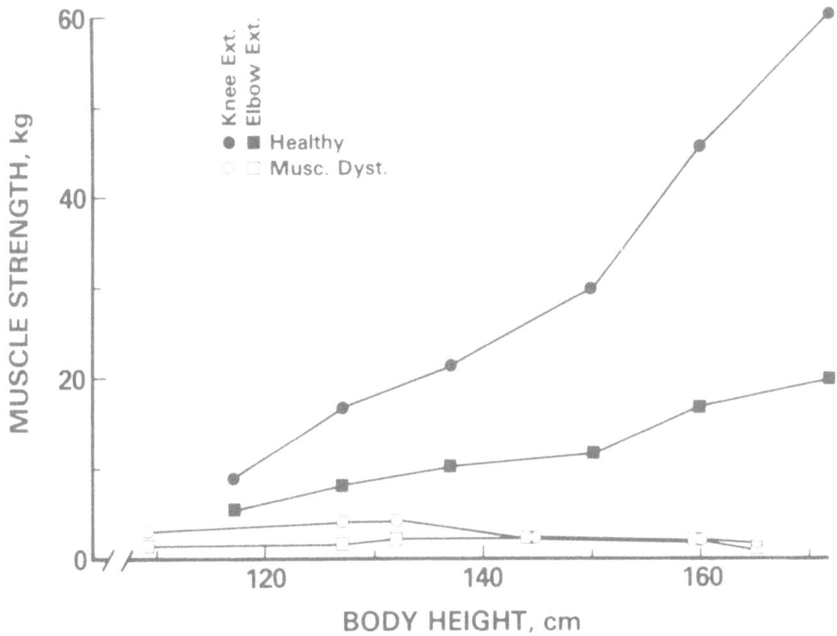

Figure 7.6. Isometric muscle strength vs. body height in muscle dystrophic patients (n = 43) and healthy boys (n = 45), 5–16 years old. Measurements were taken by cable tensiometry. Mean cross-sectional data on knee and elbow extensors (Ext.) from Fowler and Gardner.[27]

objectively by measuring the time for which a supine child can hold his neck or leg at 45° from the ground.[32] Ninety-two percent of the patients scored below the 5th percentile of healthy children. The drawback of such a measurement is its low reproducibility and questionable standardization.

We have begun to use the 30-sec Wingate anaerobic test (see Appendix II) to assess the muscle endurance of children with PMD. The score of leg pedaling of a boy with Becker's dystrophy who performed the test is shown in Fig. 7.8. While the peak and mean power outputs are much lower in this patient than in a healthy control, the rate of fatigue (the decrease in power as a percentage of peak power) is similar in both boys (i.e., about 40% of peak power). Similar results were obtained for arm cranking.

Maximal Aerobic Power. Peak power output and maximal O_2 uptake are markedly low in patients with PMD[61] and other myopathies.[18] This is due to the small functional muscle mass of these patients as well as to their compromised pulmonary[34] and cardiac[29] function. Mechanical effi-

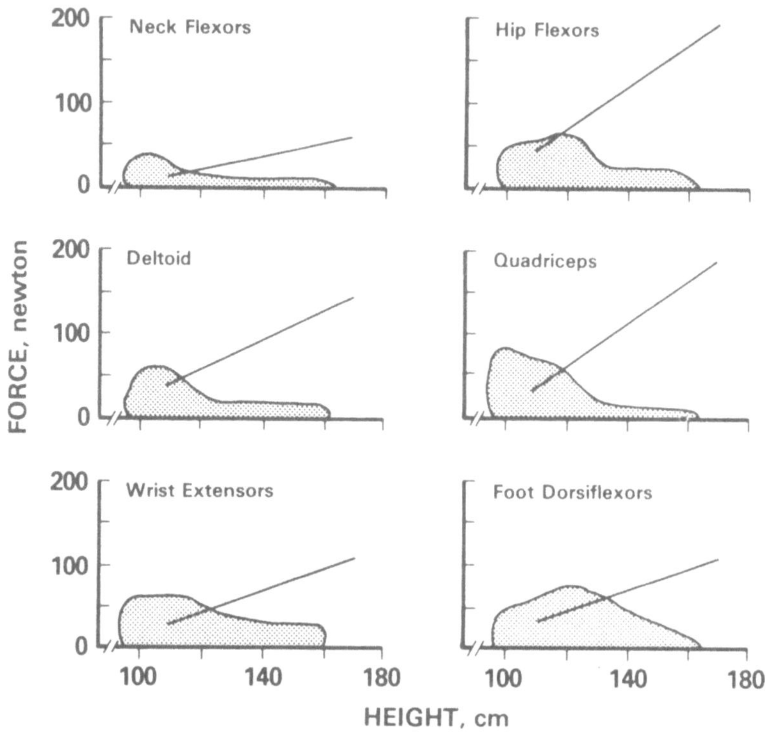

Figure 7.7. Muscle strength of children with Duchenne muscular dystrophy (shaded area) compared with lower limits of normal in healthy children. Measurements were made with the Hammersmith myometer. Regression lines are "near fit" 5th percentile of 215 healthy children. Adapted with permission from Hosking et al.[32]

ciency and anaerobic threshold are not different for patients with myopathies and healthy controls.[18]

Habitual Activity

The physical activity of patients with PMD parallels the natural history of their disease. Duchenne muscular dystrophy, which almost invariably is first manifested in early childhood, is a rapidly progressing disease. Although mobile at first, the dystrophic child will be bedridden or wheelchair-bound by the early teens and often even earlier. During a transitional phase, the child can still retain a certain degree of ambulation, but only with the support of leg braces. With more slowly developing diseases, such as Becker's muscular dystrophy, the pattern is similar, although loss of function occurs more slowly.

The activity level of the ambulatory PMD patient can be divided into four stages[69]:

1. Normal ambulation pattern, with only a slight functional deficit.
2. Definite decrease in ability to perform strenuous tasks and a concomitant reduction in habitual activity.
3. Decrease in all types of physical activity; daily duration of walking and standing, less than 2 hours.
4. Daily duration of walking and standing, less than 30 minutes.

Such a division is clinically important. Admittedly, one would also like to obtain an objective rating of muscle strength and fatigability. But it is the overall ability of the child to ambulate, rather than the characteristics of a discrete muscle group, which determines the stage of the disease.

The decline to wheelchair status is a result of such factors as residual muscle strength (especially the knee and hip extensors); extent of contractures in the lower limb joints (especially the knee); injury and bed rest; development of obesity, which causes an excessive load on the residual muscles; psychic factors such as anxiety due to fear of falling, or withdrawal from the outside world; and the decision by well-meaning relatives or teachers to "make it easy" for the child by providing him with a wheelchair. Additional factors that enhance the loss of walking ability include: prolonged sitting with bent knees; plantar flexion contracture,

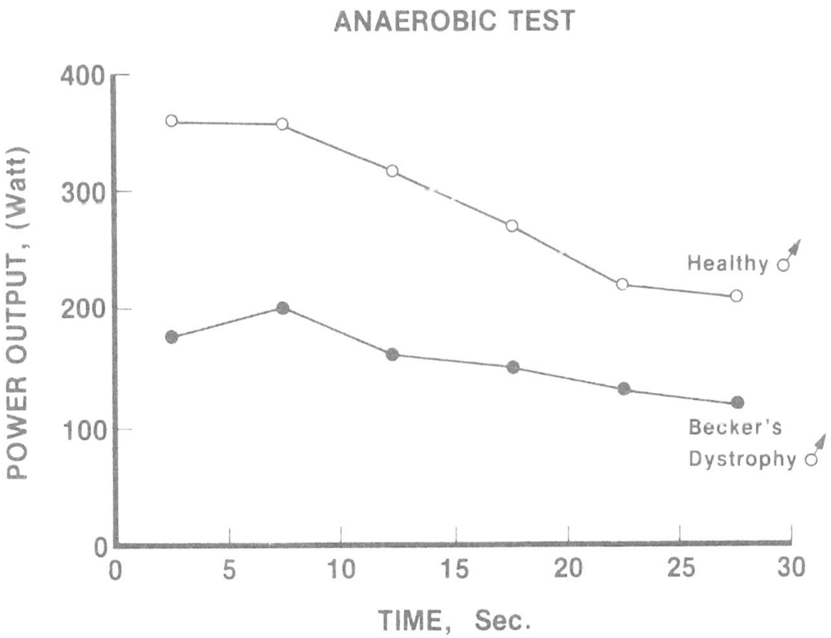

Figure 7.8. Muscle endurance in muscular dystrophy. A 14-year-old boy with Becker's muscular dystrophy and a 13-year-old weight-matched healthy boy performed the Wingate anaerobic cycling test.

which forces the child to walk on his toes; asymmetry of lower limb deficit, which induces "favoring" of one side and unilateral contractures. Whatever the reason for nonambulation, a child with muscular dystrophy who is put to bed for a few weeks may never be able to walk again. Contractures and atrophy of residual muscle develop so fast during bed rest that even surgery or braces may not reambulate such a child.[17]

In conclusion, although transition from an ambulatory to a more sedentary life is inevitable, a *premature* decline in activity status may well reflect some failure of management.

Exercise in Management of the Child

Residual, unaffected muscle tissue is trainable. If the strength and endurance of the residual muscle can be improved, one can also expect some arrest of functional deterioration and even functional *improvement.* The end result may be a prolongation of the period in which the child can still maintain an upright position and be ambulatory. The effects of conditioning have been evaluated in various studies.[1,22,28,50,70]

An example of the benefits of physical muscle strength training is shown in Fig. 7.9. Six- to 10-year-old ambulatory boys with Duchenne dystrophy underwent a one-year program that included resistive and assistive exercise of the hip abductors, hip extensors, knee extensors, arm flexors, and abdominal muscles.[70] As seen in Fig. 7.9, there was a decrease in the strength index of about 15% during the year preceding

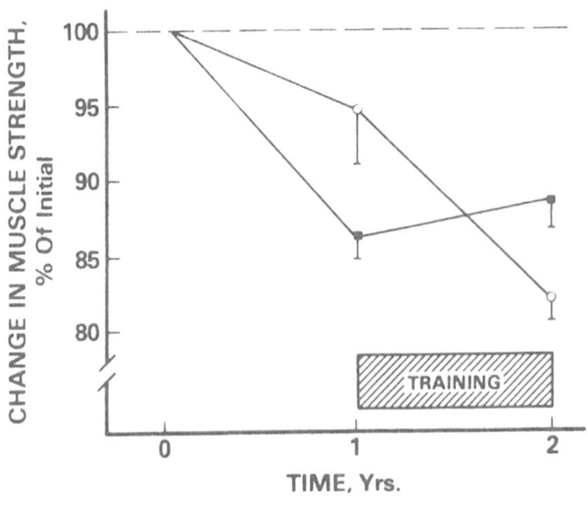

Figure 7.9. Changes in muscle strength of Duchenne dystrophic children (n = 14) before and during a one-year strength conditioning program (■). Comparison with a sedentary group of dystrophic children (n = 14) (○). "Muscle strength" represents a composite index, based on the strength of each muscle group weighted according to its assumed mass. Vertical lines denote 1 S.E.M. Adapted from Vignos and Watkins.[70]

the program. But there was no further decline in strength during the year of conditioning. This is in contrast to age-matched patients who did not train and whose strength kept declining. It thus seems that the residual muscle is indeed trainable and the progressive loss of strength can be slowed down. An interesting, as yet unexplained, finding is the improvement in short-term associative learning of boys with PMD following a bout of exercise.[22,24]

More information is needed on the optimal training methods for children with PMD. Special attention must be given to the improvement of muscle *endurance*. Based on current knowledge, one can draw the following conclusions:

1. Muscle strength in children with Duchenne dystrophy, and especially in more slowly progressing muscular dystrophies, *can* be increased by strength training of a few months' duration.
2. The degree of improvement is a function of residual muscle mass. Therefore, the more advanced the disease, the less apparent the improvement in strength.
3. Strength training *per se* is ineffective in slowing down the loss of ambulation. There is a need for additional activities that specifically make the patient rise, stand, and walk, even if helped by assistive devices or by others.
4. Specific attention should be given to the prevention and management of contractures. A muscle that is strong enough to move a limb may prove too weak when a contracture is superimposed.
5. One should guard against obesity. The weak muscles may not be able to carry an overweight individual.

Even though wheelchair status is inevitable for most patients, it always represents a major setback to child and parents alike. In giving himself up to physical passivity, the patient is inviting obesity, cardiorespiratory dysfunction, scoliosis, lower extremity edema, accelerated muscular atrophy, weakness, contractures, and emotional crises.[60] It is therefore the duty of the therapist to turn such a change of status into an opportunity for *enhanced* activities, rather than to let the patient resign himself to a passive lifestyle. Examples of specific wheelchair-based exercises are available in textbooks of physiotherapy or adapted physical education.[21]

Scoliosis

Functional and Physiologic Deficiencies

Children and adolescents with minor or moderate scoliosis have virtually no exercise-related complaints nor functional deficit. When scoliosis is advanced, however, exertional dyspnea is common and the patients have a low maximal aerobic power.[14,58,59] Maximal O_2 uptake as low as 11 ml/kg × min and usually not exceeding 25 ml/kg × min has been found among severely affected patients (compared with 40–55 ml/kg × min in healthy children).

The low maximal aerobic power is related to the chest deformity, undersized lung, hypoactivity, and generalized detraining. The main pulmonary functional abnormalities at rest and exercise are summarized in Table 7.1. Chest deformity of the child with scoliosis results in: reduced total lung and vital capacities at rest (even when allowance is made for the reduced body stature); increased work of the respiratory muscles, which contract at a mechanical disadvantage due to distorted position of the ribs;[13,14] ventilation-perfusion imbalance in some lung regions, which is manifested by a widened alveolar-arterial PO_2 gradient, particularly during exercise (gradients as high as 35–40 mm Hg have been reported); high physiologic dead space during exercise; and a high pulmonary artery blood pressure (but not pulmonary wedge pressure) during rest and exercise.[57] The O_2 cost of walking is high among patients with advanced scoliosis,[37] probably due to mechanically distorted gait but also because of the high O_2 cost of breathing.

The child with severe scoliosis is often reluctant to take part in games and other physical activities because of his unesthetic appearance and

Table 7.1. Pulmonary Function During Rest and Exercise in Moderate or Severe Scoliotics (Compared with Healthy Individuals)

Function	*Compared with Healthy Individuals*
Rest	
Total lung capacity	Lower
Vital capacity	Lower
Work of breathing	Higher
Alveolar ventilation	Lower
Pulmonary artery pressure	Higher
Submaximal Exercise	
Minute ventilation	Higher
(A-a) PO_2 gradient	Higher
Pulmonary artery pressure	Higher
Maximal Exercise	
O_2 uptake	Lower
Minute ventilation	Lower
Pulmonary artery pressure	Higher

exercise-induced dyspnea. A vicious circle of further detraining and functional deterioration may ensue. Such detraining may be more of a limiting factor than are the direct respiratory deficiencies, which become limiting only when the deformity has reached an advanced stage—quite uncommon in children or young adolescents with idiopathic scoliosis.[14] According to another opinion,[56] ventilation is the limiting factor for the majority of exercising children and adolescents with scoliosis.

The exercise-related physiologic benefit of spinal fusion is marginal. Only a slight increase in maximal ventilation[66] and a small decrease in submaximal ventilation[58] have been noted. There has also been some decrease in O_2 cost per kilogram body weight during a treadmill walk.[37] Such changes (or their absence, in other functions) are hard to interpret as there are differences in body dimensions, level of maturity, activity patterns, and fitness status between pre- and postsurgical patients.

Is Conditioning Beneficial?

Exercise therapy is recommended in scoliosis for two reasons. First is the mobilization and strengthening of those trunk and leg muscles that are related to posture. Training is expected, in conjunction with other modes of therapy, to slow down and possibly reverse the progress of the spinal curvature. Second, it is hoped that through exercise some pulmonary functions and physical working capacity will be improved.

It is not within the scope of this book to discuss in detail those "remedial exercises" that are meant to arrest the postural deterioration of the spine. This topic is covered in textbooks of physiotherapy and adapted physical education. Suffice it to say that the results are equivocal, especially when a control group is also observed.[67]

An increase in maximal aerobic power and in ventilatory efficiency (determined by the ratio pulmonary ventilation : O_2 uptake) can be obtained by conditioning. This has been shown for patients with various degrees of scoliosis.[14,65,66,68] As is true for other conditioning programs, compliance is better when the sessions are supervised. When the program is home-based there may be no appreciable conditioning effect.[59]

Do some children with scoliosis respond physiologically better than others to a conditioning regimen? There is no definitive study comparing large enough groups of patients, selected according to the severity of their lesion. It does seem, though, that the least trainable are those with very advanced scoliosis (e.g., 130–150° curvature). Some of these patients also develop high pulmonary arterial blood pressure during exercise, and for these intense exertion is not advisable.[57] Among the less severely affected patients there has been no correlation between the improvement in maximal aerobic power and the initial spinal curvature.[14,65]

References

1. Abramson AS, Rogoff J: Physical treatment in muscular dystrophy (abstract). Proceedings of the 2nd Medical Conference of the Muscular Dystrophy Association 123–124, 1952.
2. American Academy of Pediatrics. Committee Report: The epileptic child and competitive school athletics. Pediatrics 42:700–703, 1968.
3. American Medical Association Committee on the Medical Aspects of Sports and the Committee on Exercise and Physical Fitness: Convulsive disorders and participation in sports and physical education. Joint statement. JAMA 206:1291, 1968.
4. American Medical Association Committee on the Medical Aspects of Sports: Epileptics and contact sports. Position statement. JAMA 229:820–821, 1974.
5. Anderson AD, Levine SA, Gellert H: Loss of ambulatory ability in patients with old anterior poliomyelitis. Lancet 2:1061–1063, 1972.
6. Angel RW, Hofmann WW: The H reflex in normal, spastic and rigid subjects. Arch Neurol 8:591–596, 1963.
7. Aptekar RG, Ford F, Bleck EE: Light patterns as a means of assessing and recording gait. II. Results in children with cerebral palsy. Dev Med Child Neurol 18:37–40, 1976.
8. Bar-Or O, Inbar O, Spira R: Physiological effects of a sports rehabilitation program on cerebral palsied and post-poliomyelitic adolescents. Med Sci Sports 8:157–161, 1976.
9. Berg K: Effect of physical training of school children with cerebral palsy. Acta Paediatr Scand Suppl 204:27–33, 1970.
10. Berg K: Heart-rate telemetry for evaluation of the energy expenditure of children with cerebral palsy. Am J Clin Nutr 24:1438–1445, 1971.
11. Berg K, Bjure J: Methods for evaluation of the physical working capacity of school children with cerebral palsy. Acta Paediatr Scand Suppl 204:15–26, 1970.
12. Berg K, Olsson T: Energy requirements of school children with cerebral palsy as determined from indirect calorimetry. Acta Paediatr Scand Suppl. 204:71–80, 1970.
13. Bergofsky EH, Turino GM, Fishman AP: Cardiorespiratory failure in kyphoscoliosis. Medicine 38:263–317, 1959.
14. Bjure J, Grimby G, Nachemson A: The effect of physical training in girls with idiopathic scoliosis. Acta Orthop Scand 40:325–333, 1969.
15. Boucharlat J, Maitre A, Ledru J: Sport et épilepsie de l'enfant. Ann Med Psychol (Paris) 1:392–401, 1973.
16. Bower BD: Epilepsy and school athletics. Dev Med Child Neurol 11:244–245, 1969.
17. Bowker JH, Halpin PJ: Factors determining success in reambulation of the child with progressive muscular dystrophy. Orthop Clin North Am 9:431–436, 1978.
18. Carroll JE, Hagberg JM, Brooke MH, Shumate JB: Bicycle ergometry and gas exchange. Measurements in neuromuscular diseases. Arch Neurol 36:457–461, 1979.
19. Chase D: With epilepsy they take the medicine and play. Physician Sportsmed 2:61, 1974.

20. Cragg S: Patty's Magnificent Marathon. Readers' Digest April 1978, pp. 75–78.
21. Cratty BJ: Adapted Physical Education for Handicapped Children and Youth. Love Publishing Co., Denver, 1980.
22. Dowben RM: Treatment of muscular dystrophy with steroids. A preliminary report. New Eng J Med 268:912–916, 1963.
23. Eickelberg WWB, Less M: The effects of passive exercise of skeletal muscles on cardiac cost, respiratory function and associative learning in severe myopathic children. J Hum Ergology 3:157–162, 1975.
24. Eickelberg WWB, Less M, Engels WC: Respiratory, cardiac and learning changes in exercised muscular dystrophic children (abstract). Percept Motor Skills 43:66, 1976.
25. Ekblom B, Lundberg A: Effects of physical training on adolescents with severe motor handicaps. Acta Paediatr Scand 57:17–23, 1968.
26. Eriksson BO, Hansson O, Karlsson J, Piehl K: Muscle metabolic studies of a girl with McArdle-like syndrome. In: Borms J, Hebbelinck J (eds.) Pediatric Work Physiology. Karger, Basel, 1978, pp. 65–71.
27. Fowler WM Jr, Gardner GW: Quantitative strength measurements in muscular dystrophy. Arch Phys Med Rehab 48:629–644, 1967.
28. Fowler WM Jr, Pearson CM, Egstrom GH, Gardner GW: Ineffective treatment of muscular dystrophy with an anabolic steroid and other measures. N Engl J Med 272:875–882, 1965.
29. Gailani S, Danowski TS, Fisher DS: Muscular dystrophy. Catheterization studies indicating latent congestive heart failure. Circulation 17:585–588, 1958.
30. Gotze W, Kubicki St, Hunter M, Teichmann J: Effect of physical exercise on seizure threshold. Dis Nerv Syst 28:664–667, 1967.
31. Gündel L: Empfehlungen für die Befreiung anfallskranker Kinder vom Schulsport. Deutsch Med Wochensr 100:491–494, 1975.
32. Hosking GP, Bhat US, Dubowitz V, Edwards RHT: Measurements of muscle strength and performance in children with normal and diseased muscle. Arch Dis Child 51:957–963, 1976.
33. Huberman G: Organized sports activities with cerebral palsied adolescents. Rehab Lit 37:103–107, 1976.
34. Inkley SR, Oldenburg FC, Vignos PJ: Pulmonary function in Duchenne muscular dystrophy related to stages of the disease. Am J Med 56:297–306, 1974.
35. Korczyn AD: Participation of epileptic patients in sports. J Sports Med Phys Fitness 19:195–198, 1979.
36. Landin S, Hagenfeldt L, Saltin B, Wahren J: Muscle metabolism during exercise in hemiparetic patients. Clin Sci Mol Med 53:257–269, 1977.
37. Lindh M: Energy expenditure during walking in patients with scoliosis. The effect of surgical correction. Spine 3:122–134, 1978.
38. Livingston S: Should physical activity of the epileptic child be restricted? Clin Pediatr 10:694–696, 1971.
39. Livingston S: Physical activity for the epileptic child. In: Livingston S (ed.) Comprehensive Management of Epilepsy in Infancy, Childhood and Adolescence. Charles C. Thomas, Springfield, 1972, pp. 143–148.
40. Livingston S: Should epileptics be athletes? Sports Med 3:67–72, 1975.

41. Livingston S, Pauli LL, Pruce I: Epilepsy and sports (letter). JAMA 239:2, 1978.
42. Lundberg A: Changes in the working pulse during the school year in adolescents with cerebral palsy. Scand J Rehab Med 5:12–17, 1973.
43. Lundberg A: Mechanical efficiency in bicycle ergometer work of young adults with cerebral palsy. Dev Med Child Neurol 17:434–439, 1975.
44. Lundberg A: Oxygen consumption in relation to work load in students with cerebral palsy. J Appl Physiol 40:873–875, 1976.
45. Lundberg A: Maximal aerobic capacity of young people with spastic cerebral palsy. Dev Med Child Neurol 20:205–210, 1978.
46. Lundberg A, Ovenfors CO, Saltin B: Effect of physical training on schoolchildren with cerebral palsy. Acta Paediatr Scand 56:182–188, 1967.
47. Lundberg A, Pernow B: The effect of physical training on oxygen utilization and lactate formation in the exercising muscle of adolescents with motor handicaps. Scand J Clin Lab Invest 26:89–96, 1970.
48. McArdle B: Myopathy due to a defect in muscle glycogen breakdown. Clin Sci 10:13–35, 1951.
49. Molbech S: Energy cost in level walking in subjects with an abnormal gait. In: Evang K, Andersen KL (eds.) Physical Activity in Health and Disease. Universitets Forlaget, Oslo, 1966, p. 146.
50. Nesvadba Z, Hoskova L, Rennerova A: Rehabilitation of children with muscular dystrophy at the state spa of Jansko Lazne. In: Walton L, Canal N, Scarlato G (eds.) Muscle Diseases. Excerpta Medica, Amsterdam, 1970, pp. 555–557.
51. Pernow BB, Havel RJ, Jennings DB: The second wind phenomenon in McArdle's syndrome. Acta Med Scand Suppl 472:294–307, 1967.
52. Ricker K, Hertel G: Influence of local cooling on the muscle contracture and paresis of McArdle's disease. J Neurol 215:287–290, 1977.
53. Rieckert H, Bruhn L, Schwalm U, Schnizer W: Ein Ausdauertraining im Rahmen des Schulsports bei worweigend spastisch gelähmten Kindern. Med Welt 28:1694–1701, 1977.
54. Rose KD: Should epileptics be barred from contact sports? AMA changes position. Med World News 62B–63B, 1974.
55. Rotzinger H, Stoboy H: Comparison between clinical judgment and electromyographic investigations of the effect of a special training program for CP children. Acta Paediatr Belg 28[Suppl.]:121–128, 1974.
56. Shneerson JM: The cardiorespiratory response to exercise in thoracic scoliosis. Thorax 33:457–463, 1978.
57. Shneerson JM: Pulmonary artery pressure in thoracic scoliosis during and after exercise while breathing air and pure oxygen. Thorax 33:747–754, 1978.
58. Shneerson JM, Edgar MA: Cardiac and respiratory function before and after spinal fusion in adolescent idiopathic scoliosis. Thorax 34:658–661, 1979.
59. Shneerson JM, Madgwick R: The effect of physical training on exercise ability in adolescent idiopathic scoliosis. Acta Orthop Scand 50:303–306, 1979.
60. Siegel IM: Muscular dystrophy: interdisciplinary approach and management. Postgrad Med 69:125–133, 1981.

61. Sockolov R, Irwin B, Dresseuderfer RH, Bernauer EM: Exercise performance in 6- to 11-year-old boys with Duchenne muscular dystrophy. Arch Phys Med Rehab 58:195–201, 1977.
62. Sommer M: Improvement of motor skills and adaptation of the circulatory system in wheelchair-bound children in cerebral palsy. In: Simri U (ed.) Sports as a Means of Rehabilitation. Wingate Institute, Natanya, 1971, pp. 11/1–11/11.
63. Spira R: Contribution of the H-reflex to the study of spasticity in adolescents. Dev Med Child Neurol 16:150–157, 1974.
64. Spira R, Bar-Or O: An investigation of the ambulation problems associated with severe motor paralysis in adolescents. Influence of physical conditioning and adapted sport activities. Final report. Project No. 19-P-58065-F-01, U.S. Dept. HEW, Social Rehabilitation Services, 1975.
65. Stoboy H: Pulmonary function and spiroergometric criteria in scoliotic patients before and after Harrington Rod surgery and physical exercise. In: Borms J, Hebbelinck M (eds.) Pediatric Work Physiology. Karger, Basel, 1978, pp. 72–81.
66. Stoboy H, Speierer B: Lungenfunktionswerte und spiroergometrische Parameter während der Rehabilitation von Patienten mit idiopathischer Skoliose. Arch Orthop Unfall-Chir 81:247–254, 1975.
67. Stone B, Beekman C, Hall V, et al: The effects of an exercise program on change in curve in adolescents with minimal idiopathic scoliosis. A preliminary study. Phys Ther 59:759–763, 1979.
68. Sünram F, Götze HG, Scheele K: Arterielle Blutgase und Säure-Basen-Verhältnisse nach dosierter Ergometerbelastung bei 12-18-jährigen Mädchen mit idiopathischen Thorakalskoliosen vor und nach einem vierwöchigen Training. Sportarzt Sportmed 25:6–12, 3–38, 1974.
69. Vignos PJ Jr, Archibald KC: Maintenance of ambulation in childhood muscular dystrophy. J Chron Dis 12:273–290, 1960.
70. Vignos PJ, Watkins MP: The effect of exercise in muscular dystrophy. JAMA 197:843–848, 1966.

8

Hematologic Diseases

Anemia

The O_2-carrying capacity of blood and the blood flow to any tissue determine the capability of this tissue to raise its aerobic metabolism. The same principle applies to the body as a whole. Changes in hemoglobin (Hb) concentration, unless accompanied by a variation in blood volume or in an affinity of Hb to O_2, closely reflect changes in the O_2-carrying capacity of the blood. For practical purposes, therefore, we shall use the reduction of Hb concentration as an index of the severity of anemia and of a decrease in O_2-carrying capacity.

Compensatory Mechanisms

One gram of Hb can carry 1.36 ml of O_2. Low Hb concentration results in a low O_2 content of arterial blood and a low arterio-venous O_2 difference at all levels of exercise. Two compensatory mechanisms serve to increase the availability of O_2 to the tissues: an increase in cardiac output and a "shift to the right" in the O_2 dissociation curve.

Cardiac output is higher in anemic patients than in healthy individuals, both at rest[5] and during exercise.[26] This is facilitated by an increase in heart rate and, to a lesser extent, in stroke volume. The increase in exercise heart rate is proportional to the degree of anemia. In a child with severe anemia (e.g., 6 g Hb/100 ml blood) it can be some 30–40 beats/min higher than in a healthy child. Such a difference occurs at various exercise intensities, as shown in Fig. 8.1. While compensation through high cardiac output is adequate to provide O_2 to body tissues during low and moderate exercise intensities, it becomes insufficient once cardiac output has reached maximum.

Another compensatory mechanism in the anemic patient is a "shift to

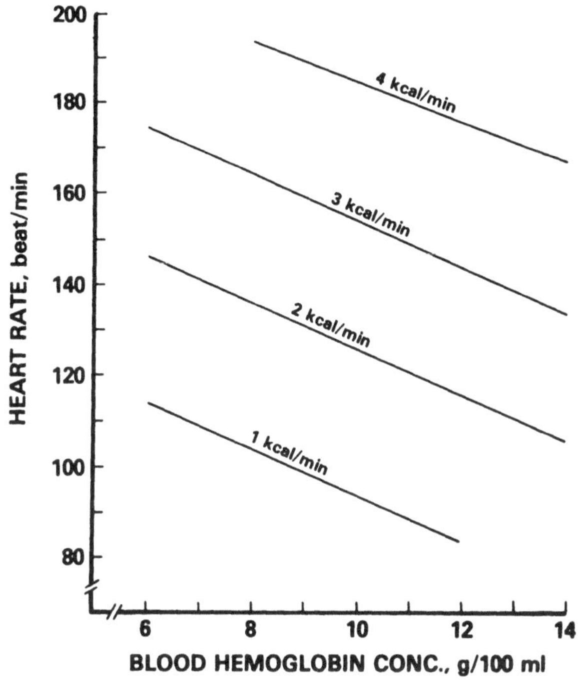

Figure 8.1. Anemia and the heart rate response to exercise. Effect of blood hemoglobin concentration on heart rate at four metabolic levels. Schematic adaptation for a 25 kg child. Based on data by Gandra and Bradfield.[15]

the right" in the O_2 dissociation curve.[26] This facilitates a greater release of O_2 to the tissue at a given Po_2. The exact causes of such a shift are not clear. It could result from an increase in Pco_2, a decrease in pH, or an increase in the erythrocyte 2,3 DPG content.[3] These, however, have not been evaluated in anemic children during exercise.

Physical Working Capacity

Thanks to the above compensatory mechanisms, an anemic child can perform well at low and moderate exercise intensities. His maximal aerobic power, however, is low, in proportion to the degree of anemia.[23,27] It also depends on the maximal cardiac output of each child. These relationships are shown in Fig. 8.2, which is based on theoretical considerations (see also Ref. 29). One should realize, however, that anemia is seldom found as an isolated entity. For example, it is just one manifestation of undernutrition, which itself can reduce the physical working capacity.[8] In some children, severe anemia accompanies chronic renal failure, which contributes to deficiency in exercise performance.[27] Hypoactivity is another potential cause of reduced exercise performance

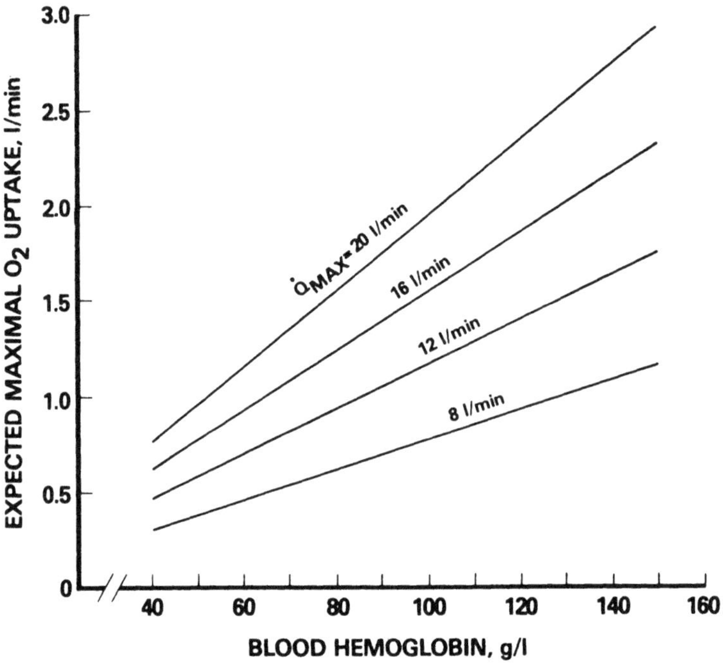

Figure 8.2. Maximal O_2 uptake as a function of hemoglobin concentration and maximal cardiac output ($\dot{Q}$ max). Regression lines were constructed based on the following assumptions: 1 g hemoglobin carries 1.36 ml O_2; arterial Hb is 90% saturated at maximal exercise; O_2 extraction at maximal exercise is 80% of arterial O_2 content.

in anemia. Only scant information is available, though, on the habitual activity of children with anemia. In a study from Brazil[15] there was *no difference* in the activity patterns of anemic and nonanemic school children. Adult anemics, on the other hand, tend to adopt a sedentary lifestyle so that they can function well even though their maximal aerobic power is reduced.[29]

Conditioning-Induced Iron Deficiency

Even though conditioning is accompanied by some increase in total Hb, endurance athletes sometimes suffer from iron deficiency, which has been termed "sports anemia." For a review of this topic see Pate.[22] In a carefully conducted study, adolescent swimmers who trained extensively (14–20 hours/week) without an iron supplement were followed for 2–4 months.[11] Their Hb, hematocrit, and serum iron levels were normal. However, the bone marrow was depleted of iron. While these young swimmers had normal bone marrow cellularity, a group of older endurance runners, who had been practicing for some years, had hypocellular bone marrow. In addition to subnormal production, endurance athletes also have an above normal rate of *elimination* of iron from the body. One avenue for such elimination is sweat.[28] Another mechanism of excessive elimination of iron is by intravascular hemolysis[31] (see also Hemoglobinuria—Exertional, below). The above data suggest a need for dietary iron supplementation for young endurance athletes. This, however, should be limited only to those who pursue a vigorous regimen on a continuous basis, for months or years.

Hemoglobinuria—Exertional

The Phenomenon

Exercise-induced hemoglobinuria, also called "march hemoglobinuria," was first documented in 1881[12] and has since been described among long-distance runners and others who perform prolonged, intense activity. It is manifested by red or dark urine, hemoglobinuria, and high levels of plasma Hb concentration. It is a benign, reversible disorder that does not lead to renal or other complications.

Although rare among the general population, exercise-induced hemoglobinuria is common among athletes. In one survey, 10% of those who completed a long-distance race (10–42 km) had exercise-induced hemoglobinuria.[16] Exertional hemoglobinemia, without spillage into the urine, appears in the majority of long-distance runners. This is particularly prevalent among young adults but has been described also in adolescents.[2,4,6,21]

Possible Mechanism

Intravascular hemolysis is an obligatory stage for exercise-induced hemoglobinuria. The Hb released into the plasma is first bound by haptoglobin. The newly formed haptoglobin-Hb complex is removed by the reticuloendothelial system, which results in a decrease in plasma Hb-binding capacity and accumulation of free plasma Hb. Hemoglobinuria will ensue if the renal "tubular maximum" (i.e., the maximal amount of Hb that can be reabsorbed in one minute by the tubuli) has been exceeded.

The cause of intravascular hemolysis during exercise is not clear. The accepted theory is that repeated pounding of the feet or other parts of the body causes mechanical damage to red blood cells in the area of impact, with eventual hemolysis. The disappearance of hemoglobinuria once the runner uses padded shoes or substitutes soft running surfaces for hard ones speaks in favor of this "mechanical" theory.[2,6,7] A mechanically caused hemolysis has been confirmed by *in vitro* experiments.[6] An interesting case is that of a 14-year-old boy with exercise-induced hemoglobinuria and ventricular septal defect. The authors[4] suggest that, during exercise, the increase in turbulent flow through the defect may have caused mechanical damage to the red blood cells and eventual hemolysis.

An alternative explanation[12,20,21] assumes some transitory red cell abnormality in susceptible individuals—for example, an increase in autohemolysis, increased mechanical fragility, or presence of heat-labile Hb.

Management

An individual who complains of red or dark urine following exercise should be examined to differentiate exercise-induced hemoglobinuria from hematuria, paroxysmal cold hemoglobinuria, paroxysmal nocturnal hemoglobinuria, and myoglobinuria. Upon diagnosis of exercise-induced hemoglobinuria, the patient should be reassured that this is a benign, reversible phenomenon. The most effective treatment is a reduction of mechanical impact, for instance, padding of soles or choosing soft running surfaces. Some athletes may have to change their running style to reduce impact. Often, remission is spontaneous.

Hemophilia

Rationale for Conditioning

The main role of conditioning in the hemophiliac child is the prevention or reduction of muscle atrophy. Such atrophy is often secondary to joint bleeds, pain, guarding, and limited mobility. It can also result from direct bleeds into the muscle.

A recent approach has recognized that exercise may be useful for the *prevention* of bleeds.[14,18] This notion is based upon the assumption that a joint supported by strong musculature and ligaments (which also serve as shock absorbers) is more stable and less liable to bleeds. Furthermore, a supple, fit child can better avoid accidents during daily activities than can a clumsy, unfit one. Although the preventive value of conditioning in hemophilia is plausible, studies that support it[14] are based on uncontrolled clinical trials.

One should not overlook the psychologic impact of conditioning. The inclusion of sports in the daily life of a child afflicted by an incurable disease, like hemophilia, is a major boost to his morale and self-esteem.

This may be the only component of therapy that will bring him closer to his peers rather than label him as different.

Recommended Activities

When commencing rehabilitation of an affected limb, the key principle is a *gradual* increase of resistance and range of movement. Beyond physiotherapeutic exercises, the best sport to start with is swimming. Water resistance helps develop strength at various ranges of motion and provides cushioning to protect a tender site. Movements in this sport are relatively slow and therefore easily controllable.

Swimming does not effectively strengthen the quadriceps, a muscle often affected in hemophilia, which requires knee extension against resistance for its rehabilitation. Isokinetic devices (e.g., a Cybex machine) have been found most useful as the patient can determine the force he will exert at a fixed speed. This device, however, is expensive and not easily accessible. Pedaling on a cycle ergometer, in which seat height has been adjusted to ensure an almost full knee extension, is an alternative. A regular bicycle can also be used, providing the child is a proficient rider. At a more advanced stage, when the leg can tolerate greater impact, running can be undertaken. The same principles apply for muscles that are in proximity to other "target joints." In general, any noncollision activity is allowable to the patient with hemophilia.

Risk of Bleeds and Their Prevention

An ambivalent approach has been taken by clinicians toward physical exercise in hemophilia.[30] On the one hand, conditioning is an important element in rehabilitation; on the other, exertion has been considered conducive to injury and bleeding.

There are no randomized prospective studies that evaluate the risk of intense exertion to the hemophiliac. Our own experience and that of others[9,18] has been that even extremely strenuous noncollision activity does not induce any increase in the incidence of bleeds. Not one of 20 young patients with hemophilia, even at severe levels, had any immediate or delayed bleeding following an all-out progressive continuous test on the cycle ergometer, or a 30-sec exhausting pedaling task. We do not, however, recommend participation in such collision sports as boxing, ice hockey, American football, or rugby, in which intense impact and contusion are integral to the activity.

Adolescents and young adults with hemophilia seem to have a greater sense of daring than do younger patients.[24] Indeed, we have seen hemophiliac adolescents and young adults who insisted on taking part in ice hockey or boxing. It will take a great deal of tact to dissuade a youngster from participating in a sport of his choice. A knowledgeable and compassionate physician should be able to suggest alternatives that are challenging but less trauma-inducing.

With the advent of home-based management, where replacement therapy is given by a family member, the objections against participation in sports of the child with hemophilia are no longer justified. Parents and children should be taught that bleeds due to minor traumata can be easily prevented by use of cryoprecipitate or Factor VIII prior to the activity. When pain or tightness appear following the activity, a bleed can still be aborted by the immediate infusion of replacement therapy. The above precautions have been found successful among patients with mild, moderate, and severe hemophilia.

Sickle-cell Anemia

In addition to their anemia, sickle-cell homozygotes have other pathologic changes that may adversely affect their response to exercise. These include cardiomyopathy, microscopic pulmonary infarctions and shunting, and isosthenuria (inability to concentrate urine by the kidney), which may lead to hypohydration. O_2 saturation of blood at rest and during exercise is subnormal in these patients, even when compared with other anemic individuals with an identical Hb concentration.[25] In addition, patients with sickle-cell anemia have exceedingly high blood viscosity, which may increase resistance to blood flow.

Few data are available to show that children with sickle-cell disease[1,17] or with sickle-cell trait[13] have a deficient work performance. Peak mechanical power in such patients is about 30–40% lower than among healthy (black) controls.

Ischemic ECG response is common during exercise among children with sickle-cell anemia. In one study among 5- to 18-year-old patients, 15% had ischemic changes and another 34% responded with "questionable" ischemic changes.[1] These ST abnormalities do not denote permanent myocardial damage, as confirmed by post-mortem studies. The exercise-induced ischemic changes are usually not accompanied by chest pain, dysrhythmia, or a vaso-occlusive crisis. It is possible, however, that some of the chest- or abdominal-pain typical to this disease does result from exercise-induced myocardial ischemia.[17]

Based on the reversible nature of exercise-induced changes in sickle-cell anemia, we recommend that the patients be allowed to exercise at will. Their advanced anemia and low maximal aerobic power will serve as a "natural brake" on their activities. Those who complain of exercise-induced chest or abdominal pain should be given an exercise test with ECG monitoring.

Thalassemia Major

Exercise performance of patients suffering from this genetic disease is subnormal, possibly due to the following reasons:

1. Low hemoglobin concentration.
2. Myocardial involvement due to multiple blood transfusions.
3. Skeletal abnormalities typical to the disease.
4. A sedentary lifestyle.

Not enough data are available to determine which of the above causes is paramount in the low working capacity of patients with thalassemia. Their low motivation to exert[10] has made it hard to interpret the results of all-out stress tests. An attempt to use isometric hand-grip as a stressor[19] did not elicit a high enough stress on the cardiovascular system to obtain sufficient information on its response to exercise. Preliminary data (Mansell, personal communication) show that the "anaerobic threshold" of adolescents with thalassemia is low, suggesting that they must resort to anaerobic metabolic pathways at low levels of exercise.

References

1. Alpert BS, Gilman PA, Strong WB, et al: Hemodynamic and ECG responses to exercise in children with sickle cell anemia. Am J Dis Child 135:362–366, 1981.
2. Attlee WHW: Haemoglobinuria following exertion. Lancet 1:1400, 1937.
3. Austin PL, Stegink LD, Gisalfi CV: The effect of exercise on red blood cell 2,3-diphosphoglycerate in children. J Pediatr 83:41–45, 1973.
4. Chaplin H Jr, Perkoff GT, Frisbie JH, et al: March hemoglobinuria associated with asymptomatic congenital heart disease. JAMA 208:1700–1702, 1969.
5. Cropp GJA: Cardiovascular function in children with severe anemia. Circulation 39:775–784, 1969.
6. Davidson RJL: Exertional hemoglobinuria: a report on three cases with studies on the haemolytic mechanism. J Clin Pathol 17:536–540, 1964.
7. Davidson RJL: March or exertional hemoglobinuria. Semin Haematol 6:150–161, 1969.
8. Davies CTM, Chukweumeka AC, Van Haaren JPM: Iron deficiency anemia—its effect on maximum aerobic power and responses to exercise in African males aged 17–40 years. Clin Sci 44:555–562, 1973.
9. Dietrich S: Hemophilia: a total approach to treatment and rehabilitation. Los Angeles Orthopedic Hospital, 1968.
10. Ehlers KH, Levin R, Klein AA, et al: The cardiac manifestations in thalassemia major: natural history, non-invasive cardiac diagnosis studies and results of cardiac catheterization. In: Engle MA (ed.) Pediatric Cardiovascular Disease. Davis, Philadelphia, 1980, pp. 171–186.
11. Ehn L, Carlmark B, Hoglund S: Iron in young sportsmen. In: Eriksson B,

Furberg B (eds.) Swimming Medicine IV. University Park Press, Baltimore, 1978, pp. 85–88.
12. Flatmark T: Studies on the hemolytic mechanism in March hemoglobinuria. Acta Med Scand 173:307–313, 1963.
13. Flood NL, Alpert BS, Strong WB, et al: Exercise in children with sickle cell trait (abstract). Med Sci Sports Exercise 14:123, 1982.
14. Gandini S, Panicucci F, Bastianini C: L'attivitá sportiva como fattore preventivo, terapeutico e riabilitativo delle artropatie negli emofilici. Med Sport 32:43–44, 1979.
15. Gandra YR, Bradfield RB: Energy expenditure and oxygen handling efficiency of anemic schoolchildren. Am J Clin Nutr 24:1451–1456, 1971.
16. Gilligan DR, Altschule MD, Katersky EM: Physiological intravascular hemolysis of exercise. Hemoglobinemia and hemoglobinuria following cross-country runs. J Clin Invest 22:859–869, 1943.
17. Hamilton W, Rosenthal A, Berwich D, Nadas AS: Angina pectoris in a child with sickle-cell anemia. Pediatrics 61:911–914, 1978.
18. Ireland T: The role of exercise in the management of hemophilia (personal communication), 1978.
19. Levin AR, Klein AA, Ehlers KH, et al: Hemodynamic and left ventricular volume and function characteristics in thalassemia major. Pediatr Res 12:386, 1978.
20. Martin H, Kilian P: Marschhämoglobinurie. Folia Haemat 4:92–117, 1959.
21. Ohno Y, Sato M, Kurokawa I, et al: Exertional hemoglobinuria. Tohuko J Exp Med 117:187–191, 1975.
22. Pate RR: Sports anemia: a review of the current research literature. Physician Sportsmed 11:115–131, 1983.
23. Parsons EC, Wright FH: Circulatory function in the anemias of children. I. Effect of anemia on exercise tolerance and vital capacity. Am J Dis Child 57:15–28, 1939.
24. Russ K, Bartlett GS: Risk-taking behavior in hemophiliac and non-hemophiliac adolescent boys. Pediatr Res 12:366, 1978.
25. Sproule BJ, Halden ER, Miller WF: A study of cardiopulmonary alterations in patients with sickle cell disease and its variants. J Clin Invest 37:486–495, 1958.
26. Sproule BJ, Mitchell JH, Miller WF: Cardiopulmonary physiological responses to heavy exercise in patients with anemia. J Clin Invest 39:378–388, 1960.
27. Ulmer HE, Griener H, Schuler HW, Scharer K: Cardiovascular impairment and physical working capacity in children with chronic renal failure. Acta Paediatr Scand 67:43–48, 1978.
28. Vellar OD: Studies on sweat losses of nutrients. I: Iron content of whole body sweat constituents, serum iron levels, hematological indices, body surface area, and sweat rate. Scand J Clin Lab Invest 21:157–167, 1968.
29. Viteri FE, Torun B: Anaemia and physical working capacity. Clin Haematol 3:609–626, 1974.
30. Weigel N, Carlson BR: Physical activity and the hemophiliac: yes or no? Am Correct Ther J 29:197–205, 1975.
31. Yoshimura H: Anemia during physical training (sports anemia). Nutr Rev 28:251–253, 1970.

9

Climate and the Exercising Child

Introduction

In the preceding chapters, physical exertion has been discussed in isolation from other stressors. For the sake of simplicity, we analyzed children's responses to exercise in a "neutral" climate: neither too cold nor too warm, neither very humid nor very dry. Yet such conditions are quite hypothetical. In many geographic regions climate is not neutral and may impose upon the child an added environmental stress, especially during outdoor exercise.

Climate can play a major, even crucial, role in performance, subjective comfort, and health. Abundant data are available on the combined effects of exercise and climate on adults, with relevance to industry, sports, and the military. Even though children are habitually more active than adults and perform many of their recreational activities out of doors, their response to exercise in hostile climates has not been thoroughly studied.

In this chapter we shall first introduce basic physical and physiologic concepts of thermoregulation and acquaint the reader with the nomenclature used by the environmental physiologist.

Theoretically—according to geometric and functional principles—children are less efficient thermoregulators than are adults, especially in extreme climatic conditions. An analysis of this topic follows, with data on heat acclimatization, heat and cold tolerance, and water and electrolyte balance in exercising children.

Certain children are at special risk of acquiring such heat-related illnesses as heat stroke or heat exhaustion. The problems encountered by such high-risk groups are discussed, as well as the precautions to be taken for their prevention. The chapter concludes with recommendations for the conduct of athletics in a hot climate.

Some Concepts in Thermoregulation

Heat Stress and Heat Strain

Climatic "heat stress" denotes a combination of environmental conditions that stress the thermoregulatory system. "Heat strain," on the other hand, is the physiologic and mental response to heat stress. Two individuals exposed to an identical heat stress may respond with a different heat strain.

Ambient temperature is just one component of climatic heat stress, and not necessarily the most important. Other components are air humidity, air movement (wind), and radiant heat. The main source of the latter is solar radiation, but such heated objects as artificial turf can also generate marked radiant heat.

Various *heat stress indices* have been constructed that include one or more of the above components. A popular index, originally designed for the military but also utilized in industry and sports, is the Wet Bulb Globe Temperature (WBGT). This incorporates air temperature, humidity, and radiation as measured by three thermometers: a dry bulb (DB), a wet bulb (WB), and a black globe (G):

$$\text{WBGT} = 0.7\ \text{WB} + 0.2\ \text{G} + 0.1\ \text{DB}$$

In this index, air temperature accounts for only 10% of heat stress, while humidity is taken as 70%! The implication is that a highly humid and mildly warm day can be more stressful than a very hot but dry one. For indoor use, when radiant heat is less important, the index is called WBT, taken as 0.7 WB + 0.3 DB. Psychrometers that monitor WB, DB, and G can be operated in the field by the coach, teacher, or team physician.

Another index, the "effective temperature," combines humidity, ambient temperature, and air velocity and is especially suitable for indoor use. Others take into account the type of clothing (which may interfere with sweat evaporation and heat dissipation) and the type and intensity of physical activity. A detailed description of these indices is available.[64]

Heat strain components include sweating rate, rectal and skin temperatures, and heart rate. Other functions such as skin blood flow, cardiac output, or ventilation play a role in thermoregulation but are less often measured. Mental functions considered to reflect heat strain are mental acuity and perception of heat intensity.

Heat Production and Heat Exchange

Living cells continuously generate heat. A major component of this metabolic heat (*M*) is generated by muscle, in proportion to the intensity and duration of muscular activity. Some 75–80% of the chemical energy used for muscle contraction is converted into heat. Dissipation of this heat (which may exceed 10 times the resting metabolic rate) is a major challenge to the thermoregulatory system during exercise.

Heat can penetrate the body from the environment through conduction (C_D), convection (C_V), and radiation (R). It can also be dissipated from the body by these three avenues. The direction and intensity of such heat transfer depends on the temperature gradient between skin temperature and ambient temperature (for C_D and C_V) and on the gradient between skin temperature and the surrounding objects (for R). Another avenue for heat dissipation is evaporation (E) of sweat, or of water from the epidermis and from the respiratory mucosa. Evaporation of one liter of water requires 580 kcal (2430 kJ) at 33°C. It is especially important during intense exertion or whenever ambient temperature is high and dissipation by C_D, C_V, or R becomes ineffective. The efficiency of evaporative cooling depends largely on sweating rate, but also on water vapor pressure in the air (high humidity attenuates evaporation), wind velocity, and air temperature (stronger wind and warmer air both enhance evaporation). Because of the insulative quality of air, heat transfer by C_D between air and skin is negligible. During water immersion, however, C_D is a major avenue for heat exchange: thermal conductivity of water is 25 times that of air.

The above components of heat generation and transfer can be put into a "heat balance" equation:

$$M \pm C_D \pm C_V \pm R - E = S$$

in which S denotes heat storage within the body. When S is zero, heat generation and penetration equal heat dissipation and the body is in thermal balance. A negative S indicates heat loss. Ordinarily, this will not occur during exercise (even on a cold day) unless it is performed in water.

Physiologic and Behavioral Means of Thermoregulation

There is a narrow range of climatic conditions, termed "neutral zone," in which metabolic heat is passively dissipated to the surroundings. The body is then kept in thermal balance without active participation of the thermoregulatory apparatus. This zone is not constant. It will vary among and within individuals depending on their activity level, clothing, body surface area, and amount of subcutaneous insulating fat. In a resting naked individual of average adiposity, the neutral zone is 25–27° C, 50–60% relative humidity. Any deviation from this neutral zone will induce, via hypothalamic centers, a physiologic response. When heat stress rises mildly, vasodilatation takes place in the skin, thus increasing convection from body core to the periphery. Skin temperature then rises, facilitating better heat dissipation by convection, conduction, and radiation. When heat stress further increases, sweat is produced by the eccrine glands and spreads over the skin so that heat is also dissipated by evaporation. This is the single most important means for heat dissipation during exercise. When the climate is mildly cool, peripheral vasoconstriction takes place, decreasing convection from core to skin and reducing heat dissipation. In addition, the resting metabolic rate rises. Further environmental cooling will induce shivering (i.e., rhythmic, involuntary, and uncoordinated muscle contractions), which generates heat and compensates for heat loss.

Thermoregulation is achieved not only by physiologic processes but also by behavior. Migration of birds, wetting of fur, and daytime shelter-seeking of desert animals are just a few examples from the animal kingdom. Humans resort extensively to behavioral means when confronted by heat or cold stress. These include the use of clothes to increase insulation; the seeking of shade or the wearing of a hat to reduce solar radiation; using fans to increase evaporation; or curling up in bed to reduce the effective surface area of the body. Behavioral means of thermoregulation allow us to widen the range of climates in which we can function and are instrumental in the proper conduct of athletic activities and for prevention of heat-related illness.

Characteristics of Children Relevant to Thermoregulation

Figure 9.1 is a diagram comparing heat production, penetration, and dissipation of an 8-year-old child and a young adult. Although the child has a smaller absolute surface area, his surface area per unit mass is some 36% *greater*. Because heat flux between two objects depends on their area of contact (or, in the case of R, the effective surface area), heat transfer to and from the body is greater in the child than in the adult for a given unit mass. The difference is depicted in the diagram by the length of the arrows for conduction, convection, and radiation. The higher the temperature gradient between air and skin, the greater the difference in heat flux. In practical terms, the geometric difference is an asset to the child in mildly warm environments (when ambient temperature is still lower than skin temperature) or during intense exercise performed in mildly cool weather, when greater heat dissipation is advantageous. However, it becomes a liability in climatic extremes, when heat transfer should be minimized. In spite of their larger surface area, children have a lower sweating rate and their evaporative capacity is deficient.

Children expend more chemical energy per unit mass than do adults while performing similar tasks (see Figs. 1.5 and 1.6 and Chapter 1). As a result they produce greater metabolic heat per kilogram body weight

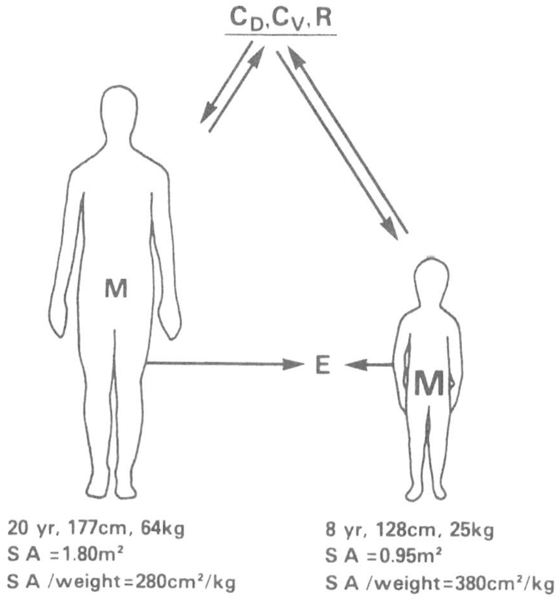

Figure 9.1. Schematic presentation of heat production and transfer in a child and an adult. C_D = conduction, C_V = convection, R = radiation, E = evaporation, M = metabolic heat production, SA = surface area of body. Length of arrows represents rate of heat transfer per unit mass. Reprinted with permission from Bar-Or (Ref. 26 in Chapter 1).

which, during intense exercise, subjects their thermoregulatory system to a greater strain.

Another potential handicap for the child is his lower cardiac output at a given metabolic rate (see Chapter 1, section entitled Cardiac Output and Stroke Volume). This may cause insufficient blood flow to internal organs[36] and intolerance to prolonged activity.[31] Furthermore, during intense activities (when cardiac output approaches maximum), blood flow to the skin will be insufficient, with a resultant deficit in heat transfer from core to periphery.

The above theoretical considerations suggest that in mild climatic stress one need not anticipate age- or size-related differences in thermoregulation. In climatic extremes, however, children are *a priori* at a disadvantage.

Sweating Pattern

Sweating Rate

The sweating apparatus is apparently fully developed by the third year of life,[60] but even so children perspire less than adults.[6,31,36,47,49,90,97,98]

The lower sweating rate of children is apparent not only in absolute terms but also when normalized for a unit surface area, as shown in Fig. 9.2. While prepubescents seldom produce more than 400–500 ml sweat per square meter per hour, adults exposed to identical conditions produce as much as 700–800 ml/m^2 × hr. Young women perspire more than girls,[36,37] but the difference is smaller among males. A relationship between sweating rate and the concentration of circulating androgens has been suggested[55] but not confirmed.

Heat-activated Sweat Glands

The total number of eccrine sweat glands in humans seems to become fixed at age 2, at about 2–2.3 million.[55] Their recruitment during exercise or exposure to heat is cyclical,[60,79] with some 1–1.7 million glands being activated at any given time.[8]

Do children perspire less because they activate fewer glands or because of lower sweat production by each gland? As seen in Fig. 9.3, the population density of active glands is *greater* in children than in adolescents or adults. It is, therefore, the *output per gland* rather than the number of glands that limits the sweating rate of children. Whether at rest,[55] during exercise-in-the-heat,[49] or during pilocarpine ionthophoresis,[48] sweat excretion per gland is 2.5 times higher in adults than in children. Figure 9.4 is an example of such a comparison.

It is not clear why sweat output per gland is smaller in children. In part, it may result from a smaller duct length and cross-sectional area of the coil,[61] but neural and hormonal causes cannot be excluded. Sweat output per gland in females is less dependent on age than in males.[8] This again suggests that circulating androgens may have some influence on the control of sweating.

Functional Implications

Is the lower sweating rate in children an advantage or a handicap? Children's greater reliance on heat dissipation via convection, conduction, and radiation and their lesser dependence on evaporation is an economical thermoregulatory pattern that minimizes water loss. On the other hand, their lower sweating rate can be looked upon as a functional handicap because of a lower evaporative capacity. Insufficient evaporation results in high skin temperature and a less favorable temperature gradient for convection of heat from body core to periphery, as seen in Fig. 9.5. One can further see in this figure that children require a greater increase in core temperature to start perspiring, which suggests a higher hypothalamic threshold.

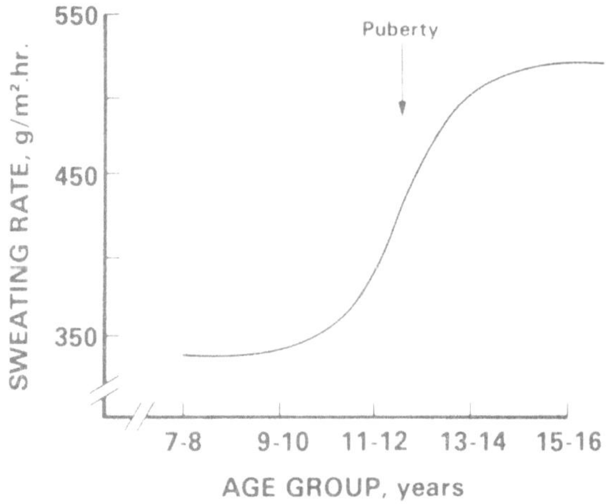

Figure 9.2. Development of sweating rate. Forty boys 7–16 years old exercised at moderate intensity (heart rate 160–170 beats/min) on the cycle ergometer, at 29°C, 60% relative humidity. Exercise time was 15–35 min. The arrow indicates the age at which pubertal changes were first noted. Data by Araki et al.[6] Reproduced with permission from Bar-Or[9]; copyright American Academy of Pediatrics, 1983.

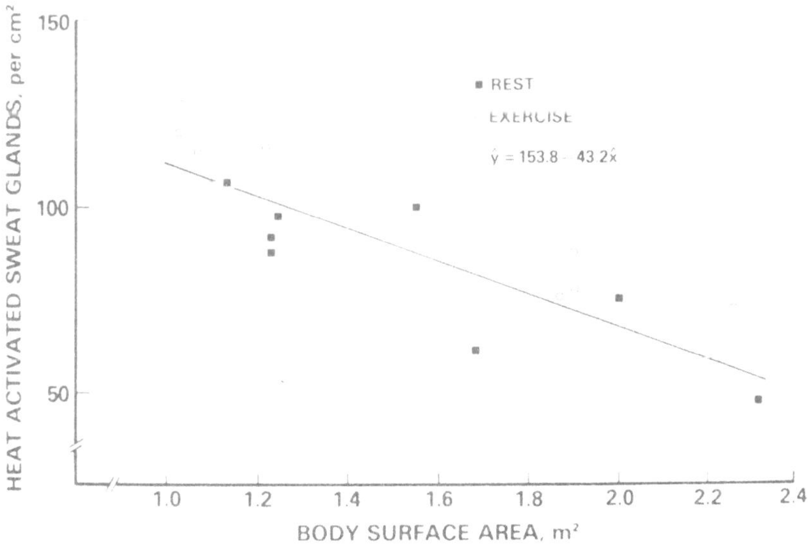

Figure 9.3. Relationship between population density of heat-activated sweat glands and body surface area, at rest or following exercise in 165 children, adolescents, and young adults exposed to a dry climate. Reproduced by permission from Bar-Or.[8]

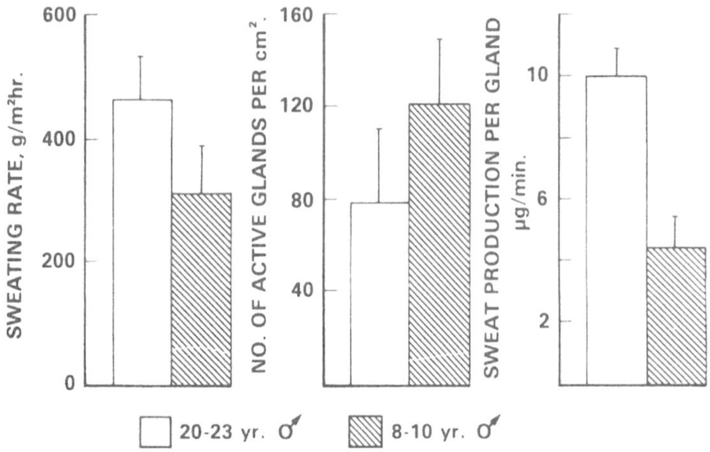

Figure 9.4. Sweating characteristics and age. Sweating rate, population density of heat-activated sweat glands, and calculated sweat production of a single gland. Data on 15 8- to 10-year-old boys and 16 20- to 23-year-old men who exercised intermittently at 50% of maximal O_2 uptake in dry heat (43°C, 20% RH). Adapted from Bar-Or,[8] based on data by Inbar.[49]

In conclusion, although their low sweat production helps children conserve water, it hampers their ability to sustain high metabolic production in a hot climate.

Heat Tolerance

What Is Heat Tolerance?

Heat tolerance is the ability to sustain optimal function during exposure to heat stress. Various criteria have been developed for heat tolerance as related to industry, the military, or space-oriented research (see Leithead and Lind[64]).

One criterion used for exercising children is the climatic heat stress beyond which a prescribed task cannot be completed. Incompletion is determined either by the appearance of dizziness, aggressiveness, apathy, disorientation, nausea, exhaustion, marked headache, or abdominal cramps or by heat strain indicators (e.g., rectal temperature higher than 39.4°C, heart rate exceeding 90% of maximum).

Heat Tolerance—Children vs. Adults

Table 9.1 summarizes studies in which heat tolerance was compared between exercising children and adults. These consistently show the

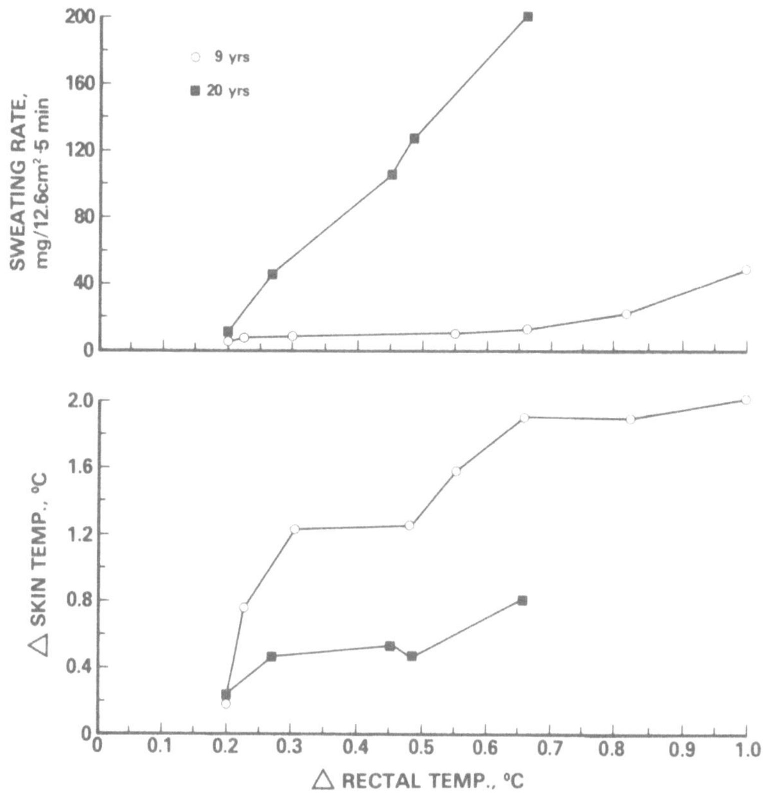

Figure 9.5. Sweating rate and increase in mean skin temperature in relationship to increase in rectal temperature. Comparison between seven 9-year-old boys and seven 20-year-old men who performed a continuous exercise task on the cycle ergometer at 29°C, 60% RH. Adapted from Araki et al.[6] Reproduced with permission from Bar-Or (Ref. 26 in Chapter 1).

lower ability of the children to complete their task in a hot climate. An example is presented in Fig. 9.6. While children have low tolerance to extreme heat (air temperature exceeding 45°C, effective temperature exceeding 30°C), they thermoregulate as effectively as adults when exercising in neutral[31,43] or in moderately warm[36,46,47,49] climates. The possible causes for children's relative intolerance to exercising in extreme heat are discussed in the section entitled Characteristics of Children Relevant to Thermoregulation, above.

Little information is available on tolerance levels to heat when exercise is intense and short. Performance of a 30-sec supramaximal cycling task (the Wingate anaerobic test) was not adversely affected by warm-humid (39°C, 90% RH) or hot-dry (39°C, 25% RH) climate in 10- to 12-year-old girls and boys.[35]

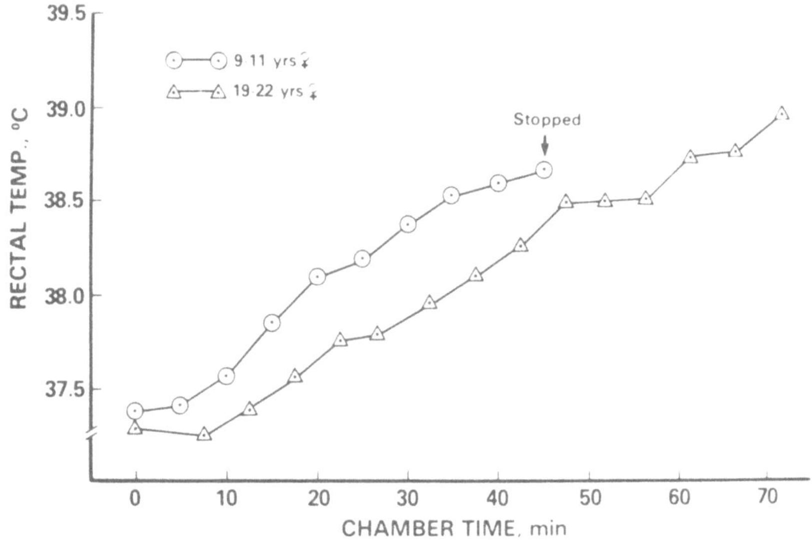

Figure 9.6. Heat tolerance of girls (n = 12) and women (n = 29). Changes in rectal temperature during an exposure to extremely hot climate (50°C, 15% RH). Subjects walked intermittently at 4.8 km/hr. Reproduced by permission from Bar-Or.[8] Based on data by Haymes et al[46] and Bar-Or et al.[13]

Cold Tolerance

Exposure to the cold may result in local skin cooling and in generalized heat loss. Local cooling with an eventual cold-related injury (such as frostbite) may occur both at rest and during exercise. Skin sites commonly affected are the chin, cheeks, ears, fingertips, and toes. In contrast, generalized body cooling is seldom a problem during exercise, because metabolic heat production, if intense enough, more than compensates for heat loss from the skin. Generalized cooling becomes a problem whenever activities are not intense metabolically, such as in mountain hikes, or snow-shoeing (see discussion below for aquatic activities).

In this context, one should be aware of the "wind-chill" concept: the effect of cold on the skin, as well as on overall heat loss, is greatly dependent upon the wind. For example, during exposure to 0°C, a wind of 24 km/hr will affect the exposed skin, causing heat loss as though the actual air temperature were minus 10.5°C. The equivalent wind-chill factor at 40 km/hr mind will be minus 16°C. A wind effect is created not only by the blowing wind but also by the moving individual. Thus, in calm air of minus 10°C the wind-chill factor is minus 31°C for a skater or a skier who advances at 40 km/hr. A young cross-country skier should be

Table 9.1. Heat Tolerance of Children vs. Adults

Age, years (Sex)	Prescribed Task	Findings	Author
10–17 vs. 21–23 (M)	Completion of 1-hr desert walk	Syncope—2 boys Not in men	Van Beaumont[97]
11–14 vs. 25–30 (M)	Completion of 90-min walk (5.6 km/hr) at 40°C, 22% RH	Boys stopped after 50 min Men completed task	Wagner et al[98]
9–11 vs. 19–22 (F)	Completion of 3 × 20-min walk (4.8 km/hr, 0–5% grade) at 50°C, 13% RH	Girls stopped after 43 min Women completed task	Haymes et al[46] vs. Bar-Or et al[13]
9–11 vs. 39.5 (M)	Completion of 3 × 20-min walk (4.8 km/hr) at 48°C, 22% RH	Completed by all boys and men	Haymes et al[47] vs. McCormick and Buskirk[71]
12 vs. 20 (F)	Completion of 2 × 50-min walk (30% max $\dot{V}O_2$) at 35°C, 65% RH or 48°C, 10% RH	Most girls could not complete either task Women completed both	Drinkwater et al[36]

advised to cover his head, including his face, when the wind-chill factor drops to minus 20–23°C.[73]

When activity is intense, core temperature will not decline and will often rise, even if exercise is performed on a cold day. In such strenuous prolonged races as marathon runs, core temperature may reach 40–42°C even when the effective temperature is as low at 5°C.[94] Another example is cross-country skiing in which one sweats profusely due to body heating even when clothing is light and the ambient temperature is below freezing. In such intermittent activities as ice hockey, there is an overall high level of metabolic heat production and, even while off the ice, the child is in no danger of body cooling.[65,75]

An exception to the above occurs when exercise is performed in water. During land exercise heat is dissipated primarily by evaporation of sweat, whereas during swimming it is lost mostly by conduction. Heat conductance of water is 25 times and specific heat 1,000 times that of air. Thus, heat loss from the skin may be as much as 30 times higher during swimming than it is during cycling.[74]

Figure 9.7 presents an experiment in which 8- to 19-year-old club swimmers of both sexes swam in 20.3°C water at a speed of 30 m/min. This corresponded to a metabolic level of four to five times their resting metabolic rate. While most of the older participants managed to main-

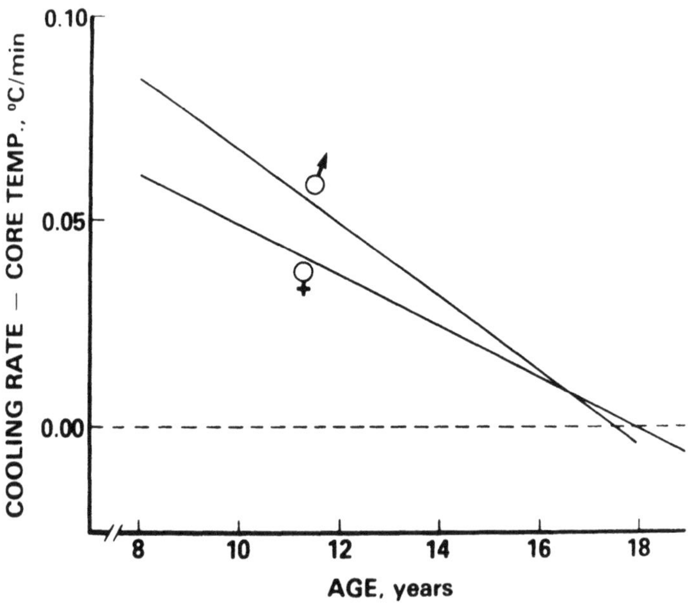

Figure 9.7. Rate of cooling of body core during swimming (20.3°C water) in relationship to age. Subjects were 16 eight- to 19-year-old female and male trained swimmers. Adapted by permission from Sloan and Keatinge.[89]

tain their core temperature at pre-swim levels, the younger swimmers had a 2–3°C drop in core temperature. Furthermore, the older swimmers stayed in the water for about 30 min, whereas the youngest had to leave after 18–20 min due to a marked subjective distress.[57,89] The inability of the young swimmers to sustain their activity in cool water was related to their higher surface area per heat-producing unit mass and especially the lower thickness of their subcutaneous fat. Adipose tissue has the highest insulative capacity of any body tissue. The thicker the fat layer, the better insulation it provides to the swimmer,[15,56] especially when blood flow to the skin is reduced.

Water temperature in swimming pools is usually kept at above 25°C, but there are still many children who swim outdoors at below 20°C. The question of excessive cooling in young swimmers is therefore not merely academic. It is possible that the extreme cold-induced distress forces swimmers out of the water before hypothermia becomes dangerous. Still, a potential risk exists for a lean, overzealous, small-sized swimmer who may be reluctant to leave the water in spite of impending danger. Coaches and health practitioners should be aware of such a possibility. Early warning signs to look for are euphoria and disorientation.

Acclimatization to Exercise in the Heat

Heat Acclimatization—What and How?

Acclimatization to heat comprises those adaptive changes that result from continuous or repeated exposures to heat stress. It is manifested by improved thermoregulation and physical and mental performance, and thermal comfort. This process is particularly important to those individuals who are abruptly confronted with a warmer environment, whether due to a heat wave or when traveling to a warmer geographic region. During the unacclimatized state, physical and mental tasks are performed at the cost of a greater physiologic strain. Attempts to perform intense prolonged activities during this stage increase the risk for heat-related illness (see also Insufficient Acclimatization, below).

The major physiologic changes that take place progressively during acclimatization are: decrease in heart rate and rectal and skin temperature at a given metabolic level; increase in sweating rate and in the sensitivity of the sweating apparatus to increments in core temperature; and a drop in electrolyte (especially Na^+ and Cl^-) concentration in the sweat. Rating of perceived exercise intensity is reduced and there is a general increase in thermal comfort.

In adults, a reasonable degree of acclimatization can be achieved following four to seven exposures to the combined stresses of heat and exercise. Effective exposures should last 1–4 hours each, at a rate of three to seven per week. The intensity of exercise should be gradually increased such that by the end of acclimatization one can perform at par with one's performance in the cooler climate. The acclimatized state can be retained for up to 7–10 days without further exposure. It is then gradually lost.

Acclimatization—Children vs. Adults

Young teenagers acclimatize to exercise in the heat, but to a distinctly lesser degree than older adolescents or young adults.[98] The ability of boys 8–10 years old to acclimatize to exercise in dry heat, compared with 20- to 23-year-old men, has been extensively studied in the author's laboratory in Israel.[12,49,50,51,52] Body temperatures (Fig. 9.8) and heart rate declined and sweating rate increased to the same degree in both groups during a 2-week acclimatization program. The main age-related difference was in the *rate* of acclimatization: while the adults reached a certain level of acclimatization within two sessions, the children needed four to five sessions for a similar result. This relatively sluggish response is depicted in Fig. 9.9. A similar "delay" in response was found for the sensitivity of the sweating apparatus to changes in core temperature.[49]

Children can acclimatize, to some extent, when they exercise in *neutral* environments, and when they *rest* in hot climates.[50,52] In adults such protocols are only partially effective.

It seems, therefore that children lag behind adults in the rate of physiologic acclimatization and therefore require a longer and more gradual program. On the other hand, they acclimatize (although slowly) under less stringent protocols than are commonly recommended for adults.

Perceptual Changes with Acclimatization

There is a subjective component to acclimatization. Concurrently with a decrease in physiologic strain, lassitude disappears and there is an improvement in general well-being. To gauge such subjective improvement, children in the above project were asked to rate the intensity of exercise that they were performing,[12] using the rating of perceived exertion (RPE) of Borg (see Chapter 1, section entitled Perceptual Change and Age). Even though power load and environmental conditions were identical in all exposures, the rating markedly declined from one session to another, indicating that the same task seemed gradually easier. Fig. 9.10 is a comparison of such changes between children and adults. The ratio RPE/HR is taken to represent subjective difficulty at a given physiologic strain. The rate of decline in RPE/HR during the 2-week program is faster in the children, reaching lower final levels than in the adults.

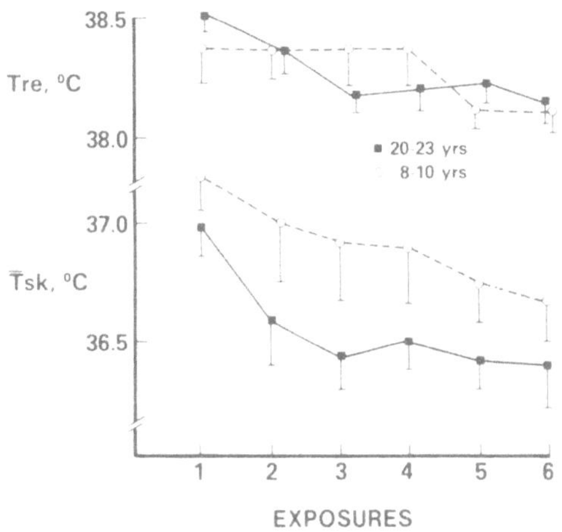

Figure 9.8. Heat acclimatization of boys (n = 9) and young men (n = 9). Changes in rectal (Tre) and mean skin ($\overline{T}sk$) temperatures during a 2-week acclimatization program. Values are as obtained at the end of each six 80-min exposures to 43°C, 21% RH. Subjects cycled intermittently at 40–50% of their maximal O_2 uptake. Vertical lines denote 1 S.E.M. Reproduced by permission from Inbar.[49]

This phenomenon implies that even though children's physiologic acclimatization is slow, their subjective improvement is faster than adults'.

While subjective well-being may be looked upon as an advantage, it may also signify a potential hazard: whereas the insufficiently acclimatized adult will be reluctant to exert in the heat, a child who is not yet acclimatized may be more daring, in spite of a marked objective strain.

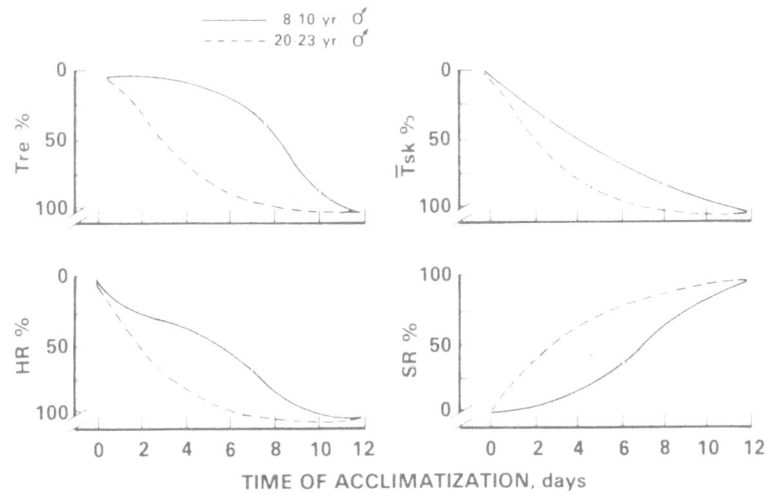

Figure 9.9. Rate of heat acclimatization in boys (n = 9) and young men (n = 9). Changes in rectal temperature (Tre), mean skin temperature ($\overline{T}sk$), heart rate (HR), and sweating rate (SR) during a 2-week acclimatization program. Values are presented as percent of final acclimatization, baseline being 0%. Conditions and protocols as in Fig. 9.8. Schematic adaptation from Bar-Or.[8] Reproduced with permission from Bar-Or.[9] Copyright American Academy of Pediatrics, 1983.

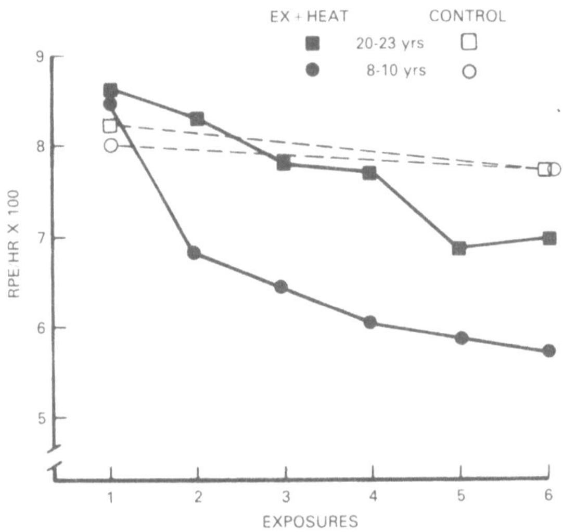

Figure 9.10. Changes in the ratio of rating of perceived exertion/heart rate (RPE/HR) during a 2-week acclimatization program in boys and adults. The "ex + heat" groups are as shown in Figure 9.8. The controls were seven boys and seven adults who were tested on the first and last days of the program but not exposed to exercise and heat in between. Reproduced by permission from Bar-Or.[8] Based on data by Bar-Or and Inbar.[12]

Fluid and Electrolyte Balance

Water Shifts During Exercise

At the start of strenuous exercise, plasma volume drops by about 10–15%. Some decline is evident as early as 10 sec following the onset of a highly intense exercise[83] and is probably due to an increase in intracapillary hydrostatic pressure. Another mechanism that follows and causes further escape of water from the intravascular compartment is hyperosmolarity of the interstitial compartment, secondary to efflux of K^+ and metabolites from the contracting muscle fibers. Exercise of prolonged duration (30 min or more), if not accompanied by sufficient fluid intake, results in a further, slow decline in plasma volume.[27] This decline reflects total body dehydration, mostly due to sweat loss. It may also result from further osmotic drive. In dry climates, or during exposure to high altitude, large amounts of water are evaporated from the upper and lower airways. If dehydration is prevented by adequate fluid replenishment, there will be no further drop in plasma volume. When fluid is fully replenished during a 90–120 min exercise, plasma volume rises gradually and eventually reaches its pre-exercise level.[28]

Urinary output declines during exercise, which partially compensates for sweat loss. Such a decline results from a reduced renal plasma flow and glomerular filtration rate. Urinary output is further reduced with dehydration secondary to increased antidiuretic hormone activity. In contrast, sweating rate is *not* reduced during progressive dehydration (Fig. 9.11) as long as the fluid deficit does not exceed 5–6% of initial body weight. It seems that to support heat dissipation and thermal balance the body surrenders its fluid balance.

Electrolyte Loss During Exercise

Human sweat contains more than 99% water. Its electrolyte concentration is invariably lower than in the extracellular fluid. Sweat osmolality of adults seldom exceeds 180 mOsm/liter[28] as compared with about 300 mOsm/liter in their body fluids. In children and prepubertal adolescents sweat is even more hypotonic,[6,34] as shown in Fig. 9.12. Sweating therefore is always accompanied by an increase in osmolarity of body fluids. Concentration in the sweat of Na^+ and Cl^-, but not of K^+, increases with the increase in sweating rate. In contrast, ionic concentration of Mg^{++} and Ca^{++} decreases at high sweating rate.[27] Conditioning and acclimati-

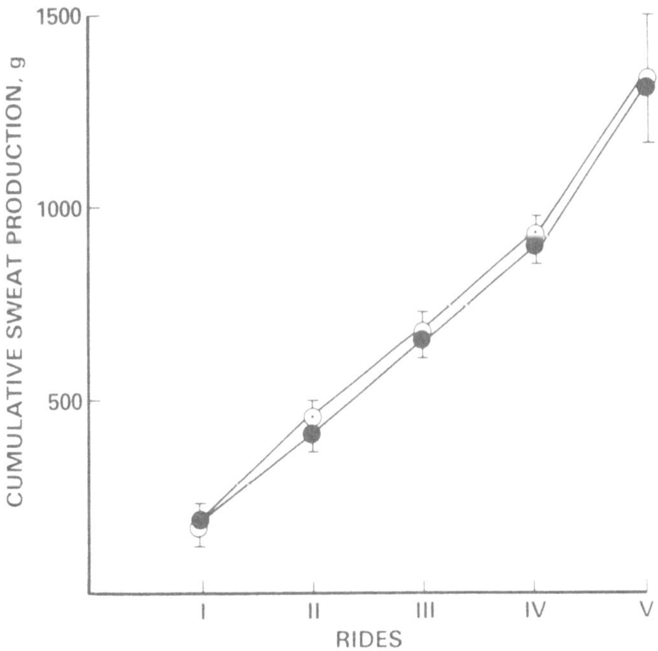

Figure 9.11. Sweat production and fluid deficit. Eleven 10- to 12-year-old boys rode intermittently on a cycle ergometer for 3.5 hr. In one session (○) they fully replenished their fluid losses, while in the other (●) they progressively dehydrated. For details see text section entitled Voluntary Dehydration. Mean values ±1 S.E.M. Reproduced by permission from Bar-Or et al.[10]

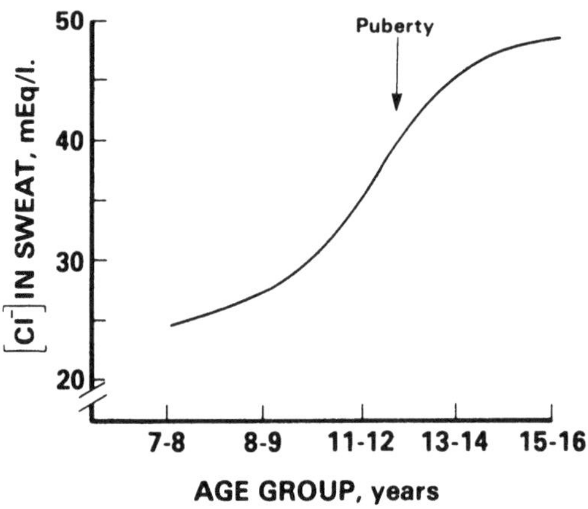

Figure 9.12. Development of salt concentration in sweat. Subjects and conditions are as in Fig. 9.2. Sweat was collected from the pectoralis area. Data by Araki et al.[6]

zation to heat both induce a decrease in ionic concentration of sweat, in spite of an increase in sweating rate. Even though sweat is hypotonic, prolonged and repeated exercise in hot weather can induce marked salt loss, particularly in NaCl.

Hypohydration

The marked fluid shifts that occur with exercise, especially when sweating rate is increased, may result in a fluid deficit. Such a state of fluid deficit will be referred to as "hypohydration." The actual process during which negative fluid balance takes place will be called "dehydration." In this section we shall first describe two types of dehydration that are common in physically active children and then discuss the implication of hypohydration to the performance and well-being of the child. Hypohydration, which accompanies several diseases, will be discussed later.

Voluntary Dehydration. It has long been recognized[76] that people who exercise in hot climates do not drink sufficient amounts to replenish fluid loss, even when allowed to drink *ad libitum*. The term "voluntary dehydration" has been coined for such a phenomenon.[82] In adults, fluid deficits due to voluntary dehydration ranged between 1.5 and 7% of initial body weight, depending on the climate, duration and intensity of the activity, and the type of fluids used for replenishment. Some striking examples are available from marathon races: among 63 runners who completed a race, mean sweating rate was 0.96 liter/hr and fluid intake only 0.13 liter/hr. The result was a fluid loss of 5.2% of initial body weight. The winner of the race had a 6.9% fluid loss.[77]

Voluntary dehydration has been also documented in children who exercised in dry heat (39°C, 45% RH) without being forced to drink.[10] Figure 9.13 presents a comparison of the fluid loss of these boys when

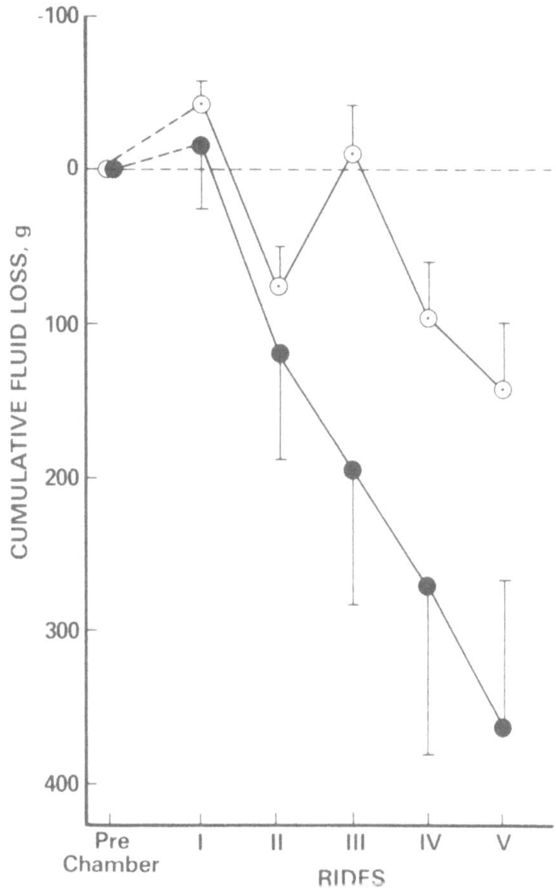

Figure 9.13. Voluntary dehydration in children cycling intermittently in dry heat. Cumulative fluid loss during 3.5 hr in 11 ten- to 12-year-old boys. ○ denotes forced drinking regimen. ● represents voluntary drinking regimen. Vertical lines = 1 S.D. Reproduced by permission from Bar-Or et al.[10]

forced to drink periodically and when allowed to drink *ad libitum*. Although sweating rate was virtually identical in both exposures (Fig. 9.11), fluid intake in the voluntary drinking exposure was only 66% of the intake needed to replenish sweat and urinary losses. Urinary output in that session was only 68% of the respective output during the forced-drinking session, but this was not enough to prevent hypohydration. We can assume that the fluid deficit would have been greater had the session lasted longer.

Deliberate Dehydration. Another type of dehydration occurs when individuals deliberately induce a negative fluid balance, either by depriving themselves of fluid intake or by inducing excessive fluid loss. Deliberate dehydration is common among athletes who wish to "make weight" just prior to competition (wrestlers, weight lifters, boxers, judo competitors, and horse jockeys). The most common procedure is to induce sweating, but some individuals reduce fluid and food intake, take diuretics or laxatives, and even induce vomiting! In North America deliberate dehydration is often found among school-age wrestlers.

In a survey among 10% of all interscholastic high school wrestlers in Iowa,[96] body weight was measured repeatedly until "weighing in" time. Most wrestlers lost 5–7% initial body weight during the last 10 days prior to competition. Eight percent of them lost 10% or more. It is safe to assume that most of the weight loss represented fluid rather than calorie deficit. Especially disheartening was the marked loss among the very lightweight categories (47 kg or less), which in some youngsters reached 15%. Most of these athletes consulted a friend or a coach, but not a physician, as to the best way of reaching their target body weight.

Implications for Performance and Health. Hypohydration results in physiologic dysfunction[1,5,10,26,27,28,101] and is often detrimental to performance and health. There is a reduction in plasma volume, stroke volume, cardiac output, renal blood flow, glomerular filtration rate, and liver glycogen content. With exercise, water depletion is proportionately greater in the extracellular than in the intracellular compartment. Heart rate at rest and in submaximal exercise is elevated. Rehydration within one hour following 4–5% hypohydration can effect a return to normal in the hemodynamic function of high school wrestlers.[1]

There is an electrolyte deficit, especially Na^+ and Cl^-, but also K^+, Ca^{++}, and Mg^{++}. Such losses notwithstanding, plasma concentration of these ions may be elevated due to hemoconcentration. Body fluid osmolality is at first high but, with intake of water, may eventually become normal.

Thermoregulation may become inefficient when the triad heat stress, exercise, and hypohydration is in effect. Convection by blood of heat from body core to skin is particularly disrupted. This results in a rise of core temperature, which is proportional to the fluid deficit. Figure 9.14 is a comparison of changes in rectal temperature between children and adults who progressively dehydrated during exercise in the heat. For each 1% weight loss, the adults had a 0.15°C rise in temperature, compared with 0.28°C in the children.[10] It is not clear whether this greater increase in body heating has clinical significance. At mild to moderate hypohydration (up to 5%), sweating rate remains fairly constant,[11,26] (see Figure 9.11) but at higher levels of hypohydration, or when plasma osmolality is markedly increased, there may be a reduction in sweat output, which further impedes heat dissipation.

Decrements in performance that accompany hypohydration include reduction in muscular strength,[16] in the time that strenuous activity can

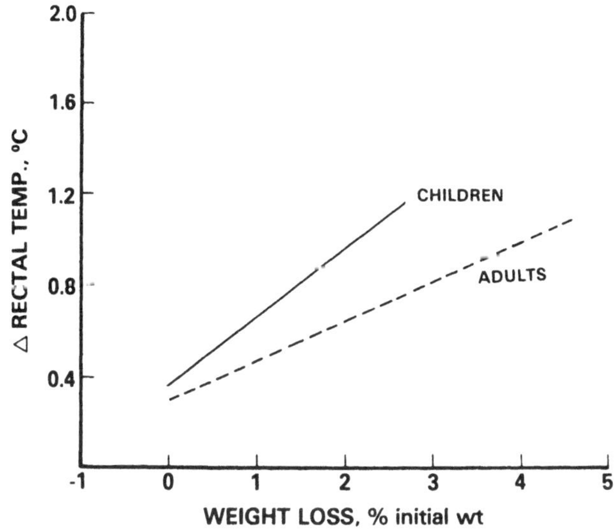

Figure 9.14. Relationship between rise in rectal temperature and level of hypohydration in children and adults exposed to exercise in dry heat. Adapted by permission from Bar-Or et al.[10]

be sustained,[26,84] and in mental alertness.[63] Maximal O_2 uptake is not reduced[84] unless hypohydration is extreme. Reaction time to visual cues is not prolonged.[63] Anaerobic capacity, as measured by the 30-sec Wingate anaerobic test, is not affected by hypohydration of up to 5%.[53]

To the clinician, hypohydration should always signal danger as a disruptor of thermoregulation (see Pediatric Health Hazards in Hot Climates, below). There are no data on the effect upon growth of repeated hypohydration, but such a possibility cannot be ignored in youngsters who repeatedly disrupt their fluid and mineral balance. In addition, chronic loss of K^+ in sweat and urine may contribute to muscle fatigue and cramps.

Water and Electrolyte Replenishment

To minimize hypohydration and electrolyte losses, fluid replacement must become part and parcel of the conduct of any prolonged physical activity. This axiom has been applied in industry, in the military, and in sports. Any inhabitant of warm climates soon recognizes the necessity for proper drinking habits.[30,95]

In an attempt to replenish fluid and electrolyte losses, one must consider the following:

1. Fluids must be selected that do not quench thirst but, instead, stimulate further drinking.

2. In the exercising individual, gastric emptying should be rapid in order to avoid fluid stasis and gastric distension.
3. Fluid quantity must be sufficient to replenish any previous deficit and, preferably, to anticipate further losses.
4. Mineral content in the fluid should not be exaggerated. It must be coordinated with dietary intake.

Although water is an adequate fluid for immediate replenishment, it quenches thirst and does not stimulate further drinking. In contrast, flavored drinks promote thirst and are therefore preferred. A common fallacy in nutrition is that cold water should not be taken when an individual is tired and sweaty. There are no documented reports on any detrimental effects of cold water before, during, or after exercise. Cold drinks have the advantage over tepid or warm drinks because they are emptied faster from the stomach[29] and are more palatable. The direct cooling effect of a cold fluid is minimal when its dilution in the total water compartment is calculated. Even so, such cooling may sometimes give the edge on a very hot day.

Although large volumes of fluid cause faster emptying from the stomach than smaller amounts, the former can cause temporary abdominal "heaviness," which is not well tolerated during sports. Smaller volumes of fluid are therefore recommended, even though they must be taken more often. An 11-year-old child, for example, who loses 400 ml water/hr should have 100 ml/15 min rather than 200 ml/30 min. Once dehydration starts, its reversal during exercise is hard. It is therefore essential to ascertain adequate hydration *prior* to the beginning of exercise and then not to wait more than 15–20 min with the first drink. For children who participate in long-distance road races, the official drinking stations are often too far apart and start too late in the race. Special drinking arrangements may have to be made for these young runners.

The osmolarity of ingested fluid may determine gastric motility and emptying. A solution containing 20 mEq/liter of salt, for example, will be emptied faster than water. Fluids with high salt content, however, will retard gastric emptying, as will sugar solutions which exceed 200 mosm/liter (i.e., 25 g/liter glucose).[29] Addition of salt in liberal amounts to the regular diet is practiced by many athletes, often in the form of salt tablets. Such a habit is not beneficial for children and may even be detrimental, for the following reasons:

1. Plasma aldosterone activity is markedly increased during acute exercise such that Na^+ is conserved by the kidney.[27] In heat-acclimatized people, it is also conserved by the sweat glands. Thus, habitual exercisers who are heat acclimatized often have a positive Na^+ balance, even when not adding salt to their regular balanced diets.
2. For occasional exercisers, salt content in most diets is high enough to balance sweat and urinary losses.

3. Sweat is invariably hypotonic, and body fluids during dehydration, hypertonic. Ingestion of too concentrated solutions will increase the hypertonicity of the interstitial space and induce further depletion of the intracellular compartment.
4. Children's sweat is more dilute than is adults'.
5. Studies of K^+ balance, tallying intake, sweat, and urine losses, have shown that K^+ can be fully preserved by a regular balanced diet, even in people who exercise often. There is therefore no need for special KCl additives.

The American College of Sports Medicine in its "Position Statement on Prevention of Heat Injuries during Distance Running" recommends that adult runners periodically drink fluids that include not more than 10 mEq/liter Na^+, 5 mEq/liter K^+ and 2.5 g/liter glucose.[4] We recommend that children, because of the low salt content of their sweat, have drinks that do not exceed 5 mEq/liter Na^+ (0.3 g/liter NaCl), 4 mEq/liter K^+ (0.28 g/liter KCl) and 25 g/liter sugar.

There are various commercially prepared drinks that contain minerals and sugar and are palatable to most athletes. Such drinks, with proper dilution, can be suitable for children. A less expensive alternative are homemade drinks prepared according to the above recommendations. These can be flavored to individual taste. For children who exercise sporadically, plain water replacement is sufficient, providing the child is given a balanced diet.

How much fluid is enough? For athletic children a simple and reliable gauge is their body weight. One should strive to attain a post-activity (practice or competition) body weight identical to that at pre-activity level. This routine is based on the assumption that the child is at an adequate hydration level at the start of the activity. To secure such a level he should drink 300–400 ml fluid 20–30 min prior to warm-up time. This will leave enough time for voiding any extra fluid. Nonathletic children will not comply with such a body weighing routine. Such children must be taught to drink until they are no longer thirsty—and then to add 150–200 ml.

In conclusion, thirst is an inadequate gauge for sufficient drinking during prolonged activities, or in hot climates. It is a prime educational responsibility of the physician to impress this fact upon parents, teachers, coaches, and children.

Pediatric Health Hazards in Hot Climates

Introduction

Previous sections in this chapter have dealt with response to exercise in the heat from a physiologist's vantage point. In this section we shall discuss the health implications of such a response.

While this book focuses on the exercising child, some information is included on the health hazards of hot climates to resting children. We include such information on the assumption that health hazards to the inactive child can only be magnified when a high metabolic load is superimposed on climatic heat stress. The concept of "heat-related illness" will be introduced first, followed by epidemiologic data on the vulnerability to hot climates of infants and children. The section will end with a closer look at those groups of children who are at an especially high risk for heat-related illness.

Heat-related Illness

"Heat-related illnesses" are pathologic conditions that result from exposure to heat, at rest or during exercise. Other terms commonly used are "heat disorders" or "heat injuries." With the worldwide exponential increase in jogging, popular races such as "fun runs," and marathon races with thousands of competitors, health practitioners are being called upon to treat increasing numbers of victims of heat-related illness.[92] Heat stroke ranks second to head injury among documented causes of death in high school sports.[59] While a detailed discussion of therapy belongs in medical texts, we cannot overemphasize one common denominator shared by these conditions: *they are all preventable.* The health practitioner should therefore become involved *before* a race starts or a team commences its preseason summer training.

Table 9.2 outlines the etiology, manifestations, and principles of prevention of the major heat-related illnesses. A more detailed classification can be found elsewhere.[64] Conditions range in severity from mild heat cramps and heat syncope to the often fatal heat stroke. Their presentation is not always clear-cut, and there is an overlap among them. For example, heat exhaustion due to salt depletion is often accompanied by water depletion which, in itself, may present first as syncope rather than exhaustion. Heat stroke can be manifested in ways other than the classical triad of hyperpyrexia, dry skin, and neurologic deficit. One notable variation is that the person still perspires even though his thermoregulatory control is disrupted. Such a person may mistakenly be diagnosed as having heat exhaustion rather than heat stroke. A sound therapeutic approach is to assume the worst whenever in doubt. Heat-related illnesses can occur even on mildly warm or climatically neutral days. A prolonged, intense activity such as a marathon race, bicycle road race, or soccer match, especially when accompanied by hypohydration, can induce any of the heat-related illnesses irrespective of climate.[93]

The two most important preventive measures are prior acclimatization and adequate hydration prior to and throughout the activity. Details are provided in previous sections of this chapter. Guidelines for Con-

Table 9.2. Heat-related Illnesses: Classification, Etiology, Clinical Presentation, and Prevention*

Illness	*Etiology*	*Presenting Symptoms and Signs*	*Prevention*
Heat cramps	Intense, prolonged exercise in heat; negative Na^+ balance	Tightening, cramps, and involuntary spasms of active muscles; somewhat low serum Na^+	Replenish salt loss; ensure acclimatization
Heat syncope	Peripheral vasodilatation and pooling of blood; hypotension; hypohydration	Giddiness, syncope; mostly in an upright resting or exercising person; pallor; high T_{re}†	Ensure acclimatization and fluid replenishment; reduce exertion on hot days; avoid standing
Heat exhaustion (water-depletion type)	Continuous and accumulating negative water balance	Exhaustion; symptoms and signs of hypohydration; flushed skin; reduced sweating in extreme dehydration; syncope; high T_{re}; hemoconcentration	Ascertain proper hydration before effort and adequate replenishment during effort; ensure acclimatization
Heat exhaustion (salt-depletion type)	Negative Na^+ balance accumulating during a few days	Exhaustion; nausea; vomiting; muscle cramps; giddiness. More insidious than the water-depletion type	Replenish electrolytes lost, based on type and duration of effort and on climate; ensure acclimatization
Heat stroke	Extreme hyperthermia leading to thermoregulatory failure; enhanced by dehydration	Acute medical emergency that classically includes: hyperpyrexia ($T_{re} \geq 41°$), lack of sweating (not always), and neurologic deficit (disorientation, twitching, seizures, coma). Variations of the above also exist	Ensure acclimatization; identify and exclude individuals at risk; adapt activities to climatic constraints

* Based in part on Buskirk and Grasley.[22] Reproduced with permission from Bar-Or.[9] Copyright American Academy of Pediatrics, 1983.
† T_{re} = rectal temperature.

duct of Athletic Events in the Heat, below, outlines additional suggestions for the prevention of heat-related injury.

Epidemiologic Studies on Children's Health in Hot Climates

There are no prospective epidemiologic studies of the relationship between climatic heat stress and children's performance or health. All available reports are retrospective, usually following severe climatic heat waves. Some are summarized in Table 9.3.

In spite of methodologic constraints, these reports do convey a clear message: infants, young children, and the elderly are more affected by climatic heat waves than are adolescents or young adults. Most susceptible are those children who arrive in hospital with hypohydration or who become hypohydrated during admission secondary to diarrhea, vomiting, etc. Mothers of these patients often are ignorant of the need for added fluid intake on hot days.[17,24,30,39,58,86,95]

Populations at High Risk for Heat-related Illness

The thermoregulatory response to exercise in the heat is highly variable among individuals. Variability is to some extent hereditary,[66] but it de-

Table 9.3. Epidemiologic Studies of Clinical Responses of Infants and Children in Hot Climates

Event	*Location*	*Findings*	*Author*
Death charts 1900–1930	Three U.S. states	Heat-related death mostly prevalent in infants and aged people	Shattuck and Hilferty[88]
5-day heat wave, 1948	New York City	All hospitalized infants developed hyperpyrexia	Cardullo[24]
Heat wave, 1959	Melbourne	25% fatalities among 0- to 6-year-old patients were due to heat illness	Danks et al[30]
Summer 1965	Baghdad	Dehydration preceded most cases of childhood heat exhaustion	Taj-Eldin and Falaki[95]
5-day heat wave, 1966	St. Louis	Infants and old men mostly susceptible	Ellis[39] and Ellis et al[40]
Heat waves, 1936–1966	St. Louis	Infants and old men mostly susceptible	Bridger et al[17]

pends largely on acquired differences. This section will highlight those groups of children who respond with high heat strain to any given heat stress. Such a response may impede their exercise performance and may also be detrimental to their health. The early identification of children who are at high risk of developing heat-related illness is of obvious relevance for the physician.

Table 9.4 outlines those conditions and diseases that put a child at a potentially high risk. A possible mechanism is offered for the thermoregulatory deficiency.

Anorexia Nervosa (AN). Adolescents with AN often like to exercise, which fits in with their "strategy" of burning up calories. The thermoregulatory capability of these patients is often deficient.[33,72,99] At rest they complain of cold, especially in their extremities, which is accompanied by cyanosis (acrocyanosis). These changes reflect a low skin temperature secondary to peripheral vasoconstriction. Such vasoconstriction may represent a compensation for deficient insulation due to the paucity of subcutaneous fat. Core temperature is lower than normal in the resting AN patient (about 36°C). It drops further when the patient is exposed to the cold and rises during exposure to heat. Such lack of stability of core temperature could reflect hypothalamic dysfunction.[72] Even though AN patients have a sluggish vasodilatory response to heat, they dissipate more heat by convection and radiation than by evaporation. The reverse is true for healthy individuals exposed to a similar environment.

During prolonged exercise in a thermoneutral climate, the rise of core temperature to a new plateau is slow compared with that of healthy controls. This could reflect a greater heat capacity,[33] the specific heat of lean tissues being higher than that of adipose tissue.[13]

No reports are currently available on the response to hot or cold climates of the exercising AN patient. Preliminary data suggest, however, that the environmental zone in which they can maintain a constant core temperature is narrower than normal.[32]

Whether their deficient thermoregulatory capability is due to insufficient insulation or suboptimal hypothalamic control, these patients may be at risk for heat- or cold-related illness. Some AN youngsters practice vomiting, which may induce hypohydration and electrolyte disturbances.

Congenital Heart Disease (CHD). Clinical experience has shown that some infants and children with CHD sweat excessively. This phenomenon has been confirmed in controlled indoor observations.[2,69,78] Especially affected are those suffering from congestive heart failure or right-to-left shunts. Sweat glands of some CHD patients more than 6 months old have longer ducts than normal.[62] The physiologic implication of such an anatomic difference is not clear.

Table 9.4. Conditions and Diseases that Predispose the Exercising Child to Thermoregulatory Insufficiency*

Condition or Disease	*Possible Mechanism*				
	Reduced Heat Convection to Periphery	*Insufficient Sweating*	*Excessive Sweating*	*Potential Hypo-hydration*	*Other*
Anorexia nervosa	X			X	Reduced subcutaneous insulation
Congenital heart disease	X		X	X	
Cystic fibrosis	X		X	X	
Diabetes (mellitus, insipidus)	X			X	
Diarrhea and vomiting	X			X	
Excessive eagerness				X	High heat production
Fever	X		X	X	Regulatory insufficiency
Hypohydration	X	X (if extreme)			
Insufficient acclimatization	X	X			
Insufficient conditioning	X	X			
Malnutrition					Reduced subcutaneous insulation
Mental deficiency					Insufficient drinking
Obesity					High heat production, low specific heat and surface area
Prior heat-related illness	Various (depends on illness)				
Sweating insufficiency syndromes		X			

* Reproduced with permission from Bar-Or.[9] Copyright American Academy of Pediatrics, 1983.

When exposed to temperatures ranging from 22 to 32°C, children with severe congenital defects lose more heat by evaporation and less by convection and radiation than do children with mild or asymptomatic CHD.[69] The reason for such an aberration is not clear. One might speculate that low cardiac output in these patients causes insufficient heat convection to the periphery and, as a result, a greater demand for sweating and evaporative cooling. Although such a mechanism may be operative, the excessive sweating of these cardiac patients was found not only to compensate for reduced convection and radiation, but to induce core temperatures that are *lower* and more labile than among healthy children.[78] As the metabolic level of infants with CHD is not lower than that of healthy controls,[54] and is perhaps even higher,[69] the low core temperature of the former suggests an "overshoot" on the part of the sudomotor apparatus. Excessive sweating should be looked on as an uneconomical thermoregulatory pattern that may lead to hypohydration.

There are no reports as to whether hot and humid climates actually interfere with the physical ability and well-being of the ambulatory cardiac child. If one extrapolates from information on adults,[20] however, there is enough evidence to recommend that the outdoor activity level of the child with severe heart defects be reduced whenever climatic heat stress is high.

Cystic Fibrosis (CF). Infants and children with CF often suffer from marked heat prostration during climatic heat waves.[30,58,100] It has been estimated that 15% of all New York City children with CF were hospitalized for heat prostration during the 1948 heat wave.[58] One of the initial signs is profuse sweating. Within 2–3 days they develop hypohydration, hyperpyrexia, and circulatory insufficiency. Serum concentration of Na^+ and Cl^- are low and may reach 125 mEq/liter and 80 mEq/liter, respectively. Rehydration with an adequate salt supplement is effective. Although the mechanisms for such heat prostration have not been systematically studied, it seems that an excessive production of hypertonic sweat leads to hyponatremic hypohydration, which in turn interferes with heat dissipation. Such increased sweating rate may be related to the greater density of active sweat glands, as shown by pilocarpine iontophoresis.[48]

Recent studies (see Chapter 3) have shown the clinical benefits to CF children of conditioning programs. Caution should be exercised on hot days with children who take part in such activities.

Diabetes Mellitus or Insipidus. Regular physical activity should become integral to the management of the diabetic child. On hot or humid days, however, caution should be exercised and special attention paid to adequate hydration. This is especially important in patients with polyuria, who may dehydrate and suffer excessive electrolyte loss. A similar risk confronts the child with diabetes insipidus.

Diarrhea and Vomiting. Any condition accompanied by diarrhea and vomiting, irrespective of etiology, can easily induce hypohydration and electrolyte imbalance in a child. Sweating will aggravate such disturbances, predisposing the child to heat exhaustion and heat stroke.

Excessive Eagerness. The discomfort experienced by people who are exposed to heat or cold stress is of a protective nature. As a rule, an individual will try to terminate such an experience by, for example, seeking shade, stopping exercise, or leaving the cold water.

Children's perception of exercise while exposed to heat stress seems to underestimate the physiologic strain that they are undergoing.[8,12] It would seem, therefore, that children do not experience the above protective discomfort that is effective in adults. A young, ambitious, overzealous athlete may decide to conduct a practice session regardless of the hazards of the prevailing climate. Such a child should be advised about the harmful consequences.

Fever. Irrespective of its etiology, fever indicates some disruption of thermoregulatory control. It will therefore predispose further hyperpyrexia in a child who is generating high metabolic heat, especially when exposed to climatic heat. Intensive activities, as a rule, should be curtailed in a febrile child even if the cause is "obvious"—such as following vaccination.

Hypohydration. Hypohydration is the link between most "high risk" conditions and thermoregulatory insufficiency. As discussed above, hypohydration can also occur in healthy children who do not replenish fluid losses before, during, or after activity.

Hypohydration is accompanied by reduced plasma volume. This impedes convection of heat from body core to skin and reduces the central blood volume so that circulatory insufficiency may ensue. Hypohydration is almost invariably present in heat exhaustion (water depletion type), and in heat stroke. It can also lead to heat syncope. "Thirst fever" is a syndrome seen in febrile young children with no apparent infectious etiology. The level of fever is proportional to the fluid deficit and disappears once the child is fully hydrated.[25,86]

Insufficient Acclimatization. For a full discussion of the physiologic and perceptual phenomena of children's acclimatization to exercise in the heat, see the earlier Section on that topic. Insufficient acclimatization is the single most important cause of heat-related illness. The clinician must bear in mind the sluggish rate of acclimatization in children on the one hand and their fast subjective improvement on the other. This is especially important during climatic heat waves, whenever a practice season starts in the summer (as in American football), or where athletes travel to hot geographic regions. Retrospective studies of heat stroke strongly suggest a relationship to insufficient acclimatization. Among eight heat stroke fatalities in high school American football, six happened following the first or second session of preseason practice, before the players had a chance to acclimatize.[7,42] Another young football player was lucky to survive a heat stroke that occurred following the first practice session in the summer.[80] Similar occurrences have been reported for unacclimatized college football players.[91] Rural infants and preschoolers in Australia were found to have less heat-related illness than urban ones, presumably because the former were habitually more exposed to heat and therefore better acclimatized.[30]

We cannot overemphasize the need to attain acclimatization in children and adolescents who are newly exposed, or likely to be exposed, to a hot climate. Guidelines are available in other sections of this chapter.

Insufficient Conditioning. As discussed above, conditioning *per se* is a means for heat acclimatization in children. One can therefore assume that the insufficiently conditioned child is also less heat-acclimatized and may encounter difficulties when abruptly confronted with hot weather.

Low heat tolerance is related to low level of fitness. The elevation of rectal temperature during exercise is proportional to the *relative* metabolic level (e.g., percentage of maximal O_2 uptake) rather than to the absolute level. A less conditioned child will therefore have a greater rise in core temperature while performing a certain task.[33] Furthermore, a fit child can perform a certain task more economically—i.e., at a lower metabolic cost.

Another potential disadvantage of the less fit individual is his or her inexperience in performing strenuous tasks. There is some evidence that novice adult long-distance runners are more prone to heat stroke because they select a running pace incompatible with their exercise capacity.[45] This issue of inexperience and heat-related illness has not been studied in children.

Malnutrition. Malnutrition of the hypocaloric type is associated with low subcutaneous insulation. As in anorexia nervosa (see above), this may lead to excessive heat loss in cold climates[18] and to an excessive rise in core temperature in hot climates (when ambient temperature exceeds skin temperature).[19] Malnourished children also were found to have a reduced sweating rate, perhaps due to local changes in the sweat glands.[19]

Although the above data are based on experiments with infants, it is likely that older malnourished children will respond in a similar way. One should bear in mind that, unless severely affected, the malnourished child is physically active and may therefore be adversely affected in hot climates.

Mental Deficiency. The mentally deficient child does not necessarily have any thermoregulatory deficiency. It has been found, however,[30] that such children are more prone to heat-related illness because they can neither understand nor verbalize their need for extra drinking on hot days.

Obesity. Since subcutaneous fat provides insulation to the body core, increased adiposity is an asset in cold climates.[21,56] In contrast, obesity is a distinct liability in hot climates.

In response to a questionnaire, overweight women reported a lesser tolerance to heat and a greater tolerance to cold than did normal weight women. Men had a similar tendency.[44] Tolerance time of obese adults is indeed short when they exercise in a heated climatic chamber.[13,71] They respond to exercise in the heat with higher heart rate, cardiac output, and body temperatures, and their post-exercise cardiovascular adaptation to orthostatic stress is less efficient than that of lean controls.[13,21,70] Among children, the obese respond to a given heat stress with a higher strain than do the lean. However, in a study among 9- to 12-year-old boys, there was no difference in the ability of the lean and the obese to complete 60 min of exercise in 46–48°C, 22% RH.[47]

Figure 9.15 is a comparison in the rise of rectal temperature and heart rate between obese (31.2% fat) and lean (15.5% fat) 9- to 12-year-old boys. Both groups walked on a treadmill in dry heat. While baseline values were similar, rectal temperature and heart rate rose faster and reached higher levels among the obese.[47]

In any given individual, an inverse relationship exists between the population density of heat-activated sweat glands and skinfold thickness. As a group, the obese have a lower density of active glands, especially in the trunk,[14] even though their sweating rate, corrected for surface area, is the same or even higher than in that of the lean.[13,21] Sweat output per gland is therefore higher in the obese. They can acclimatize well to exercise in the heat,[23] but during progressive dehydration their core temperature rises excessively with any given body fluid deficit.[11] Obesity can mask the clinical presentation of fluid deficit, especially in young children.[38] One should therefore be alert to the possibility of fluid deficit in any obese child with recent exposure to heat.

There are a number of causes that underlie the handicaps of the obese in hot climates:

1. Specific heat of fat is 0.4 kcal/g × °C, as compared with 0.8 kcal/g × °C in the fat-free mass. Thus a given amount of heat will raise the temperature of 1 g fat twice as much as it will raise the temperature of 1 g fat-free tissue.

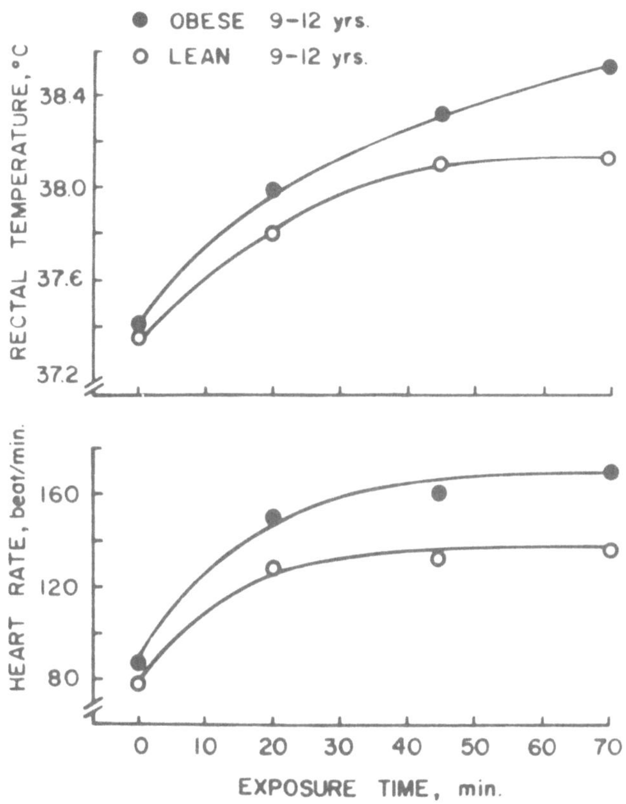

Figure 9.15. Obesity and body heating. Five obese and seven lean 9- to 12-year-old boys walked intermittently (4.8 km/hr, 5% grade) in dry heat (40–42°C, 25% RH). Values are group means. Reproduced by permission from Bar-Or.[8] Based on data by Haymes et al.[47]

2. Obese children often have a large body mass and a small surface-area-to-mass ratio. In addition, accumulating fat deposits change the contour of the body such that the above ratio is further decreased. This will result in a lower rate of heat transfer (see above) to and from the body. This is a disadvantage in moderately warm climates when skin temperature is higher than air temperature. Only in extreme heat will the small surface-area-to-mass ratio be an advantage in reducing heat influx from the environment. It will still be a handicap as far as sweating is concerned.[13,46]
3. The rise in core temperature during exercise is proportional to the *relative* metabolic rate (i.e., percentage of maximal O_2 uptake). Obese children as a group have low maximal aerobic power and, therefore, when performing a certain task, they operate at a relatively high percentage of their maximal O_2 uptake. This results in a greater rise of core temperature as well as greater fatigability.

4. Fat has less water content than do most other tissues so that obese individuals have a low total body water content per unit mass. As a result a certain hypohydration level, determined as percentage of initial body weight, represents for the obese a greater water deficit relative to total body water.

The above physiologic liabilities have clinical implications. A relationship has been found between overweight and the risk of heat stroke and death.[85] Among 125 heat stroke victims, "most of the patients were somewhat overweight and actually obese."[68] High school boys who died of heat stroke following football practice were markedly overweight.[7,42] These 15.3-year-old boys weighed on the average 89 kg, which probably represents some obesity and not only developed musculature.

Obese children, because of their low level of fitness and high heat strain, are likely to slow down or terminate their exercise earlier than other children. One must remember to allow these individuals special consideration on hot or humid days and not to push them to their limit.

Prior Heat-related Illness. An important question is whether an individual with a history of heat-related illness is predisposed to an above-average risk in the future. Young adults with a history of exercise-related heat stroke who were reexposed to exercise in dry heat responded with reduced tolerance time, high rectal temperature and heart rate, but normal sweating rate.[81,87] It is not known whether such individuals have an inherent thermoregulatory deficiency or have suffered irreversible damage from the heat stroke episode.

Sweating Insufficiency Syndromes. These are rare and often just one manifestation of a systemic disease.[41,67] Such children must rely on convection and radiation for dissipating heat. A special risk exists on days when air temperature is high and heat dissipation by these avenues is impeded.

Guidelines for Conduct of Athletic Events in the Heat

The American College of Sports Medicine has issued a "Position Statement on Prevention of Heat Injuries During Distance Running"[4] and a "Position Stand on Weight Loss in Wrestlers."[5] The American Academy of Pediatrics, Committee on Sports Medicine has published its position on "Climatic Heat Stress and the Exercising Child,"[3] and The Canadian Association of Sports Sciences is preparing a similar document. Such a plethora of statements reflects the realization that health scientists and practitioners should play a leading role in the prevention of heat-related illness among active individuals. The following guidelines have been prepared based on the above statements, bearing in mind children's specific responses to exercise and heat.

1. Ensure acclimatization to heat. The nonacclimatized child, when first exposed to heat, must cut down the intensity and duration of exercise and then gradually increase them. Exposures can be conducted three to six times per week, for a total of six to eight exposures. Exercise during the session must be interspersed with rest periods.
2. Secure full hydration before practice or competition (300–400 ml fluid, 20–30 min prior to activity for a 10- to 12-year-old child) and encourage periodic drinking (100 ml/15 min) during prolonged activities. Remember that thirst is not an accurate guide. Encourage active children to drink above and beyond subjective repletion.
3. Fluids should be chilled and flavored. Their content should not exceed 5 mEq/liter Na^+ (0.3 g/liter NaCl), 4 mEq/liter K^+ (0.28 g/liter kcal), and 25 g/liter sugar. Attempt to provide the child with his favorite drinks.
4. Discourage "making weight" by dehydration. This is a habit strongly rooted among athletes and coaches and must be eradicated, or minimized, through education. Rubberized sweat suits, laxatives, diuretics, and emetics should be absolutely disallowed. *Never* restrict fluids as a disciplinary or "character building" measure.
5. Activities must be tailored to the prevailing climate. It is the prerogative of the team or school physician to postpone, curtail, or cancel activities or to increase rest periods because of climatic stress. Suggested policy is given in Table 9.5. Those who do not possess a psychrometer can obtain information from the local weather bureau. Rest periods in well ventilated and shaded areas are important for dissipation of stored heat during a practice.
6. Clothing should be lightweight, limited to one layer of absorbent material and tight to the skin to facilitate evaporation of sweat. Excessive taping and padding should be discouraged. A hat and light-colored clothing are recommended whenever feasible to reduce solar radiation. Discourage prolonged exposure of the skin to the sun.
7. Identify and screen out, as necessary, individuals who are at high risk for heat-related illness (see above). A preseason examination is of value for such screening.
8. Look for and teach others about early warning symptoms and signs that precede heat-related illness. These include: unexplained headache, throbbing pressure in the head, chills, nausea, piloerection on the chest and arms, disorientation, ataxia, and dry skin. Activity of children with the above should be discontinued and treatment commenced.

Table 9.5. Climatic Heat Stress and Permissible Physical Activities

*WBGT °C** (*°F*)	*WBT °C†* (*°F*)	*Changes in Activity*
<25 (<77)	<15 (<59)	All activities allowed
25–27 (77–81)	15–21 (59–70)	1. Longer breaks in the shade 2. Drinking each 15 min 3. Alert for warning symptoms of heat-related illness
27–29 (81–84)	21–24 (70–75)	As above plus: 1. Stop activities of all unacclimatized, unconditioned, and high-risk persons 2. Limit activities of all others (drastically cut down duration of each activity, increase rest periods, disallow long-distance races)
>29 (>84)	>24 (>75)	Stop all athletic activities of all participants

* WBGT (Wet Bulb Globe Temperature) = 0.7 WB + 0.2 G + 0.1 DB, where WB is wet bulb, G is black globe, and DB is dry bulb, measuring humidity, radiation, and air temperature, respectively.
† WBT = 0.7 WB + 0.3 DB (for indoor use when radiant heat is less important).

9. Coaches, athletes, and even parents may disregard heat-related risk for the sake of victory. Whether a team physician, a school physician, or a physician attending a competition, do your utmost to educate them. If other means fail, assert your authority.

References

1. Allen TE, Smith DP, Miller DK: Hemodynamic response to submaximal exercise after dehydration and rehydration in high school wrestlers. Med Sci Sports 9:159–163, 1977.
2. Alter BP, Czapek EE, Rowe RD: Sweating in congenital heart disease. Pediatrics 41:123–129, 1968.
3. American Academy of Pediatrics, Committee on Sports Medicine: Climatic heat stress and the exercising child. Pediatrics 69:808–809, 1982.
4. American College of Sports Medicine: Position Statement on prevention of heat injuries during distance running. Med Sci Sports 7(1):vii–viii, 1975.
5. American College of Sports Medicine: Position Stand on weight loss in wrestlers. Med Sci Sports 8:xi–xiii, 1976.
6. Araki T, Toda Y, Matsushita K, Tsujino A: Age differences in sweating during muscular exercise. Jpn J Phys Fitness Sports Med 28:239–248, 1979.

7. Barcenas C, Hoeffler HP, Lie JT: Obesity, football, dog days and siriasis: a deadly combination. Am Heart J 92:237–244, 1976.
8. Bar-Or O: Climate and the exercising child—a review. Int J Sports Med 1:53–65, 1980.
9. Bar-Or O: Thermoregulation, fluid and electrolytes in the young athlete. In: Smith NJ (ed.) Sports Medicine: Health Care for Young Athletes. American Academy of Pediatrics, Evanston Ill., in press, 1983.
10. Bar-Or O, Dotan R, Inbar O, et al: Voluntary hypohydration in 10- to 12-year-old boys. J Appl Physiol: Respir Environ Exercise Physiol 48:104–108, 1980.
11. Bar-Or O, Harris D, Bergstein V, Buskirk ER: Progressive hypohydration in subjects who vary in adiposity. Isr J Med Sci 12:800–803, 1976.
12. Bar-Or O, Inbar O: Relationship between perceptual and physiological changes during heat acclimatization in 8- to 10-year-old boys. In: Lavallée H, Shephard RJ (eds.) Frontiers of Activity and Child Health. Pélican, Quebec, 1977, pp. 205–214.
13. Bar-Or O, Lundegren HM, Buskirk ER: Heat tolerance of exercising obese and lean women. J Appl Physiol 26:403–409, 1969.
14. Bar-Or O, Lundegren HM, Magnusson LI, Buskirk ER: Distribution of heat-activated sweat glands in obese and lean men and women. Hum Biol 40:235–248, 1968.
15. Bergh U, Ekblom B, Holmer I, Gullstrand L: Body temperature response to a long distance swimming race. In: Eriksson B, Furberg B (eds.) Swimming Medicine IV. University Park Press, Baltimore, 1978, pp. 342–344.
16. Bosco JS, Terjung RL, Greenleaf JE: Effects of progressive hypohydration on maximal isometric muscular strength. J Sports Med Phys Fitness 8:81–86, 1968.
17. Bridger CA, Ellis FP, Taylor HL: Mortality in St. Louis, Missouri during heat waves in 1936, 1953, 1954, 1955 and 1966. Environ Res 12:38–48, 1976.
18. Brooke OG: Thermal insulation in malnourished Jamaican children. Arch Dis Child 48:901–905, 1973.
19. Brooke OG, Salvosa CB: Response of malnourished babies to heat. Arch Dis Child 49:123–127, 1974.
20. Burch GE, DePasquale NP: Hot Climates, Man and His Heart. Charles C Thomas, Springfield, 1962.
21. Buskirk ER, Bar-Or O, Kollias J: Physiological effects of heat and cold. In: Wilson NL (ed.) Obesity. F.A. Davis, Philadelphia, 1969, pp. 119–139.
22. Buskirk ER, Grasley WC: Heat injury and conduct of athletics. In: Physiological Aspects of Sports and Physical Fitness. Athletic Institute, 1968, pp. 49–52.
23. Buskirk ER, Lundegren H, Magnusson L: Heat acclimatization patterns in obese and lean individuals. Ann NY Acad Sci 131:637–653, 1965.
24. Cardullo HM: Sustained summer heat and fever in infants. J Pediatr 35:24–42, 1949.
25. Choremis K, Danelatou C, Maounis F, et al: Paper chromatography for amino-acids in thirst fever. Helv Paediatr Acta 14:44–53, 1959.
26. Claremont AD, Costill DL, Fink W, Van Handel P: Heat tolerance following diuretic induced dehydration. Med Sci Sports 8:239–243, 1976.

27. Costill DL: Sweating: its composition and effects on body fluids. In: Milvy P (ed.) The Long Distance Runner. Urizen Books, New York, 1978, pp. 290–303.
28. Costill DL, Miller JM: Nutrition for endurance sports: carbohydrate and fluid balance. Int J Sports Med 1:2–14, 1980.
29. Costill DL, Saltin B: Factors limiting gastric emptying during rest and exercise. J Appl Physiol 37:679–683, 1974.
30. Danks DM, Webb DW, Allen J: Heat illness in infants and young children: a study of 47 cases. Br Med J 2:287–293, 1962.
31. Davies CTM: Thermal responses to exercise in children. Ergonomics 24:55–61, 1981.
32. Davies CTM, Fohlin L, Thorén C: Temperature regulation in anorexia nervosa. J Physiol (London) 268:8P–9P, 1977.
33. Davies CTM, Fohlin L, Thorén C: Thermoregulation in anorexia patients. In: Borms J, Hebbelinck M (eds.) Pediatric Work Physiology. Karger, Basel, 1978, pp. 96–101.
34. Dill DB, Hall FG, Van Beaumont W: Sweat chloride concentration: sweat rate, metabolic rate, skin temperature and age. J Appl Physiol 21:99–106, 1966.
35. Dotan R, Bar-Or O: Climatic heat stress and performance in the Wingate Anaerobic Test. Eur J Appl Physiol 44:237–243, 1980.
36. Drinkwater BL, Kupprat IC, Denton JE, et al: Response of prepubertal girls and college women to work in the heat. J Appl Physiol: Respir Environ Exercise Physiol 43:1046–1053, 1977.
37. Drinkwater BL, Horvath SM: Heat tolerance and aging. Med Sci Sports 11:49–55, 1979.
38. Editorial: Dehydration and fat babies. Br Med J 1:125, 1971.
39. Ellis FP: Mortality from heat illness and heat-aggravated illness in the United States. Environ Res 5:1–58, 1972.
40. Ellis FP, Exton-Smith AN, Foster KG, Weiner JS: Eccrine sweating and mortality during heat waves in very young and very old persons. Isr J Med Sci 12:815–817, 1976.
41. Foster KG, Hey EN, O'Connell B: Sweat function in babies with defects of the central nervous system. Dev Med Child Neurol 20[Suppl.]:94, 1969.
42. Fox EL, Mathews DK, Kaufman WS, Bowers RW: Effects of football equipment on thermal balance and energy cost during exercise. Res Q Am Assoc Health Phys Educ 37:332–339, 1966.
43. Gullestad R: Temperature regulation in children during exercise. Acta Paediatr Scand 64:257–263, 1975.
44. Hadland DG, Stock JF, Hewitt MI: Heat and cold tolerance: relation to body weight. Postgrad Med 55:75–80, 1974.
45. Hanson PG, Zimmerman SW: Exertional heatstroke in novice runners. JAMA 242:154–157, 1979.
46. Haymes EM, Buskirk ER, Hodgson JL, et al: Heat tolerance of exercising lean and heavy prepubertal girls. J Appl Physiol 36:566–571, 1974.
47. Haymes EM, McCormick RJ, Buskirk ER: Heat tolerance of exercising lean and obese prepubertal boys. J Appl Physiol 39:457–461, 1975.
48. Huebner DE, Lobeck CC, McSherry NR: Density and secretory activity of

eccrine sweat glands in patients with cystic fibrosis and in healthy controls. Pediatrics 38:613–618, 1966.
49. Inbar O: Acclimatization to dry and hot environment in young adults and children 8–10 years old. EdD Dissertation, Columbia University, 1978.
50. Inbar O, Bar-Or O, Dotan R, Gutin B: Conditioning versus exercise in heat as methods for acclimatizing 8- to 10-year-old boys to dry heat. J Appl Physiol Respir Environ Exercise Physiol 50:406–411, 1981.
51. Inbar O, Dotan R, Bar-Or O, Gutin B: Heat acclimatization—a comparison between prepubertal boys and young men. Submitted for publication.
52. Inbar O, Dotan R, Bar-Or O, Gutin B: Passive vs. active exposures to dry heat as methods for heat acclimatization in prepubertal children. Submitted for publication.
53. Jacobs I: The effects of thermal dehydration on performance of the Wingate Anaerobic Test. Int J Sports Med 1:21–24, 1980.
54. Kappagoda CT, Macartney FJ: Effect of environmental temperatures on oxygen consumption in infants with congenital disease of the heart. Br Heart J 38:1–4, 1976.
55. Kawahata A: Sex differences in sweating. In: Yoshimura H, Ogata K, Itoh S (eds.) Essential Problems in Climatic Physiology. Nankodo, Kyoto, 1960, pp. 169–184.
56. Keatinge WR: Body fat and cooling rates in relation to age. In: Folinsbee LJ, et al (eds.) Environmental Stress. Individual Human Adaptation. Academic Press, New York, 1978.
57. Keatinge WR, Sloan REG: Effect of swimming in cold water on body temperatures of children (abstract). J Physiol 226:55P–56P, 1972.
58. Kessler WR, Andersen DH: Heat prostration in fibrocystic disease of the pancreas and other conditions. Pediatrics 8:648–656, 1951.
59. Knochel JP: Dog days and siriasis: How to kill a football player. JAMA 233:513–515, 1975.
60. Kuno Y: Human Perspiration. Charles C. Thomas, Springfield, 1956.
61. Landing BH, Wells TR, Williamson ML: Studies on growth of eccrine sweat glands. In: Cheek BD (ed.) Human Growth. Body Composition, Cell Growth, Energy and Intelligence. Lea and Febiger, Philadelphia, 1968, pp. 382–395, Appendix 22.
62. Landing BH, Wells TR, Williamson ML: Anatomy of eccrine sweat glands in children with chronic renal insufficiency and other fatal chronic diseases. Am J Clin Pathol 54:15–21, 1970.
63. Leibowitz HW, Abernathy CN, Buskirk ER, et al: The effect of heat stress on reaction time to centrally and peripherally presented stimuli. Hum Factors 14:155–160, 1972.
64. Leithead CS, Lind AR: Heat Stress and Heat Disorders. F.A. Davis, Philadelphia, 1964.
65. MacDougall JD: Thermoregulatory problems encountered in ice hockey. Can J Appl Sport Sci 4:35–38, 1979.
66. Mackie JM: Physiological responses of twin children to exercise under conditions of heat stress. MSc Thesis, University of Waterloo, 1982.
67. Mahloudji M, Livingston KE: Familial and congenital simple anhidrosis. Am J Dis Child 113:477–479, 1967.

68. Malamud N, Haymaker W, Custer RP: Heat stroke. A clinico-pathologic study of 125 fatal cases. Milit Surg 99:397–449, 1946.
69. McConnell CM, Rostan S, Puyau FA: Heat dissipation in children with congenital heart disease. South Med J 63:837–841, 1970.
70. McCormick RJ: Heat tolerance of exercising lean and obese middle-aged men. EdD Dissertation. Pennsylvania State University, 1973.
71. McCormick RJ, Buskirk ER: Heat tolerance of exercising lean and obese middle-aged men. Fed Proc 33:441, 1974.
72. Mecklenburg RS, Loriaux L, Thompson RH, et al: Hypothalamic dysfunction in patients with anorexia nervosa. Medicine 53:147–159, 1974.
73. Murray JJ: Pediatric aspect of Nordic skiing. Pediatr Clin North Am 29:1423–1429, 1982.
74. Nielsen B: Physiology of thermoregulation during swimming. In: Eriksson B, Furberg B (eds.) Swimming Medicine IV. University Park Press, Baltimore, 1978, pp. 297–303.
75. Paterson DH, Cunningham DA, Penny DS, et al: Heart rate telemetry and estimated energy metabolism in minor league ice hockey. Can J Appl Sports Sci 2:71–75, 1977.
76. Pitts GC, Johnson RE, Consolazio FC: Work in the heat as affected by intake of water, salt and glucose. Am J Physiol 142:253–259, 1944.
77. Pugh LGCE, Corbett JL, Johnson RH: Rectal temperatures, weight losses, and sweat rates in marathon running. J Appl Physiol 23:347–352, 1967.
78. Puyau FA: Evaporative heat losses of infants with congenital heart disease. Am J Clin Nutr 22:1435–1443, 1969.
79. Randell WC: Quantitation and regional distribution of sweat gland in man. J Clin Invest 25:761–767, 1946.
80. Redfearn JA Jr: History of heat stroke in a football trainee (question). JAMA 208:699, 1969.
81. Robinson S, Wiley SL, Bondurant LG, Mamlin S Jr: Temperature regulation of men following heatstroke. Isr J Med Sci 12:786–795, 1976.
82. Rothstein A, Adolph EF, Wills JH: Voluntary dehydration. In: Adolph EF, et al (eds.) Physiology of Men in the Desert. Interscience Publishers, New York, 1947, pp. 254–270.
83. Rotstein A, Bar-Or O, Dlin R: Hemoglobin, hematocrit and calculated plasma volume changes induced by a short, supramaximal task. Int J Sports Med 4:230–233, 1982.
84. Saltin B: Aerobic and anaerobic work capacity after dehydration. J Appl Physiol 19:1114–1118, 1964.
85. Schickele E: Environment and fatal heat stroke. Milit Surg 100:235–256, 1947.
86. Shaker Y: Thirst fever, with a characteristic temperature pattern in infants in Kuwait. Br Med J 1:586–588, 1966.
87. Shapiro Y, Magazanik A, Udassin R, et al: Heat intolerance in former heatstroke patients. Ann Int Med 90:913–916, 1979.
88. Shattuck GC, Hilferty MM: Sunstroke and allied conditions in the United States. Am J Trop Med 12:223–245, 1932.
89. Sloan REG, Keatinge WR: Cooling rates of young people swimming in cold water. J Appl Physiol 35:371–375, 1973.
90. Sohar E, Shapiro Y: The physiological reactions of women and children

marching during heat (abstract). Proc Isr Physiol Pharmacol Soc 1:50, 1965.

91. Spickard A: Heat stroke in college football and suggestions for prevention. South Med J 61:791–796, 1968.
92. Sutton JR: Heat Illness. In: Strauss RH (ed.) Medicine in Sports and Exercise: Non-Traumatic Aspects. Franklin Institute Press, Philadelphia, in press, 1983.
93. Sutton JR, Bar-Or O: Editorial—Thermal illness in fun running. Am Heart J 100:778–781, 1980.
94. Sutton J, Coleman MJ, Millar AP, et al: The medical problems of mass participation in athletic competition. The "city-to-surf" race. Med J Aust 2:127–133, 1972.
95. Taj-Eldin S, Falaki N: Heat illness in infants and small children in desert climates. J Trop Med Hyg 71:100–104, 1968.
96. Tipton CM, Tcheng T-K: Iowa wrestling study. Weight loss in high school students. JAMA 214:1269–1274, 1970.
97. van Beaumont W: Thermoregulation in desert heat with respect to age (abstract). Physiologist 8:294, 1965.
98. Wagner JA, Robinson S, Tzankoff SP, Marino RP: Heat tolerance and acclimatization to work in the heat in relation to age. J Appl Physiol 33:616–622, 1972.
99. Wakeling A, Russell GFM: Disturbances in the regulation of body temperature in anorexia nervosa. Psychol Med 1:30–39, 1970.
100. Williams AJ, McKiernan J, Harris F: Heat prostration in children with cystic fibrosis (letter). Br Med J 2:297, 1976.
101. Zambraski EJ, Foster DT, Gross PM, Tipton CM: Iowa wrestling study: weight loss and urinary profiles of collegiate wrestlers. Med Sci Sports 8:105–108, 1976.

Appendix I

"Norms"

Introduction

The inclusion of "norms" of physical working capacity in this book might imply that: 1) we know what is normal in children's performance; 2) the norms of one population apply to others, irrespective of habitual activity, body size and composition, nutritional status, health status, climate, altitude, ethnic origin, and social traditions; 3) the samples chosen to establish these norms are representative of their own populations; and 4) the variables follow a known distribution (e.g., Gaussian) such that our choice of limits of normality is correct.

None of these assumptions is entirely valid. We do not have criteria for normality in any of the components of working capacity. Furthermore, if a range of "normal" values were to exist, it would vary among populations, depending on their level of activity (which in itself cannot be categorized into "normal" or "abnormal"), state of health and nutrition, prevailing climate, altitude, and ethnic origin.

Norms of one ethnic group may not apply to others. This author observed Burmese children perform physical fitness tests constructed in Europe and North America. While most rural Burmese girls performed above the "Western" 95th percentile of flexed-arm hang (testing elbow flexor and shoulder girdle strength and endurance), many of them could not complete a single sit-up (testing abdominal wall strength and endurance). My Burmese host attributed these scores to specific traditions that let the rural girl perform heavy lifting tasks but forbid her from doing those activities that require dynamic abdominal muscle contraction.

Few of the samples included in the following graphs are representative of their populations. Most comprise subjects who happened to be tested in various laboratories for specific research or clinical purposes.

These graphs reflect fitness of selected groups of well-nourished, nonathletic, European, North American or Israeli children and adolescents, most of Caucasian origin, who had no overt manifestations of disease. They should be regarded merely as guidelines for those who cannot establish norms of their own.

The choice of chronologic age as the independent variable is not ideal. *Biologic* age would have been more appropriate. From a practical point of view, however, biologic age is seldom available to the clinician, whereas chronologic age can be easily ascertained. Whenever available, data are also presented per kilogram body weight which, to some extent, corrects for difference in growth.

References

1. Andersen KL, Seliger V, Rutenfranz J, Mocellin R: Physical performance capacity of children in Norway. Part I. Population parameters in a rural inland community with regard to maximal aerobic power. Europ J Appl Physiol 33:177–195, 1974.
2. Cumming GR: Exercise studies in clinical pediatric cardiology. In: Lavalée H, Shephard RJ (eds). Frontiers of Activity and Child Health. Pélican, Quebec, 1977, pp. 17–45.
3. Cumming GR, Everatt D, Hastman L: Bruce treadmill test in children: normal values in a clinic population. Am J Cardiol 4:69–75, 1978.
4. Roche PD: The development of norms for run-walk tests for children aged 7 to 17. J Canad Ass Health Phys Ed Recr J 46:6–13, 1980.
5. Rutenfranz J, Berndt I, Frost H, et al: Physical performance capacity determined as $\dot{W}_{170}$ in youth. In: Bar-Or O (ed.) Pediatric Work Physiology. Wingate Institute, Natanya, 1973, pp. 245–249.
6. Rutenfranz J, Mocellin R: Untersuchungen über die körpliche Leistungsfähigkeit gesunder und kranker Heranwachsender. I. Bezugsgrössen und Normwerte. Z. Kinderheilkd 103:109–132, 1968.
7. Seliger V, Bartunek Z: Mean values of various indices of physical fitness in the investigation of Czechoslovak population aged 12–55 years. CSTV Praha (CSSR), 1976.
8. Wirth A, Trager E, Scheele K, et al: Cardiopulmonary adjustment and metabolic response to maximal and submaximal physical exercise of boys and girls at different stages of maturity. Eur J Appl Physiol 39:229–240, 1978.

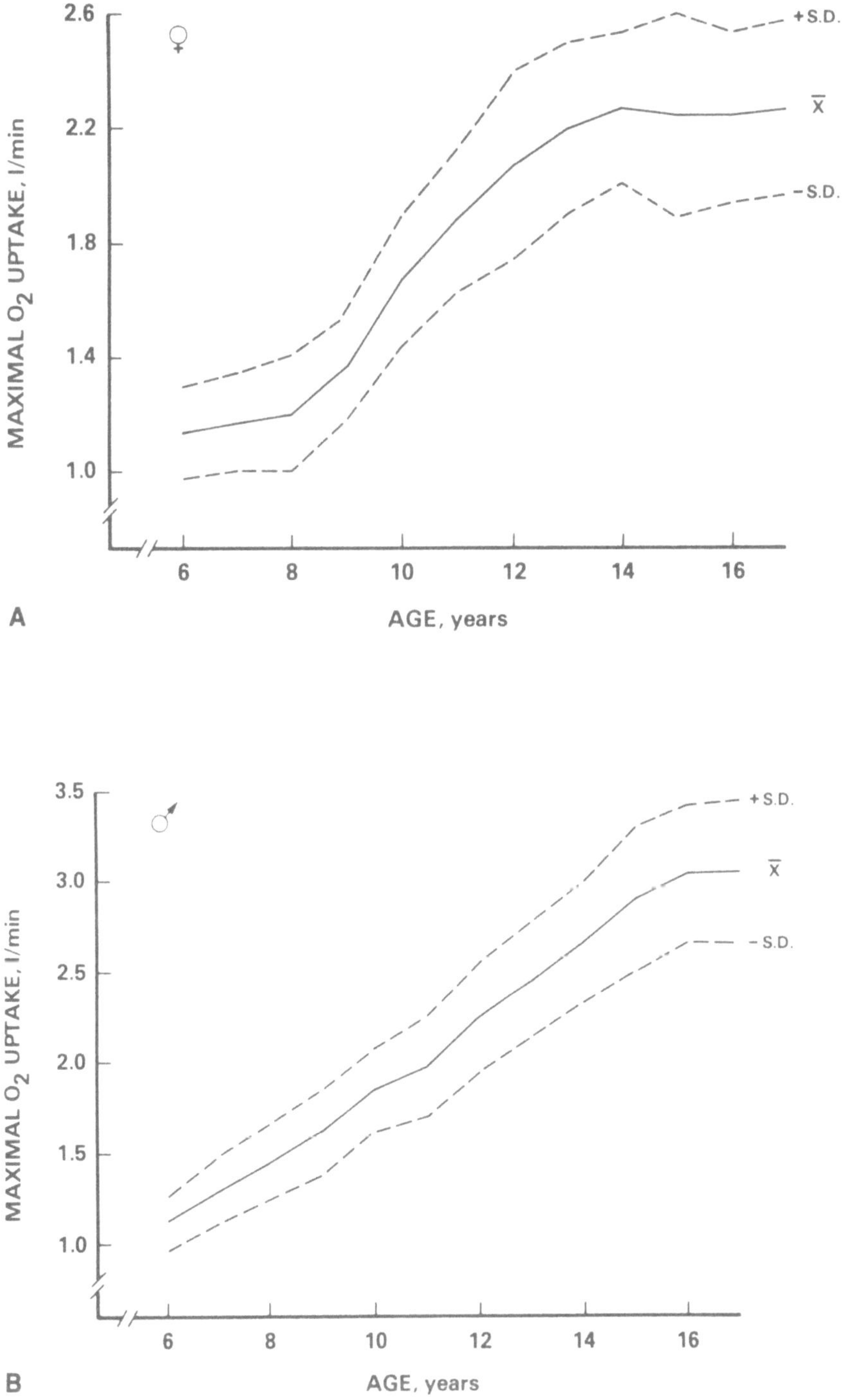

Figure I.1. Maximal O_2 uptake in children and adolescents. One hundred and seventy nine girls (**A**) and 178 boys (**B**) performed an all-out progressive-continuous protocol on a treadmill. Subjects were healthy nonathletes. Data from the author's laboratory.

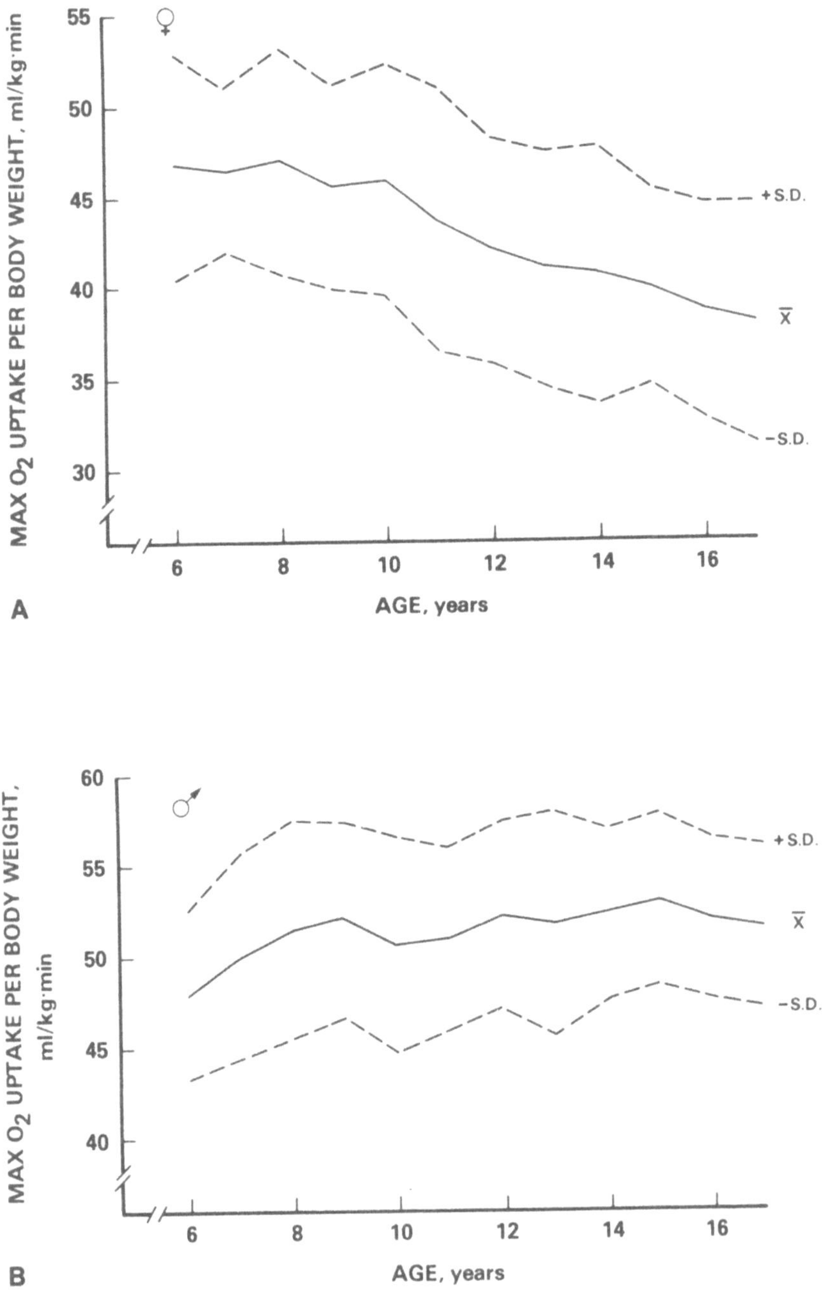

Figure I.2. Maximal O_2 uptake per kilogram body weight in children and adolescents. Subjects and protocols as in Figure I.1.

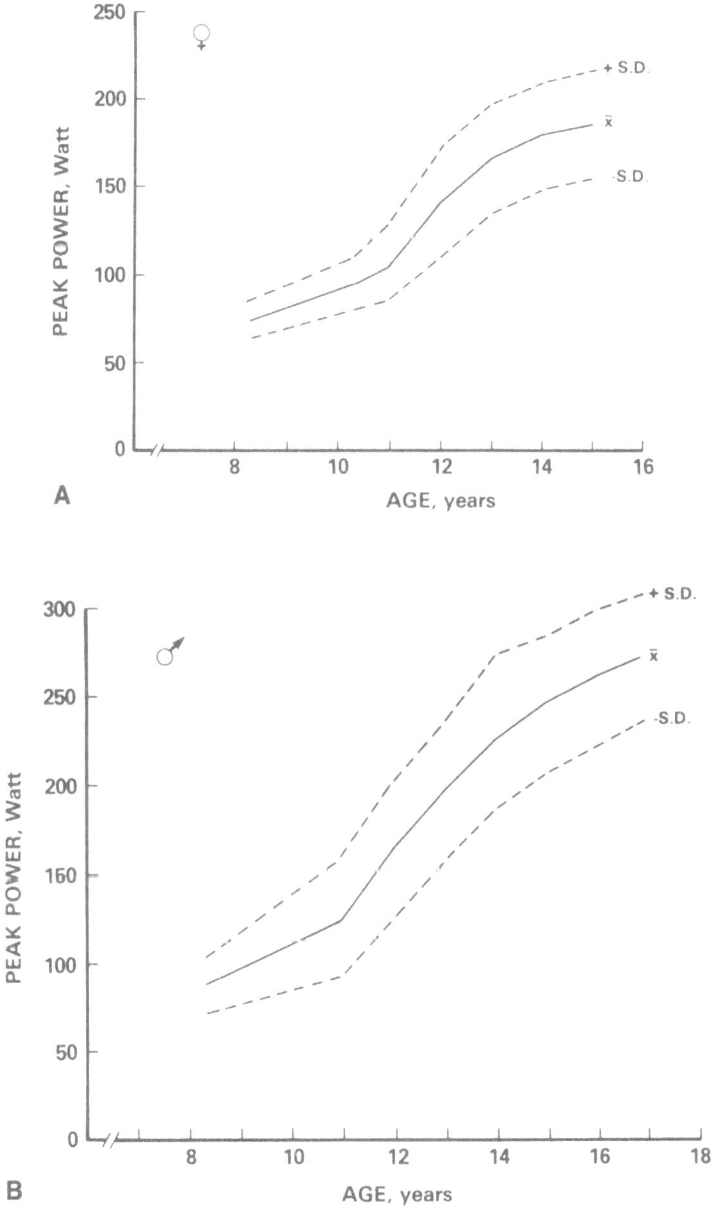

Figure I.3. Peak mechanical power in children and adolescents. The highest mechanical power output achieved during a progressive upright cycle ergometer test by healthy girls (**A**) and boys (**B**). Based on data by Andersen et al,[1] Seliger and Bartunek,[7] Wirth et al.[8]

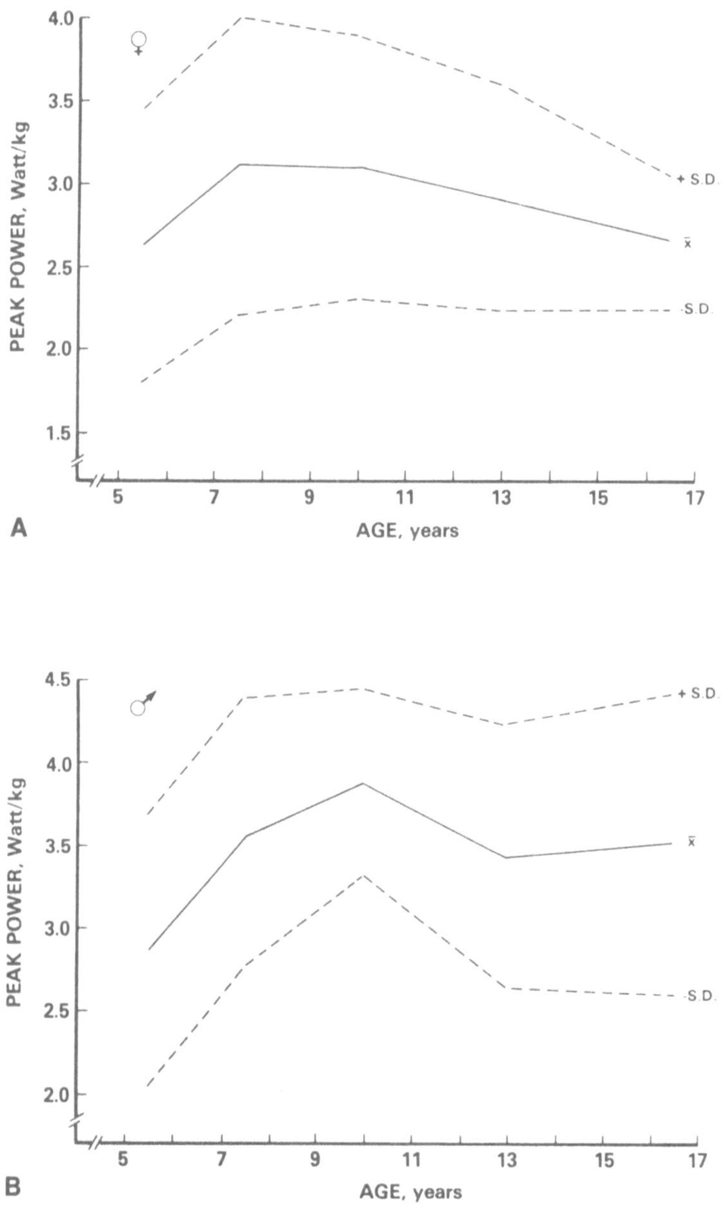

Figure I.4. Peak mechanical power per kilogram body weight in children and adolescents. The highest mechanical power output per kilogram body weight achieved during a progressive upright cycle ergometer test. Subjects were girls (**A**) and boys (**B**) of an outpatient population who had an "innocent" murmur but no organic heart disease. Data by Cumming.[2]

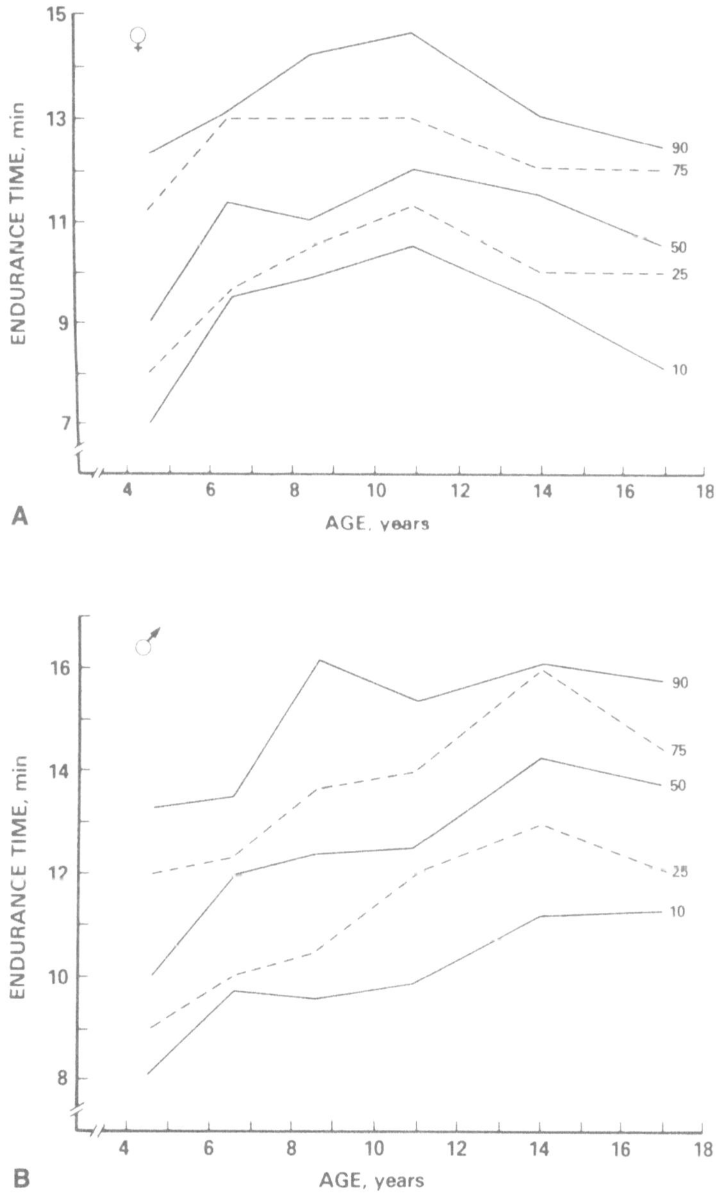

Figure I.5. Endurance time in the Bruce treadmill test in children and adolescents. Lines represent percentiles, which are based on performance of 160 girls (**A**) and 167 boys (**B**) who had "innocent" murmur, but no organic disease. Data by Cumming et al.[3]

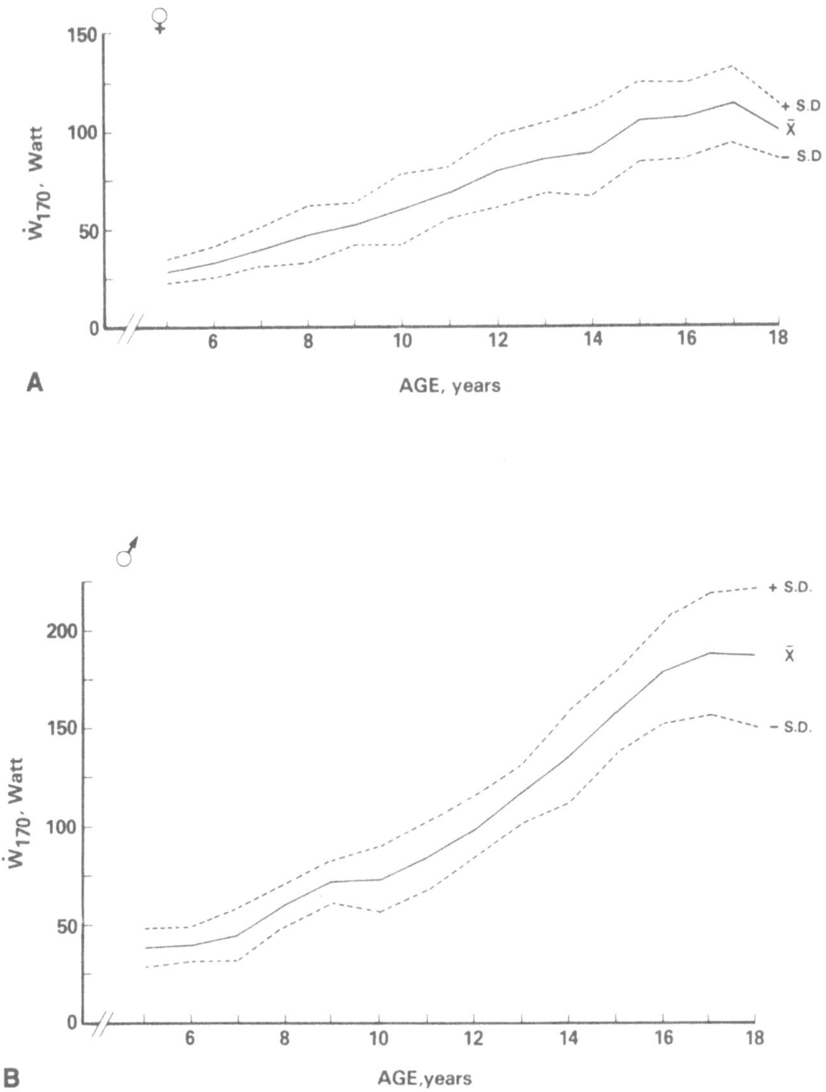

Figure I.6. Mechanical power at heart rate of 170 beats/min ($\dot{W}_{170}$) in 727 girls (**A**) and boys (**B**). Data by Rutenfranz et al.[5]

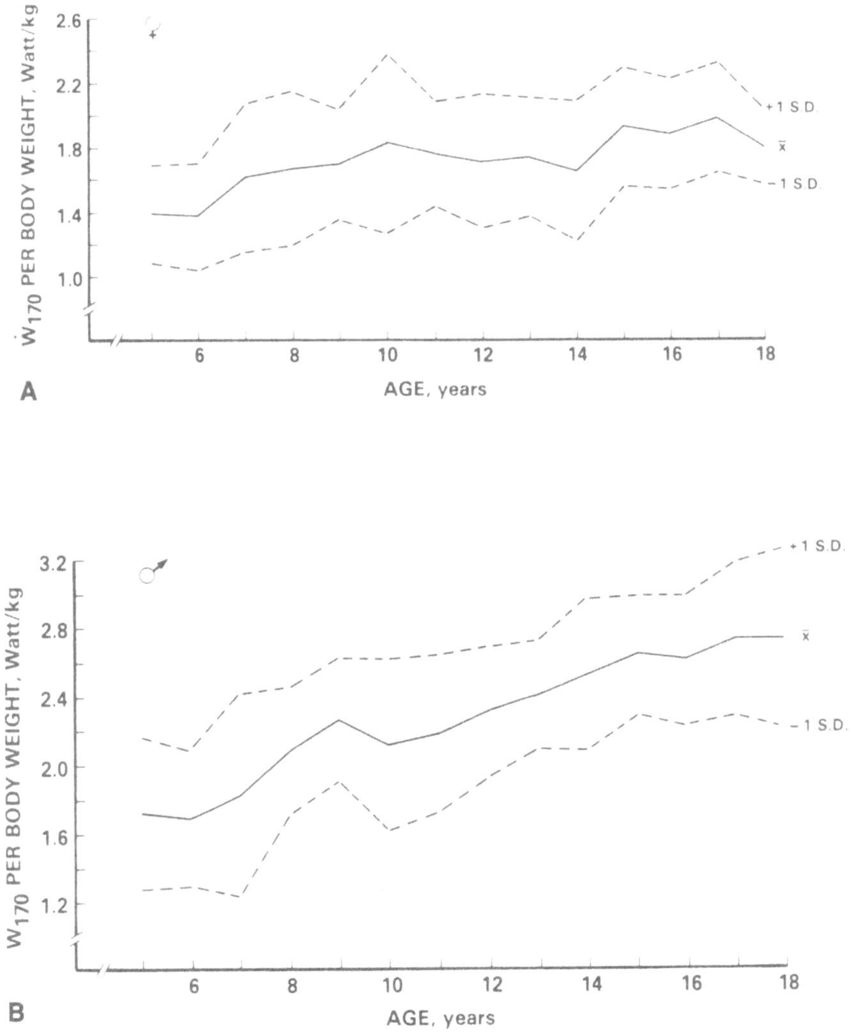

Figure I.7. Mechanical power per kilogram body weight at heart rate of 170 beats/min ($\dot{W}_{170}$) in 727 girls (**A**) and boys (**B**). Data by Rutenfranz et al.[5]

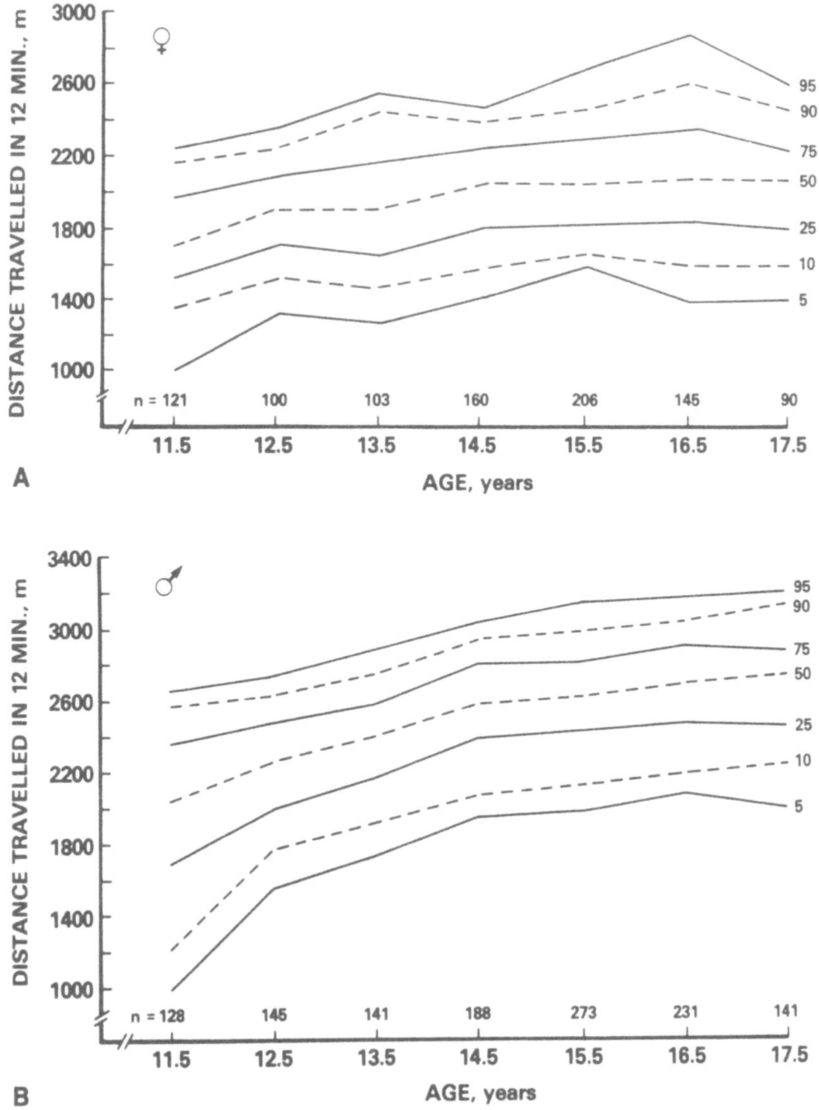

Figure I.8. Distance traveled by girls (**A**) and boys (**B**) during a 12-min run-walk test. Lines represent percentiles. Subjects were school children randomly selected from 43 schools in Halton County, Ontario, Canada. Based on data by P. Roche.[4]

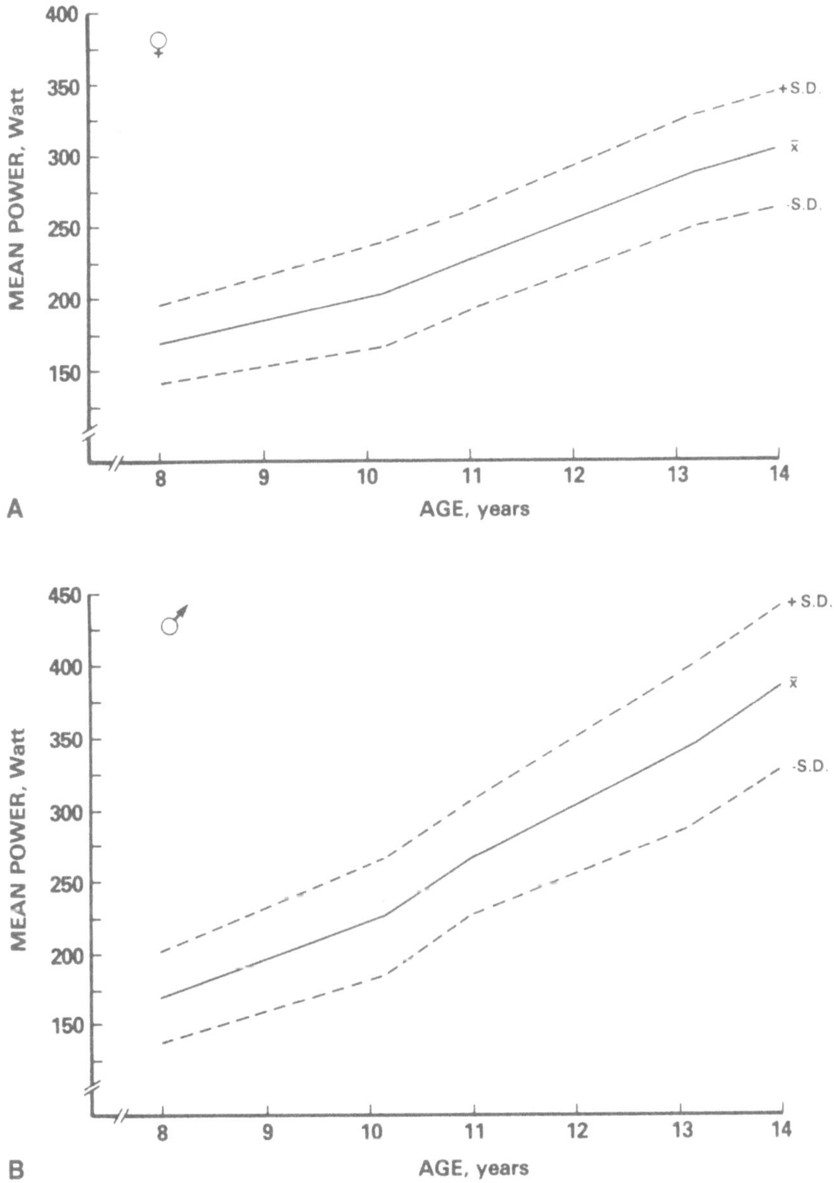

Figure I.9. Anaerobic capacity in children and adolescents. Mean power output during a supramaximal 30-sec cycle ergometer test (Wingate anaerobic test). Subjects were 144 girls (**A**) and 145 boys (**B**), all healthy nonathletes. Based on data from the author's laboratory in Israel and from the Exercise Physiology Laboratory, McMaster University.

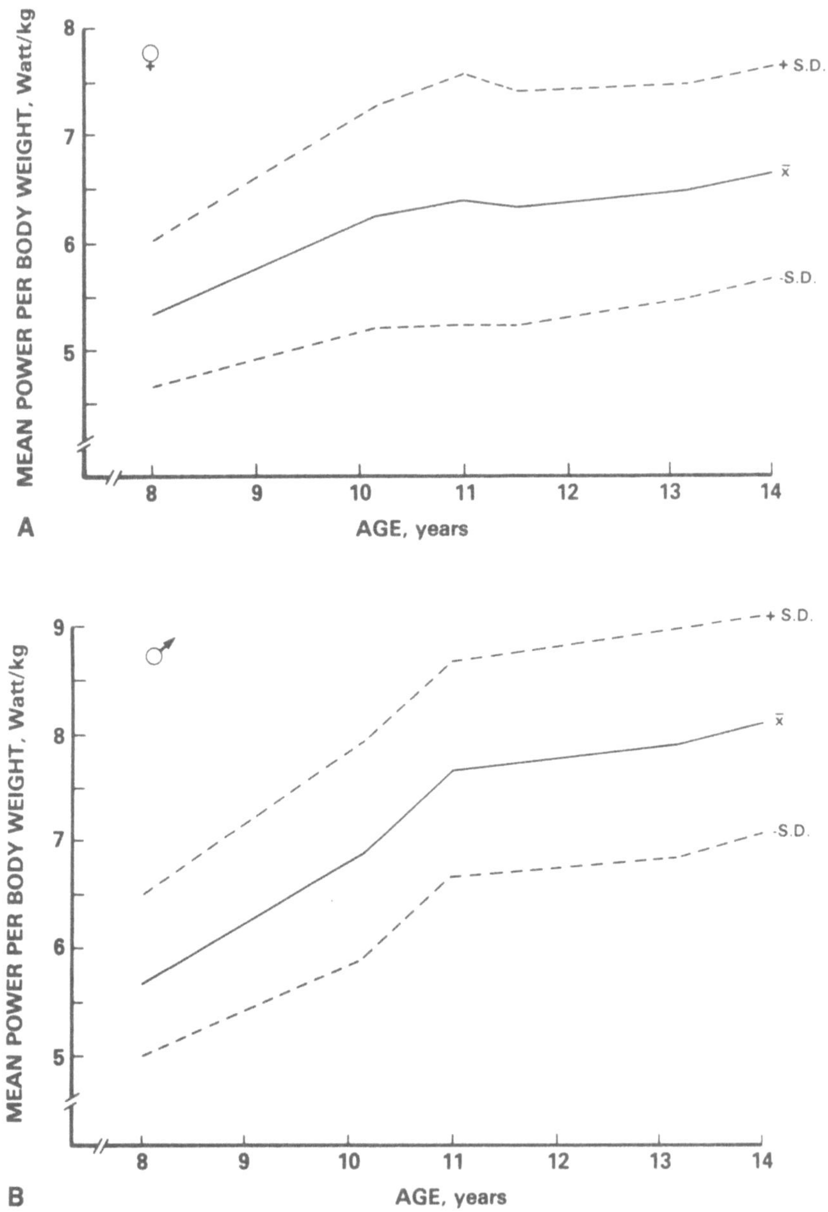

Figure I.10. Anaerobic capacity per kilogram body weight in children and adolescent girls (**A**) and boys (**B**). Mean power output during a supramaximal 30-sec cycle ergometer test (Wingate anaerobic test). See Fig. I.9 for description of subjects and source of data.

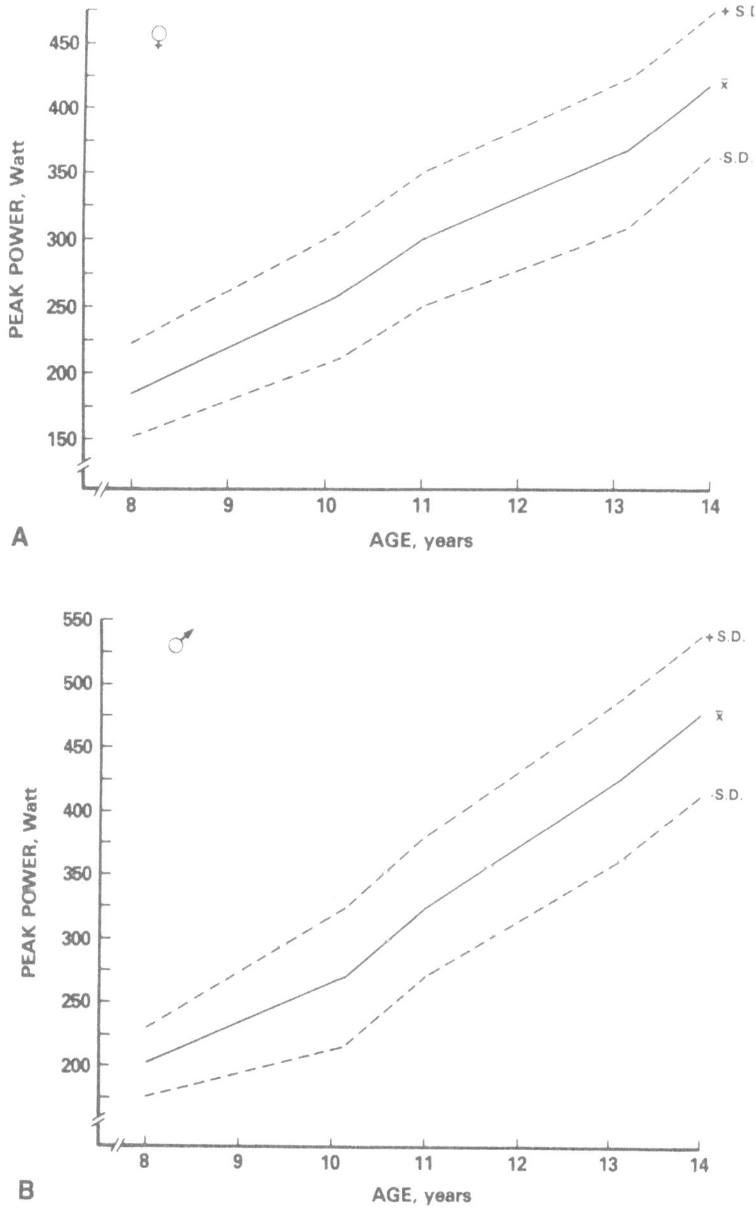

Figure I.11. Peak anaerobic power in children and adolescent girls (**A**) and boys (**B**). The highest power output at any 5-sec period during an all-out 30-sec cycle ergometer test (Wingate anaerobic test). See Fig. I.9 for description of the subjects and source of data.

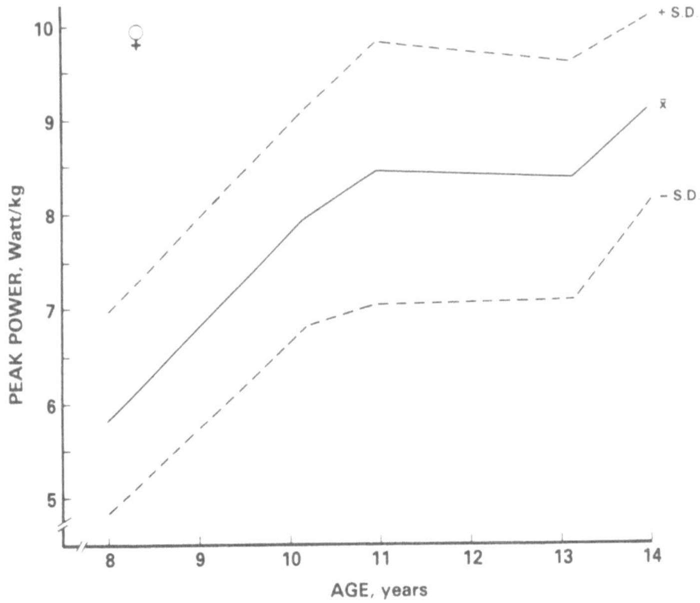

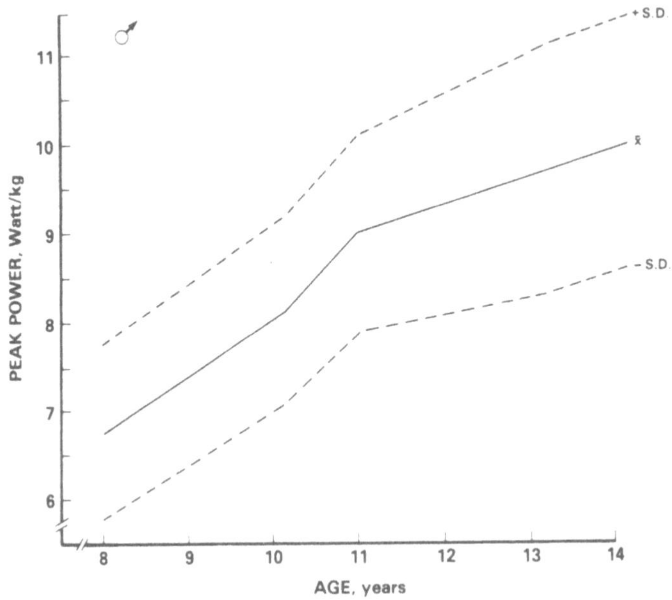

Figure I.12. Peak anaerobic power per kilogram body weight in children and adolescent girls (**A**) and boys (**B**). The highest output at any 5-sec period during an all-out 30-sec cycle ergometer test (Wingate anaerobic test). Values are normalized for body weight. See Fig. I.9 for description of the subjects and source of data.

Appendix II

Procedures for Exercise Testing in Children

An exercise test comprises three elements: the ergometer, the exercise protocol, and the physiologic or perceptual measurements.

Choice of an Ergometer

The two most commonly used ergometers in an exercise laboratory are the cycle ergometer and the motor-driven treadmill. A step can also be used, especially in a physician's office. For some neuromuscular disabilities one must resort to testing of the arms, using specially designed arm cranking ergometers or modified cycles. A comparison of the characteristics of the four ergometers is presented in Table II.1.

In spite of its greater cost and the need for safety measures and for more technicians, the treadmill is the ergometer of choice to determine maximal exercise performance of children, especially those 7 years old or younger. Children as young as three years can learn to walk and run on a treadmill.[44]

We often see a young child who cannot keep pedaling on a cycle ergometer even though his heart rate is only 160–170 beats/min. The same child can reach a heart rate of 200–210 beats/min on the treadmill, and a maximal O_2 uptake that is 20–30% higher than the peak uptake obtained on a cycle ergometer. The apparent cause of such a discrepancy is the relatively undeveloped mass of the knee extensors (which take the brunt of the load during cycling) in the young child. This results in local fatigue and premature termination of the test. In contrast, during walking or running a greater muscle mass is activated, resulting in a higher cardiorespiratory and metabolic contribution. Some children, especially the very young and the mentally retarded, cannot conform to the cadence of a metronome, which is needed for mechanically braked cycle ergometers and for step testing. Furthermore, the attention span of such children is low and they cannot maintain the required pedaling rate for the duration of the test. In contrast, a treadmill forces them to maintain a certain speed, and they can easily compensate for a momentary change of pace.

Many laboratories still prefer the cycle ergometer because of its lower cost, greater portability, and safety. In addition, certain procedures such

Table II.1. Characteristics of Ergometers in Use for Pediatric Exercise Testing

Characteristics	*Cycle Ergometer*	*Treadmill*	*Step*	*Arm Ergometer*
Cost	Low to medium	High	Low	Medium
Portability	High	No	High	High
No. of staff needed	1 to 2	2 to 3	1 to 2	1 to 2
Noise level	Low to medium	Medium to high	Low	Low to medium
Special safety measures	None	Harness, padding	None	None
Skill required by patient	Too hard for < 5 yr or mentally retarded	Little skill	Some skill	Too hard for < 5 yr or mentally retarded
Muscle mass involved	Small	Large	Large	Small
Maximal O_2 uptake	Under-estimated	Achieved	Achieved	Much under-estimated
Determination of mechanical power	Accurate	Estimated	Fairly accurate	Accurate
Feasibility of obtaining physiologic measurements	Easy	Less easy	Less easy	Easy
Anaerobic testing	Suitable	Not suitable	Not suitable	Suitable

as sphygmomanometry, rebreathing, or echocardiography are easier to perform on a sitting child than on a walking, running, or stepping one.

If a cycle ergometer is chosen, models used for adults fit most children 8 years old or older. For younger children, however, one needs special pediatric models or a modified existing cycle with the seat height and the length of the pedal crank reduced and the handlebar lengthened. The optimal (i.e., inducing the lowest O_2 cost of pedaling) pedal length for 6-year-old children is 13 cm, as compared with 15 cm for 8- and 10-year-old children[38] and 20 cm for adults. When seat height is optimal, the angle at the knee joint during extension is 15°.[38] A useful indication that the seat is too high is a lateral tilt of the pelvis, as seen from behind.

The power load in the electronically braked cycle ergometer is independent of the pedaling rate at 50–70 rev/min. In mechanically braked ergometers, the rate must be kept constant. Among 6- to 10-year-old children, 50 rev/min yield higher mechanical efficiency than 30 or 70 rev/min.[38] Recommended rates are 50–60 rev/min.

On the treadmill the child should be tested only after having practiced and gained confidence in walking, running, stepping on, and stepping off. Economy of gait depends on stride length, the most economical being that selected by the child himself: children should therefore be allowed to use their preferred stride length. A body harness is used in some laboratories as a safeguard. In our experience, this is not needed as long as an investigator stands close to the child. We do recommend, however, padding of the rear of the treadmill to prevent injury in case of a fall.

The Exercise Protocol

A variety of exercise protocols is available for children. The choice of a protocol depends primarily on the specific question(s) to be answered but also on the abilities and limitations of the particular patient.

Prototypes of Exercise Tests

Protocols are schematically presented in Fig. II.1. These can be divided into "all-out" ones, in which the patient is expected to reach his maximal aerobic power ("aerobic" peak in the diagram), and to submaximal ones, in which the peak is not reached. The latter type is further subdivided into progressive and single-stage protocols. The test depicted in Fig. II.1G is designed to determine anaerobic characeristics of a muscle group. Here the child performs a supramaximal task, which can be sustained for not more than 30–45 sec.

All the protocols, excluding the anaerobic one, are suitable for use with a cycle ergometer, arm ergometer, treadmill, or step. The anaerobic protocol is designed for leg cycling or arm cranking only. Even though running can be done at supramaximal levels, there are neither reliable nor safe protocols for anaerobic testing by treadmill.

For direct determination of maximal aerobic power the most commonly used protocol is that shown in Fig. II.1A. Resistance (cycle, arm ergometer), inclination and speed (treadmill), or height and frequency of ascents (step) are increased every 1–3 min, without interruption of the test, until the child can no longer maintain the activity. The all-out interrupted protocol (Fig. II.1B) is used whenever measurements must be taken *after* the end of each stage (e.g., blood lactate, echocardiogram) or when the investigator needs time to decide whether to continue with a higher load or to terminate the test.

Submaximal progressive protocols in which maximal aerobic power is predicted rather than directly measured (Fig. II.1C,D) are resorted to

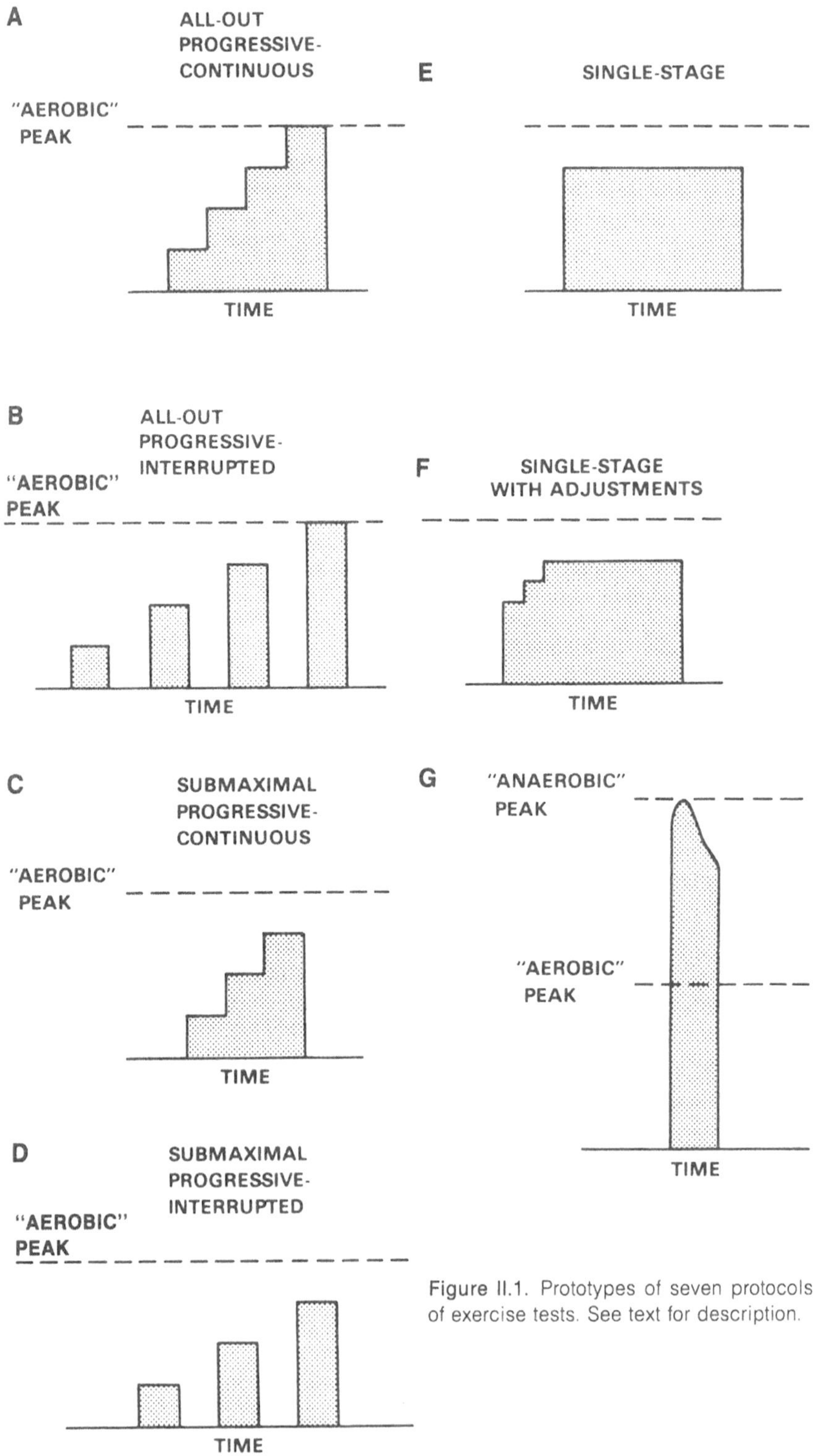

Figure II.1. Prototypes of seven protocols of exercise tests. See text for description.

when the investigator is interested only in response to submaximal exercise or is reluctant to exhaust the child by a maximal test. This is done, for example, in conjunction with heart catheterization studies.

Single-stage protocols (Fig. II.1E,F) will be chosen when the purpose of testing is not physical working capacity, but a specific pathophysiologic response to an exercise provocation—for example, diagnosis of exercise-induced bronchoconstriction or of growth hormone deficiency. The duration of a single-stage test varies from 6 to 20 min and the intensity is submaximal. Occasionally, the exact intensity cannot be determined prior to testing. The investigator then starts the test at *approximately* the required intensity and makes adjustments during the initial 1–2 min based, for example, on heart rate response. This protocol is termed "single-stage with adjustments" (see Fig. II.1F).

Examples of Exercise Protocols

The Bruce All-Out Progressive Continuous Treadmill Test (Table II.2). The end point in this test is the inability of the child to carry on walking or running in spite of verbal encouragement. Performance is graded by the number of minutes completed, or according to the calculated O_2 cost at the highest stage. The test was introduced for screening adult cardiac patients[12] and later adopted for use with children.[16] Reproducibility among children is high and the test is feasible for 4- to 18-year-old girls and boys. This test is not inherently better than other all-out treadmill tests.[46,53] Its basic drawback is that the initial load may be too strenuous for markedly unfit children. Another disadvantage is that highly fit children may exercise 18 min or more to determine their maximal aerobic power, by which time local fatigue and thermal load may interfere with their performance. In spite of the above drawbacks, the Bruce test is recommended in a clinical setup for the assessment of maximal aerobic power. Norms are available, based on the performance

Table II.2. The Bruce Treadmill Protocol

Stage	*Speed*		*Grade* (%)	*Duration* (*min*)
	km/hr	*miles/hr*		
1	2.7	1.7	10	3
2	4.0	2.5	12	3
3	5.5	3.4	14	3
4	6.8	4.2	16	3
5	8.0	5.0	18	3
6	8.8	5.5	20	3
7	9.7	6.0	22	3

of healthy children in an outpatient clinic[16] (See Fig. I.5 in Appendix I) and at a school.[18]

The McMaster All-Out Progressive Continuous Cycling Test (Table II.3). As in the Bruce protocol, the child is verbally encouraged to keep exercising until he can no longer adhere to the required pedaling rate (50 rev/min). Performance is graded by the peak mechanical power or the directly measured maximal O_2 uptake. When the child cannot complete 2 min of the final load, peak power is prorated, based on the period that he did manage to complete.

For example:	Penultimate stage	= 100 watt
	Final stage	= 125 watt
	Time completed in final stage	= 60 sec
	"Peak power"	= 112.5 watt

The protocol is constructed according to body height such that the total exercising time will range between 8 and 12 min for most children. The initial load and increments must sometimes be reduced to suit children with marked disability as in muscle dystrophy or cerebral palsy.

Other all-out progressive continuous cycling protocols are available.[24,25,31,55] They are as suitable as the McMaster protocol for obtaining peak power and maximal O_2 uptake. However, they are less suitable for obtaining submaximal data, either because the duration of each stage is too short[24] or because the progression of loads is not uniform.[25,31,55] A variation of the all-out progressive continuous protocol is the "pulse-conducted exercise test." In this test the mechanical power continuously, or almost continuously, increases to induce a continuous rise of heart rate by 5 beats/min each minute.[45,58] The increments in power, therefore, are determined by the cardiovascular response to exercise, being slower for less fit children. Although such flexibility is an advantage, the system requires a computer and other electronic devices, which are not readily available.

Table II.3. The McMaster All-Out Progressive Continuous Cycling Protocol, by Body-Height Groups

Body Height (cm)	*Initial Load (Watts)*	*Increments (Watts)*	*Duration of Each Load (min)*
≤119.9	12.5	12.5	2
120–139.9	12.5	25	2
140–159.9	25	25	2
≥160	25	♀ 25 ♂ 50	2

Table II.4. The McMaster All-Out Progressive Continuous Arm Cranking Protocol, by Body-Height Groups

Body Height (cm)	*Initial Load (Watts)*	*Increments (Watts)*	*Duration of Each Load (min)*
≤119.9	8	8	2
120–139.9	8	16.5	2
140–155.9	16.5	16.5	2
≥160	16.5	♀ 16.5 ♂ 33	2

The McMaster All-Out Progressive Continuous Arm Test (Table II.4). This procedure is similar to that described above with the exception that the child uses his arms for cranking while sitting on a wheelchair or a regular chair. The axle of the pedals is kept at shoulder height. The child should be seated comfortably, supported by the back of the seat or by a pillow. At the farthest point of the pedal the arm should be fully extended without forward bending of the trunk. Strapping in is needed to stabilize some patients (e.g., those with high paralysis) and to minimize trunk motion. As in the cycling test, loads must be tailored to the disability of the child. Some patients require very small increments, while others, such as spastic children with cerebral palsy, need a very slow rate of cranking.

The Cumming All-Out Progressive Intermittent Cycling Test (Table II.5). This protocol is used with cardiac catheterization, the patient cycling in the supine position.[15] The wide range of power loads recommended at each stage reflects the need to individualize the test according to age, disease, and functional ability. Other protocols of interrupted exercise for the catheterized child are available.[40]

Table II.5. The Cumming All-Out Progressive Intermittent Protocol

Stage	*Mechanical Power (Watts/kg body weight)*	*Duration (min)*
1	0.65–1.14	3
Rest		3
2	1.31–1.96	3
Rest	—	5–10
3	2.12–4.08	To exhaustion

Table II.6. The Adams Submaximal Progressive Continuous Cycling Protocol, by Body-Weight Groups*

Body Weight (kg)	*1st Stage Power (Watts)*	*2nd Stage Power (Watts)*	*3rd Stage Power (Watts)*
<30	16.5	33	50
30–39.9	16.5	50	83
40–59.9	16.5	50	100
⩾60	16.5	83	133

* Each stage lasts 6 min.

The Adams Submaximal Progressive Continuous Cycling Test (Table II.6). This protocol comprises three stages which, for most children, are submaximal. Each stage lasts 6 min, at the end of which heart rate is determined. Performance is graded according to the mechanical power that the child produces at a heart rate of 170 beats/min ($\dot{W}_{170}$). This value is reached by interpolation or extrapolation of the individual regression line of heart rate over power. The long duration of each stage allows for a steady state to be established. Other protocols are available for assessment of $\dot{W}_{170}$.[9,49]

The Hanne Submaximal Progressive Intermittent Step Test (Table II.7). This protocol was introduced in 1971[27] based on a feasibility study with 7- to 12-year-old girls and boys. Stepping is performed up and down a 30-cm step to a cadence given by a metronome. Each stage lasts 5 min with a 5–10 min rest in between. Grading is done according to heart rate response at a given power load, from which $\dot{W}_{170}$ can be calculated.

Calculation of mechanical work and power is done as follows: work per one ascent equals:

$$\text{Work (Joule)} = \text{body weight (kg)} \times \text{step height (m)} \times 9.80 \qquad (1)$$

Table II.7. The Hanne Submaximal Progressive Intermittent Stepping Test

Stage	*Step Height (cm)*	*Rate (ascents/min)*	*Duration (min)*	*Approx. Power (Watts/kg)*
1	30	15	5	1.0
2	30	22.5	5	1.5
3	30	30	5	2.0

Work per one descent is taken as one-third of the above. Thus the total work of one ascent and descent in a 30-cm step equals:

$$\text{Work (Joule)} = \text{body weight (kg)} \times 3.92 \tag{2}$$

Assuming N ascents and descents per minute, the mechanical power output is:

$$\text{Power (Watt)} = \frac{\text{body weight (kg)} \times 3.92 \times N}{60} \tag{3}$$

Another step test feasible for children is the Canadian Home Fitness Test. Stepping is done on a double 20.3-cm (8-in.) step in 19 ascents/min. Scoring is done according to heart rate, measured during recovery. Special records are available, which include instructions and music at the required cadence.[5]

The Wingate Anaerobic Cycling Test (Fig. II.2, Table II.8). While the tests described above are meant to measure or assess maximal aerobic power, the Wingate test is constructed to assess the ability of a muscle group to perform short supramaximal exercise.[8] In such a task the limiting factor is not the O_2 transport system but, rather, the ability to convert anaerobically chemical energy to mechanical energy.

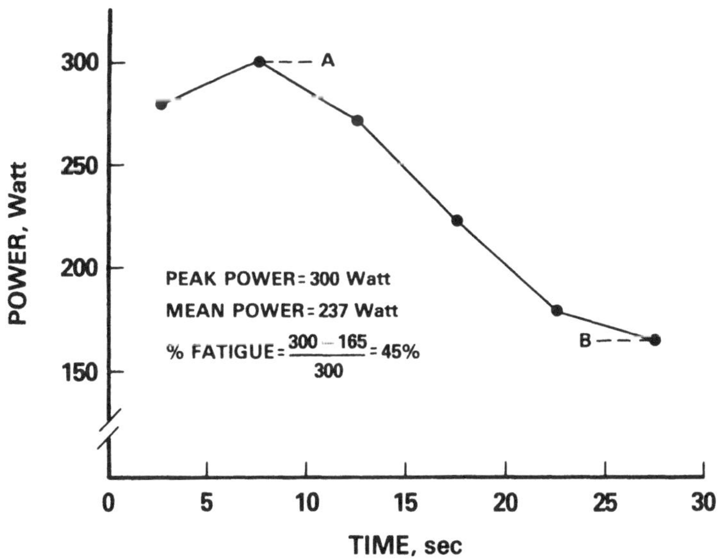

Figure II.2. The Wingate anaerobic test. Leg performance of an 11-year-old boy. The plotted power values are means for 5-sec periods. For calculation of the three indices see text.

Table II.8. Optimal Resistance for the Wingate Anaerobic Test, by Body-Weight Groups, Using the Monark Cycle Ergometer*

Body Weight (kg)	*Resistance in kp*	
	Legs	*Arms*
20–24.9	1.75	1.25
25–29.9	2.13	1.50
30–34.9	2.50	1.75
35–39.9	2.83	2.00
40–44.9	3.25	2.25
45–49.9	3.63	2.50
50–54.9	4.00	2.75
55–59.9	4.50	3.00
60–64.9	5.00	3.35
65–69.9	5.50	3.70

* Based on data from the author's laboratory at the Wingate Institute, Israel.

The test can be performed by the legs or by the arms, using an appropriate ergometer. It lasts 30 sec, during which the patient pedals at *maximal* rate against a high constant resistance. The power output is therefore a function of pedaling velocity and the mechanical work during 30 sec is a function of the total number of pedal revolutions. These can be counted by a mechanically or magnetically triggered counter.

Performance is expressed by three indices, as shown in Fig. II.2:

Peak power at any 5-sec period
Mean power (or total work) during 30 sec
Percent fatigue, which is defined as

$$\frac{A - B}{A} \times 100$$

assuming A is peak power and B the lowest power in any 5-sec period.

Ergometers suitable for this test must have a constant resistance mode. This is available in all mechanically braked ergometers and in some electrically braked ones. The following calculation of work and power is given for the commonly available Monark ergometer, in which the distance traveled by the flywheel circumference is 6 m for one pedal revolution. In our own studies we have been using the Fleisch-Metabo ergometer (Switzerland), in which the respective distance is 10 m. Work done in 1 pedal revolution is:

$$\text{Work (kpm)} = \text{resistance (kp)} \times 6 \quad (1)$$

Total work in 30 sec is therefore:

$$\text{Work (kpm)} = \text{resistance (kp)} \times 6 \times \text{total rev} \quad (2)$$

Mean power for the whole test is:

$$\text{Mean power (kpm/min)} = \text{resistance (kp)} \times 12 \times \text{total rev} \quad (3)$$

or:

$$\text{Mean power (Watt)} = \text{resistance (kp)} \times 1.96 \times \text{total rev} \quad (4)$$

Power at each 5-sec period is calculated to determine peak power and percent fatigue:

$$\text{Power (Watt)} = \text{resistance (kp)} \times 11.76 \times \text{rev in 5 sec} \quad (5)$$

The resistance setting that yields the highest mean power (or total work) for various body-weight groups is shown in Table II.8. Values are presented for the Monark ergometer. Optimal resistance for other ergometers can be calculated accordingly, based on their geometric characteristics.

The Wingate anaerobic test is feasible for use with healthy and disabled children, as young as 6 years. Warming up is recommended prior to the test[30] (for details see Chapter 1 section entitled Warm-up Effect). Unlike the aerobic tests, which should be done in a neutral environment, neither air temperature nor humidity affect performance in the Wingate test.[20] Norms for peak and mean anaerobic power are given in Appendix I (Figs. I.9–I.12).

Another test of anaerobic performance that is suitable for children is the Margaria step-running test.[19] Here the child runs up 2–4 steps at maximal speed. Power is calculated based on body weight, vertical distance, and the time needed for covering this distance. Although the test yields higher power values than peak power of the Wingate test, it requires some skill and cannot be applied to disabled children. Nor can arm performance be measured separately. Due to its short duration, performance in the Margaria test apparently does not reflect the rate of glycogen utilization.

Measurements Taken During Exercise Tests

Numerous variables can be measured during an exercise test. These range from the simple determination of peak power output, which does not require any instruments apart from an ergometer, to the measurement of intracardiac pressures or myocardial perfusion imaging, which require catheterization and sophisticated equipment. Methodologic details of such measurements are available in books on clinical exercise testing, such as Ellestad,[21] Godfrey,[24] and Jones and Campbell.[37] In this section we shall focus on those variables that can be measured in a clinic by noninvasive procedures.

Heart Rate

This variable is conveniently determined from ECG strips, by counting the number of R-R intervals during a certain time period or by measuring the time needed for a given number of R-R intervals. Special rulers are available for the latter method. Because of sinus dysrhythmia, which appears particularly at rest and during low exercise intensities, one should include at least six R-R intervals in each measurement. When using the ECG it is important to periodically calibrate the paper speed of the electrocardiograph.

Cardiotachometers are available that electronically average the R-R interval, obviating the need for paper strips. Although convenient, those instruments must be carefully calibrated. Electrocardiographic data can be taped and heart rate analyzed by computer.[3]

Heart rate is highly sensitive to environmental changes and other factors, as summarized in Chapter 1, Table 1.3. The utmost care must be taken to standardize room temperature and humidity, to reduce emotional stress, and to ensure that the child is not fatigued by prior exercise and that he is well hydrated. The reader is strongly advised to study Chapter 1, section entitled Heart Rate and Exercise, for further details.

Ventilation

Ventilatory function is often monitored to evaluate the limiting factors in performance in a dyspneic child or in a child with known pulmonary or cardiac disease (see, for example, Chapter 3, section on the child with cystic fibrosis. Ventilation is also monitored for the measurement of O_2 uptake. Expired air can be collected by use of a mouthpiece and a one-way valve system, into Douglas bags, or into a Tissot spirometer. One can also use a flow meter at the inspiratory or expiratory end, or a pneumotachograph, without collection of the air. Ventilation is presented in volume per minute, corrected for BTPS (air at body temperature, ambient pressure, and saturated with water). However, for the calculation of O_2 uptake, ventilation is corrected for STPD (air at 0°C, 760 mm Hg, dry). For children the dead-space of the valve and mouthpiece must be small—preferably not more than 50–60 ml for a sub- 10-year-old child.

Systemic Arterial Blood Pressure

The use of a sphygmomanometer during exercise is feasible with children, as it is with adults. This measurement is liable to error, resulting from lack of objectivity, insufficient skill, or nonstandardized procedures. There are two difficulties in the use of the auscultation method during exercise: movement of the arm and noises other than the Korotkow sounds. Stethoscopes are available in which the diaphragm is stabilized to the cuff. This can be helpful with a child who keeps moving his arm during exercise. Noise and vibration are transferred, during cycling, from the handlebar to the arm. It is useful therefore if the child lets go of the handlebar and keeps his arm extended during the time of measurement.

The monitoring of *diastolic* blood pressure by auscultation is quite unreliable during exercise as one can sometimes hear the Korotkow sound at zero pressure. The systolic pressure, on the other hand, can be determined with relative ease during moderate and intense activities because of the high intensity of the sounds.

Sphygmomanometers are available in which inflation and deflation of the cuff are done automatically and at a standard rate. The Korotkow sounds are detected by a microphone (some systems use electronic filtering of other noises). Such an apparatus increases the objectivity and standardization of the measurement and is recommended for clinics with a high patient load.

When testing children, special attention must be paid to the cuff size. The bladder should, as a rule, cover at least two-thirds of the arm length and completely encircle it.[46,56]

O_2 Uptake

A direct determination of O_2 uptake requires the use of O_2 and CO_2 analyzers to determine the fractional concentration of these gases in the expired air. Such analyzers are usually available in exercise physiology laboratories but not in clinics or in a physician's office. For description of the analyzers and methods of direct measurement of O_2 uptake, see Consolazio et al[13] and Jones and Campbell.[37]

O_2 uptake can be assessed *indirectly* when the mechanical power output is known, as in pedaling. Assuming that the O_2 cost of pedaling is not markedly different among individuals, one can use Table II.9 to assess the O_2 uptake equivalent of cycling.

Equations are also available for the O_2 uptake equivalents of walking, running, or stepping up and down stairs. Such equations, however, neglect to consider the marked differences that exist in the O_2 cost of these tasks among children of different ages (Chapters 1, section entitled Me-

Table II.9. Oxygen Uptake Equivalents of Mechanical Power Output During Submaximal Upright Cycling Exercise

Mechanical Power (Watts)	*Oxygen Uptake (liters/min)*
25	0.62
50	0.94
75	1.26
100	1.58
125	1.90
150	2.22

Mean values of 88 girls and 83 boys, 8–16 years old. There were no sex- or age-related differences in O_2 cost of pedaling. Based on Andersen et al.[2]

chanical Efficiency and Economy of Movement and Figs. 1.4, 1.5) and even within a certain age-group.

For children with some neuromuscular disabilities (e.g., cerebral palsy) the O_2 cost of pedaling is above normal (see Chapter 7, section entitled Mechanical Efficiency and Economy of Motion). For such children, Table II.9 should not be used.

Electrocardiogram (ECG)

The exercise ECG is an important tool in the assessment of children with a proven, or suspected, cardiovascular disease. Reviews of indications, methodology, and findings in healthy and sick children are available.[32,35,57] The following are comments on some practical aspects of exercise ECG, as well as lists of indications and major findings.

Skin Preparation. To increase the electric conductivity of the skin, hyperemia is induced through cleansing and rubbing with alcohol or acetone and the skin is then slightly abraded, using a lancet or a dental burr.

Choice of Electrodes. A straight isoelectric baseline is essential for detection and interpretation of ST-T changes. To achieve a straight baseline during exercise one must ascertain that the skin-to-electrode electrical impedance remains unchanged, in spite of body movement. This can be achieved by the use of "floating" electrodes in which the metal is not in direct contact with the skin, but is separated from it by a layer of electrolyte cream or gel. Various electrodes of this type are available commercially. Cables connecting the electrode to the ECG must be lightweight and flexible.

Choice of Leads. These can range from a single bipolar chest lead to 14 leads that combine the conventional 12-load ECG with Frank orthogonal leads.[32] Although a single CM_5 bipolar lead (reference electrode at the manubrio-sternal junction; exploratory electrode at C_5 and a ground electrode on the back) can detect dysrhythmias and most ST-T changes, one should attempt to use at least three leads simultaneously (e.g., V_3, V_5, aVF).

Indications for Exercise ECG. Exercise ECG is of value for children with known cardiovascular diseases as well as for those without any established diagnosis who complain of chest pain, palpitation, dizziness, and easy fatigability. The major indications for exercise ECG are listed in Table II.10.

Exercise-induced Electrocardiographic Changes. The most common ECG changes in children are ventricular dysrhythmias, ST depression, and atrioventricular block. These are summarized in Table II.11, with a list of conditions in which they may appear.

Exercise ECG Test vs. Long-term ECG Monitoring. One way to detect ECG abnormalities is the continuous recording of ECG for prolonged periods of time (e.g., 12 or 24 hours). Tape recorders are available that

Table II.10. Indications for Exercise ECG

Indication	*Questions to Be Answered*
History of: chest pain, palpitation, exercise-induced dizziness	Is there an underlying dysrhythmia? ischemia?
Left ventricular outflow obstruction (aortic stenosis, coarctation)	Is there ischemia? What is the severity of obstruction?
Chronic volume load, left or right (valvar incompetence, left-to-right shunt)	Is there ischemia?
Dysrhythmia at rest	Is dysrhythmia aggravated, or modified, by exercise?
Suspected ECG abnormality during long-term Holter monitoring	Can abnormality be confirmed?
Familial hypercholesterolemia	Is there premature ischemic heart disease?

Table II.11. Common ECG Changes in the Exercising Child

ECG Change	*Condition or Disease*	*Reference*
Segmental ST depression	Aortic stenosis or hypertrophic subaortic stenosis	26
	Aortic insufficiency	34
	Coarctation of aorta, pre- and postoperative	36
	Mitral incompetence	32
	Mitral valve prolapse	51
	Familial hypercholesterolemia	33
	Healthy children	33
Aggravation of heart block	Lower degree block, postsurgical atrial septal defect	32
Ventricular dysrhythmia	Complete congenital heart block	29
	Tetralogy of Fallot, postsurgical	22
	Healthy children	43
Atrial tachycardia	Sick-sinus syndrome	47

are light and can be worn by children without interfering with their spontaneous activities.

How does long-term ECG monitoring compare with a "single shot" exercise ECG test? Based on probability, long-term scanning should be a more sensitive method to detect ECG abnormalities. Recent comparisons suggest, however, that an exercise ECG test, in which the child undergoes a number of metabolic loads (maximum included), is as sensitive a method. In fact, exercise-induced dysrhythmias are more likely to be triggered, suppressed, or modified by a progressive all-out test than by a regular day's activities.[23,47,48]

A possible reason for the greater sensitivity of an all-out test is that during their spontaneous activity children may not push themselves to high-intensity exertion. Since exercise-induced ECG changes appear in any given child at a fairly constant heart rate,[48] a mild or moderate activity may not trigger any ECG change. The same child, on the other hand, when pushed to his or her maximum, will display such changes. It is also possible that the higher level of emotional stress (and sympathetic drive) in the unfamiliar laboratory setting is more conducive to dysrhythmic and other changes than is the stress in the more familiar playground.[23]

A definite advantage of long-term monitoring (whether by a tape recorder or by telemetry) is that is also yields data on the *habitual activity* pattern of the child. This information is highly valuable whenever recommendations are to be made regarding modification of the patient's daily activity.

In conclusion, long-term monitoring and laboratory exercise-ECG testing are both informative and complement each other. Whenever feasible, both should be administered, providing there are indications for assessment of electrocardiographic response to exercise. When resources are limited, a progressive all-out test in the laboratory is the method of choice.

Cardiac Output

The assessment of cardiac output in the exercising child is feasible, using invasive (direct Fick, or dye-dilution) or noninvasive techniques. For ethical reasons, a noninvasive approach is preferable. While a number of noninvasive methods are available, the indirect Fick CO_2 method has been found feasible, reliable, and valid when compared with a direct Fick determination. Such a method requires a fast (e.g., infrared) CO_2 analyzer and a bag-valve system for rebreathing. For details see Godfrey[24] and Jones and Campbell.[37]

Rating of Perceived Exertion (RPE)

The above described measures of response to exercise are all based on the *objective* physiologic strain that takes place in response to a physical stress. It is also of value, however, to determine the *subjective* strain that the child undergoes during exercise. One way of obtaining such information is by the use of an RPE scale, as devised by Borg.[10] This category-scale is shown in Table II.12.

Table II.12. The Borg Category Scale for Rating of Perceived Exertion

6	
7	Very, very light
8	
9	Very light
10	
11	Fairly light
12	
13	Somewhat hard
14	
15	Hard
16	
17	Very hard
18	
19	Very, very hard
20	

The scale is explained to the child before the start of an exercise test. It is important that the explanation be standardized. We use the following introduction:

"You are going to perform an exercise test in which the effort will be changed from time to time [one should *not* suggest to the child that the exercise test is progressive, with equal increments]. This scale describes different efforts: 6 is the lightest possible effort that you can think of and 20 the hardest possible effort. Therefore, all efforts that you will be making must be between 6 and 20.

"I shall show this scale to you during the exercise and ask you how hard you are exercising. Please answer by saying a *number* which best describes your feeling of the exercise at that moment. There are words opposite some of the numbers. These will help you remember what the numbers mean, but your answer must be by a number and not by a word. You can choose *any* number between 6 and 20 and not only those which have words opposite them. Remember—there is no 'correct' or 'incorrect' answer. We just want to know how *you* feel during the exercise."

This introduction is followed by examples and by ascertaining that the child understands the procedure. One must show the scale to the child just before the end of each stage (*after* an ECG strip has been recorded for heart rate determination) and each time ask: "How hard are you exercising?"

We have been using the RPE scale routinely in our laboratory, as part of all progressive exercise tests. In healthy children, and in many sick ones, at above 7 years of age, the subjective rating is closely correlated with physiologic indices, mostly heart rate. We have found the RPE scale of special value in children with neuromuscular disabilities, with handicapping dyspnea or with a high degree of apprehension, in whom there may be a discrepancy between exercise performance and physio-

Table II.13. The New Borg Scale for Rating of Perceived Exertion

0	Nothing at all
0.5	Very, very weak (just noticeable)
1	Very weak
2	Weak (light)
3	Moderate
4	Somewhat strong
5	Strong (heavy)
6	
7	Very strong
8	
9	
10	Very, very strong (almost maximal)
●	Maximal

logic strain. In such children the subjective rating may be high at a low power output, even when heart rate is low. The exercise performance of a child with, for example, muscular dystrophy is often more related to his subjective rating than to objective physiologic strain.

We have also been using the RPE concept for exercise prescription, to describe a certain exercise intensity that the patient should reach. This approach is useful for those 13 years or older. We have not assessed it among younger individuals.

Recently, a new RPE scale has been suggested, which is a combination of a category-scale and a ratio-scale (Table II.13). It has been found useful among adults for rating of specific symptoms during exercise, e.g., breathlessness, aches, and pains.[11] This scale has yet to be evaluated for children.

Determination of Maximal Aerobic Power

Maximal aerobic power can be determined directly by the measurement of maximal O_2 uptake or can be assessed indirectly by the measurement of heart rate during submaximal exercise; of peak power output during cycling or stepping; and of peak effort in a progressive treadmill test.

Direct Determination

The main criterion indicating that maximal O_2 uptake has been reached during a progressive protocol is that an increase in power load is accompanied by no increase, or by an increase of less than 2 ml/kg × min, in O_2 uptake. Such a plateau, although common among adults, is less often reached in children, especially during cycle ergometry.[17]

Secondary criteria are available to decide whether the child has reached his or her maximal O_2 uptake: heart rate of 195 beats/min, blood lactate concentration of 9 mmol/liter, or a respiratory gas exchange ratio (CO_2 output/O_2 uptake) that exceeds 1.0. These criteria have been used with healthy children, but they may not be valid in some diseases. A child with muscle dystrophy, for example, may reach his peak effort at a low heart rate (e.g., 130–140 beats/min), as will a child with complete heart block or cyanotic heart disease.

Unless on-line information is available, a decision that the child has reached his maximum cannot be based on physiologic variables. Experienced personnel often use such *subjective* criteria as blanching of the skin (mostly around the mouth and at the neck and shoulders), widening of the pupils, or a change in gait style, which shows that the child is struggling to keep walking or running. On the cycle ergometer, a child who is attempting to sustain the prescribed pedaling or cranking cadence and cannot do so, in spite of encouragement, has probably reached his maximum. As a rule, when using an "all-out" protocol, we do not terminate the test until the child himself cannot continue. We assign much importance to repeated verbal encouragement. Such statements as, "We know that you are terribly tired but you've almost finished" or, "We are now in the last minute (30 sec)" are most valuable and often elicit that all-important extra effort on the part of the child.

Indirect Determination—Submaximal Tests

Some investigators prefer not to bring the child to his or her self-imposed maximum. This approach has been taken for children who are sick or unmotivated; in conjunction with catheterization; or in large-scale surveys where a test must be short and simple. In submaximal tests, one measures heart rate during one or more stages and uses a derived index of maximal aerobic power.

Prediction of Maximal O_2 Uptake. Various methods are available for the prediction of maximal O_2 uptake from one or more values of submaximal heart rate. The most commonly used method is based on a nomogram of Åstrand and Rhyming.[4] To use this nomogram, heart rate must be measured at a known power load (on a cycle ergometer or by a step test), or together with a measurement of O_2 uptake (on a treadmill). This nomogram was originally designed for young adults, assuming a maximal heart rate of 198 beats/min for females and 195 beats/min for males. Because of children's higher maximal heart rate, the maximal O_2 uptake obtained from the nomogram must be multiplied by 1.1.

The use of this nomogram is based on the following assumptions:

1. Heart rate is linearly related to O_2 uptake.
2. Exercise at a certain power output requires the same O_2 uptake in all individuals.
3. Individuals of a certain age-group have the same maximal heart rate.

Even though such assumptions have some theoretical basis, none of them holds true in practice. The result is a low reproducibility and validity when the nomogram is used with children.[27,42,54,60] Moreover, the nomogram seems to underestimate the real maximal O_2 uptake by 10–25%.[28,52,54,60] To reduce this underestimation, various correction factors have been suggested. These, however, do not reduce the marked scatter in results obtained by this method of prediction. We therefore do not recommend the use of the Åstrand and Rhyming nomogram, despite its popularity, for the assessment of *individual* patients. Its use should be limited to population studies.

$\dot{W}_{170}$. This index, which is the mechanical power at which heart rate is 170 beats/min, was first introduced by Wahlund[59] and has been in use since for children[49,50] and adults. With this index one does *not* assume a certain maximal heart rate. The only assumption is that the heart rate is linearly related to power, at about 170 beats/min and less. Two, or more, measurements of heart rate are needed and the values are plotted against mechanical power, as shown in Fig. II.3. The line thus established is extrapolated (or interpolated) to heart rate = 170 beats/min and the corresponding power is $\dot{W}_{170}$. The drawback of $\dot{W}_{170}$ is that, as with other submaximal tests, it is based on heart rate, which is dependent on numerous factors in addition to the fitness of the individual.

To decrease the error in this measurement, one of the heart rates should be as close as possible to 170. Another refinement is the use of more than two values to plot the line. If one point is obviously out of line, it can be discarded. This occurs sometimes with the lowest or with the highest value. The former is often too high, because the child is emotionally excited at the start of the test. The latter may be too low if the load approaches maximum (heart rate may reach a plateau while O_2 uptake is still rising).

$\dot{W}_{0.85}$. Some sick children have a maximal heart rate *below* 170 beats/min. In such patients (e.g., those with congenital complete heart block) the use of $\dot{W}_{170}$ will *overestimate* their real maximal aerobic power. An

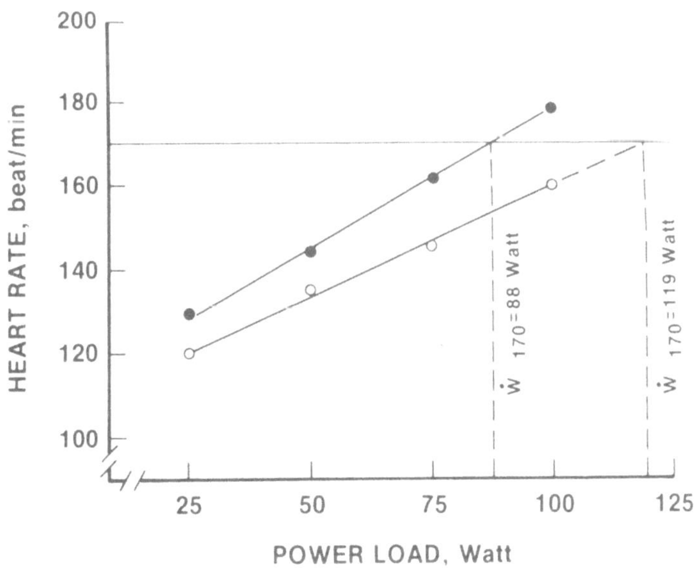

Figure II.3. A schematic diagram displaying the method by which $\dot{W}_{170}$ is derived from submaximal heart rate in two adolescents.

alternative index has been suggested,[14] which is the mechanical power that the child produces when his heart rate is 85% of his maximal heart rate ($\dot{W}_{0.85}$). To use this index, one must measure heart rate at a number of loads, including the peak load, and then interpolate to 85% of maximal heart rate.

$\dot{W}R_{17}$. An alternative index, which is not based on heart rate, is the mechanical power that the child produces when he *perceives* the exercise to be "hard," or 17 on the Berg's category-scale (Table II.12). We have found this index to correlate highly with other indices of maximal aerobic power and to serve as a valid fitness test in children.[6,7] As seen in Fig. II.4, the method of assessing $\dot{W}R_{17}$ is similar to that for $\dot{W}_{170}$, with RPE being assessed instead of heat rate.

Indirect Determination—All-Out Tests

When direct measurement of O_2 uptake is not feasible, one can assess maximal aerobic power using the *performance* of the child as a criterion. One example is the highest stage that the child can reach during the Bruce treadmill test (see above and Fig. I.5), which is highly correlated with directly measured maximal O_2 uptake. Another example is the peak power output in the McMaster cycling or arm-cranking tests (see above and Figs. I.3, I.4).

Other indices are the speed of running a certain distance[39] and the distance covered during a predetermined time period.[41]

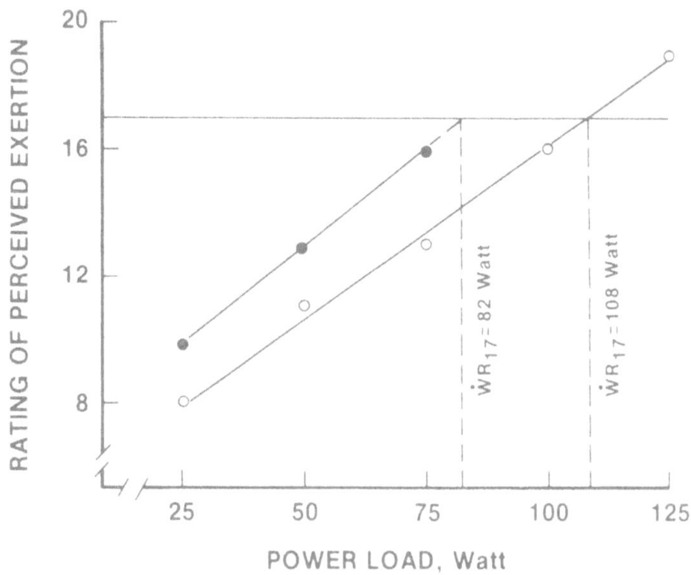

Figure II.4. A schematic diagram displaying the method by which $\dot{W}R_{17}$ is derived from the rating of perceived exertion in two adolescents.

Safety Precautions

In adults, the two main untoward effects of exercise stressing are myocardial ischemia and ventricular fibrillation. Although epidemiologic data are not available on the occurrence of such complications in the pediatric age-group, the risk seems to be extremely low.[14] It can be kept that way if one follows certain precautions.

Personnel

An exercise laboratory team comprises a physician, nurses, and exercise technicians. All should be familiar with the principles of exercise physiology, should recognize normal and abnormal response patterns to exercise, and should be proficient in cardiopulmonary resuscitation.

In some laboratories the physician takes an active part in the testing. In others, he is available nearby in case of emergency. Whenever more than one exercise protocol is used in a laboratory, the physician is the individual who should determine the protocol needed for each patient.

Contraindications for Exercise Testing

As a rule, any child can be given an exercise test. There are, however, health conditions that, on the day of testing, preclude exertion. These are summarized in Table II.14.

Individuals with the following diseases constitute a high-risk group, and special attention must be paid to them during testing: severe aortic stenosis, myocardial disease, cyanotic heart disease, advanced pulmo-

Table II.14. Contraindications for Exercise Testing in Pediatric Patients

1. Acute febrile condition
2. Acute inflammatory cardiac disease, e.g., pericarditis, myocarditis, acute rheumatic heart disease
3. Congestive heart failure—uncontrolled
4. Asthmatic child who is dyspneic at rest, or whose $FEV_{1.0}$* or PEF† are less than 60% of height-predicted value
5. Acute renal disease, e.g., acute glomerulonephritis
6. Acute hepatitis, during 3 months since onset
7. Insulin-dependent diabetic who did not take his prescribed insulin or who is ketoacidotic
8. Drug overdose affecting cardiorespiratory response to exercise, e.g., digitalis or quinidine toxicity, salicylism, antidepressants

Adapted from American Heart Association Report: Standards for Exercise Testing in the Pediatric Age Group.[1]

* $FEV_{1.0}$ = Forced expiratory volume in the first second
† PEF = Peak expiratory flow

Table II.15. Criteria for Termination of an Exercise Test in the Pediatric Age-Group

1. Clinical
   a. Symptoms—chest pain, severe headache, dizziness, chills, sustained nausea, inappropriate dyspnea
   b. Signs—sustained pallor, clammy skin, disorientation, inappropriate affect
2. Electrocardiographic
   a. Ventricular tachycardia
   b. Supraventricular tachycardia
   c. ST segmental depression, or elevation, of more than 3 mm
   d. Triggering by exercise of intracardiac block
   e. Premature ventricular contractions with increasing frequency
3. Blood Pressure
   a. Excessive levels—systolic BP 240 mm Hg
   —diastolic BP 120 mm Hg
   b. Progressive fall in systolic BP

Based in part on American Heart Association Special Report: Standards for Exercise Testing in the Pediatric Age Group.[1]

nary vascular disease, ventricular dysrhythmia (which accompanies a heart disease), coronary arterial disease, and hypercholesterolemia. While some laboratories use only submaximal protocols in these conditions, others allow the child to perform an all-out test under close supervision.

Termination of an Exercise Test

To safeguard the health of the child one must sometimes discontinue an exercise test. All personnel who administer exercise tests must be able to make such a decision, based on clinical, electrocardiographic, and blood pressure criteria. These criteria are listed in Table II.15.

References

1. American Heart Association Council on Cardiovascular Disease in the Young: Standards for exercise testing in the pediatric age group. Circulation 66:1377A–1397A, 1982.
2. Andersen KL, Seliger V, Rutenfranz J, Mocellin R: Physical performance capacity of children in Norway. Part I. Population parameters in rural inland community with regard to maximal aerobic power. Eur J Appl Physiol 33:177–195, 1974.
3. Antila K, Petäjoki M-L, Arstila M, Välimäki I. In: Stern S (ed.) Ambulatory ECG Monitoring. Year Book Medical Publishers, Chicago, 1978, pp. 69–75.
4. Åstrand PO, Rhyming I: A nomogram for calculation of aerobic capacity

(physical fitness) from pulse rate during submaximal work. J Appl Physiol 7:218–221, 1954.
5. Bailey DA, Mirwald RL: A children's test of fitness. In: Borms J, Hebbelinck M (eds.) Pediatric Work Physiology. Karger, Basel, 1978, pp. 56–64.
6. Bar-Or O: A comparison of responses to exercise and lung functions of Israeli Arabic and Jewish 12 to 17 year-old boys. In: Bar-Or O (ed.) Pediatric Work Physiology. Wingate Institute, Natanya, 1973, pp. 59–68.
7. Bar-Or O: Age-related changes in exercise perception. In: Borg G (ed.) Physical Work and Effort. Pergamon Press, Oxford and New York, 1977, pp. 255–266.
8. Bar-Or O: Le test anaérobie de Wingate. Caractéristiques et applications. Symbioses 13:157–172, 1981.
9. Bengtsson E: The working capacity in normal children, evaluated by submaximal exercise on the bicycle ergometer and compared with adults. Acta Med Scand 154:91–109, 1956.
10. Borg G: Physical Performance and Perceived Exertion. Gleerup, Lund, 1962.
11. Borg G: Psychophysical bases of perceived exertion. Med Sci Sports Exercise 14:377–381, 1982.
12. Bruce RA, McDonough JR: Stress testing in screening for cardiovascular disease. Bull NY Acad Med 45:1288–1305, 1969.
13. Consolazio CF, Johnson RE, Pecora LJ: Physiological Measurements of Metabolic Functions in Man. McGraw-Hill, New York, 1963.
14. Cumming GR: Exercise studies in clinical pediatric cardiology. In: Lavallée H, Shephard RJ (eds.) Frontiers of Activity and Child Health. Pélican, Quebec, 1977, pp. 17–45.
15. Cumming GR: Hemodynamics of supine bicycle exercise in "normal" children. Am Heart J 93:617–622, 1977.
16. Cumming GR, Everatt D, Hastman L: Bruce treadmill test in children: normal values in a clinic population. Am J Cardiol 4:69–75, 1978.
17. Cumming G, Friesen W: Bicycle ergometer measurement of maximal oxygen uptake in children. Can J Physiol Pharmacol 45:937–946, 1967.
18. Cumming GR, Hnatiuk A: Establishment of normal values for exercise capacity in a hospital clinic. In: Berg K, Eriksson BO (eds.) Children and Exercise IX. University Park Press, Baltimore, 1980, pp. 79–93.
19. Davies CTM, Barnes C, Godfrey S: Body composition and maximal exercise performance in children. Hum Biol 44:195–214, 1972.
20. Dotan R, Bar-Or O: Climatic heat stress and performance in the Wingate anaerobic test. Eur J Appl Physiol 44:237–243, 1980.
21. Ellestad MH: Stress testing. Principles and practice, 2nd ed. F.A. Davis, Philadelphia, 1980.
22. Garson A Jr, Gillette PC, Gutgesell HP, McNamara DG: Stress-induced ventricular arrhythmia after repair of tetralogy of Fallot. Am J Cardiol 46:1006–1012, 1980.
23. Ginzel H, Porkony W: Telemetrische Belastungsuntersuchungen bei Kindern mit Extrasystolen. Paediatr Paedol 14:63–67, 1979.
24. Godfrey S: Exercise Testing in Children. Applications in Health and Disease. W.B. Saunders, Philadelphia, 1974.
25. Goldberg SJ, Weiss R, Adams FH: A comparison of the maximal endurance

of normal children and patients with congenital cardiac disease. J Pediatr 69:46–55, 1966.
26. Halloran KH: The telemetered exercise electrocardiogram in congenital aortic stenosis. Pediatrics 47:31–39, 1971.
27. Hanne N: A step-test for 6- to 12-year-old girls and boys (in Hebrew). Research report, Wingate Institute, 1971.
28. Hermansen L, Oseid S: Direct and indirect estimation of maximal oxygen uptake in prepubertal boys. Acta Paediatr Scand Suppl 217:18–23, 1971.
29. Holmgren A, Karlberg P, Pernow B: Circulatory adaptation at rest and during muscular work in patients with complete heart block. Acta Med Scand 164:119–130, 1959.
30. Inbar O, Bar-Or O: The effects of intermittent warm-up on 7–9 year-old boys. Eur J Appl Physiol 34:81–89, 1975.
31. James FW: Exercise testing in children and young adults: an overview. Cardiovasc Clin 9:187–203, 1978.
32. James FW: Exercise ECG Test in Children. In: Chung EK (ed.) Exercise Electrocardiography—Practical Approach. Williams and Wilkins, Baltimore, 1979, pp. 122–145.
33. James FW, Glueck CJ, Fallat RW, et al: Maximal exercise stress studies in normal and hyperlipidemic children. Atherosclerosis 25:85–94, 1976.
34. James FW, Kaplan S: Systolic hypertension during submaximal exercise after correction of coarctation of aorta. Circulation 49, 50[Suppl. II]:27–34, 1974.
35. James FW, Kaplan S: Exercise testing in children. Primary Cardiol 3:34–40, 1977.
36. James FW, Kaplan S, Schwartz DC: Ischemic ST segments during exercise in children after coarctectomy (abstract). Am J Cardiol 37:145, 1976.
37. Jones NL, Campbell EJM: Clinical Exercise Testing. W.B. Saunders, Philadelphia, 1982.
38. Klimt F, Voigt GB: Investigations on the standardization of ergometry in children. Acta Paediatr Scand Suppl 217:35–36, 1971.
39. Krahenbuhl GS, Pangrazi RP, Petersen GW, et al: Field testing of cardiorespiratory fitness in primary school children. Med Sci Sports 10:208–213, 1978.
40. Lock JE, Einzig S, Moller JH: Hemodynamic responses to exercise in normal children. Am J Cardiol 41:1278–1284.
41. MacDougall JD, Roche PD, Bar-Or O, Moroz JR: Maximal aerobic capacity of Canadian school children. Prediction based on age-related oxygen cost of running. Int J Sports Med, in press, 1983.
42. Mocellin R, Lindemann H, Rutenfranz, J, Sbresny W: Determination of $\dot{W}_{170}$ and maximal oxygen uptake in children by different methods. Acta Paediatr Scand Suppl 217:13–17, 1971.
43. Monarrez CN, Strong WB, Rees AH: Exercise electrocardiography in the evaluation of cardiac dysrhythmias in children. Paediatrics 7:116–125, 1978.
44. Mrzena B, Máček M: Use of treadmill and working capacity assessment in pre-school children. In: Borms J, Hebbelinck M (eds.) Pediatric Work Physiology. Karger, Basel, 1978, pp. 29–31.

45. Petäjoki ML, Arstila M, Välimäki I: Pulse-conducted exercise test in children. Acta Paediatr Belg 28[Suppl]:40–47, 1974.
46. Riopel DA, Taylor AB, Hohn AR: Blood pressure, heart rate, pressure-rate product and electrocardiographic changes in healthy children during treadmill exercise. Am J Cardiol 44:607–704, 1979.
47. Rocchini A, Freed M, Rosenthal A: Detection of arrhythmias in childhood: use of treadmill and dynamic ECG (abstract). Pediatr Res 11:4, 1977.
48. Rozanski JJ, Dimich I, Steinfeld L, Kupersmith J: Maximal exercise stress testing in evaluation of arrhythmias in children: results and reproducibility. Am J Cardiol 43:951–956, 1979.
49. Rutenfranz J: Entwicklung und Beurteilung der körperlichen Leistungsfähigkeit bei Kindern und Jugendlichen. Karger, Basel, 1964.
50. Rutenfranz J: Exercise tests in children and adolescents. In: Andersen KL, Shephard RJ, Denolin H, Vernauskas E, Masironi R (eds.) Fundamentals of Exercise Testing. World Health Organization, Geneva, 1971, pp. 105–109.
51. Schwartz DC, James FW, Kaplan S: Exercise induced ST segment depression in children with mitral valve prolapse (abstract). Circulation 52 [Suppl. II]:67, 1975.
52. Shephard PJ, Allen C, Bar-Or O, et al: The working capacity of Toronto schoolchildren. Can Med Assoc J 100:560–566, 705–714, 1969.
53. Skinner JS, Bar-Or O, Bergsteinová V, et al: Comparison of continuous and intermittent tests for determining maximal oxygen intake in children. Acta Paediatr Scand Suppl 217:24–28, 1971.
54. Stewart KJ, Gutin B: The prediction of maximal oxygen uptake before and after physical training in children. J Hum Ergol 4:153–162, 1975.
55. Strong WB, Spencer D, Miller MD, Salehbhai M: The physical working capacity of healthy black children. Am J Dis Child 132:244–248, 1978.
56. Task Force on Blood Pressure Control in Children: Methodology and instrumentation for blood pressure measurement in infants and children. Pediatrics 59:800–801, 1977.
57. Thorén C: Exercise testing in children. Paediatrician 7:100–115, 1978.
58. Välimäki I, Petäjoki ML, Arstila M, et al: Automatically controlled ergometer for pulse-conducted exercise test. In: Borms J, Hebbelinck M (eds.) Pediatric Work Physiology. Karger, Basel, 1978, pp. 47–51.
59. Wahlund H: Determination of the physical work capacity. Acta Med Scand Suppl 215, 1948.
60. Waynarowska B: The validity of indirect estimation of maximal oxygen uptake in children 11–12 years of age. Eur J Appl Physiol 43:19–23, 1980.

Appendix III

Activity Questionnaire

The following is an example of a questionnaire to be filled in by a parent on the first visit of the child to the exercise clinic. Questions are designed to elicit information on the child's activities at home, at school, and at the sports club, as well as the activity habits of other family members and parental attitude toward physical activity.

Dear Parent:
The purpose of the following questions is to help us evaluate the activity habits and exercise capability of your child. Please be as accurate as possible in your answers. Feel free to add any details which seem relevant.

1. How would you compare the physical activity of your child with that of her/his friends?

   Child is as active as her/his friends ☐

   Child is more active than her/his friends ☐

   Child is less active than her/his friends ☐

   It is hard to make such a comparison ☐

   Details ____

   ____

2. How would you compare the activity of this child with that of your other children?

   There are no other children in the family ☐

   This child is as active as my other children ☐

   This child is more active than my other children ☐

   This child is less active than my other children ☐

   It is hard to make such a comparison ☐

   Details ____

   ____

3. Does this child take part in physical education classes at school?

   All activities, with no exception ☐

   Some activities only ☐

   Child does not participate in physical education classes ☐

   Child does not attend school ☐

   Details (especially type of activities forbidden) ______________

   ____

   ____

4. If child is limited in activity at school, for what reason? (You may fill in more than one answer)

   Advice of physician ☐

   Advice of teacher ☐

   Decision of parents ☐

   Child does not want to participate ☐

   Other ____________________ ☐
   please specify

   ____

   Details ____

   ____

5. Is the child a member of a sports team at school or otherwise?

   No ☐

   Yes, within school (intramural) ☐

   Yes, representing the school ☐

   Yes, other ________________ ☐

   Yes, in the past but no more ☐

6. If a member of a team, in which sport or sports? ________________

   ____

7. If child trains regularly, what is the nature of his training?

	Type of Sport	Hours per Week	Time of Year	Comments
A.				
B.				
C.				

8. Are there any other members of your family who participate in *competitive* sports?

   No, no one in this family ☐

   Yes ☐

9. If "yes," please specify

	Family Member	Type of Sport	Trains Regularly?
A.			
B.			
C.			
D.			

10. Does the child participate in any *recreational* activity that requires physical effort? (For example, skiing, canoeing, cycling, dancing, swimming). Please specify.

	Type of Activity	Time of Year	Hours per Week
A.			
B.			
C.			

11. Does any member of the family participate in recreational activities that require physical effort?

Yes ☐

No one ☐

12. If "yes" please specify

	Family Member	Type of Activity	Time of Year
A.			
B.			
C.			
D.			

13. Does this child complain of any difficulty during or after physical exertion?

No complaint ☐

Shortness of breath ☐

Pain ☐ Where? ____________

Fatigue ☐

Other ____________ ☐
please specify

Details ____________

____

14. Does your child *often* sustain bruises, injuries, or other damage when physically active?

Yes ☐

No ☐

15. If "yes," please specify ____________

____

16. In your opinion, is this child as active as she/he should be?

Yes ☐

Child is too active ☐

Not sufficiently active ☐

17. If this child is not as active as she/he should be, what in your opinion is the reason? (You can fill in more than one answer)

Lack of interest ☐

A disease ☐

Lack of suitable conditions ☐

Other ☐

I don't know ☐

Details ____
____

18. Please fill in any of the following statements that you agree with (you can check more than one statement):

Physical activity is important because it is fun ☐

Physical activity is necessary to keep fit ☐

Physical activity is good for health reasons ☐

Physical activity may be dangerous to one's health ☐

Physical activity can prevent overweight ☐

Physical activity is important mostly to those who wish to become professional athletes ☐

Appendix IV

Calorie Equivalents

Calorie Equivalent of Child's Activities. Numbers denote kcal per 10 min

Activity	*Body Weight in kg*									
	20	*25*	*30*	*35*	*40*	*45*	*50*	*55*	*60*	*65*
Basketball (game)	34	43	51	60	68	77	85	94	102	110
Calisthenics	13	17	20	23	26	30	33	36	40	43
Cross-country ski										
leisure	24	30	36	42	48	54	60	66	72	78
Cycling										
10 km/hr	15	17	20	23	26	29	33	36	39	42
15 km/hr	22	27	32	36	41	46	50	55	60	65
Field hockey	27	34	40	47	54	60	67	74	80	87
Figure skating	40	50	60	70	80	90	100	110	120	130
Horse-back riding										
canter	8	11	13	15	17	19	21	23	25	27
trot	22	28	33	39	44	50	55	61	66	72
gallop	28	35	41	48	50	62	69	76	83	90
Ice hockey (on-ice time)	52	65	78	91	104	117	130	143	156	168
Judo	39	49	59	69	78	88	98	108	118	127
Running										
8 km/hr	37	45	52	60	66	72	78	84	90	95
10 km/hr	48	55	64	73	79	85	92	100	107	113
12 km/hr	—	—	76	83	91	99	107	115	123	130
14 km/hr	—	—	—	—	—	113	121	130	140	148

Calorie Equivalent of Child's Activities (Continued)

Activity	*Body Weight in kg*									
	20	*25*	*30*	*35*	*40*	*45*	*50*	*55*	*60*	*65*
Sitting										
complete rest	8	8	9	9	10	10	11	11	12	12
quiet play	11	12	14	15	15	16	17	18	19	20
Snow shoeing	35	42	50	58	66	74	82	90	98	107
Soccer (game)	36	45	54	63	72	81	90	99	108	117
Squash	—	—	64	74	85	95	106	117	127	138
Swimming										
30 m/min										
breast	19	24	29	34	38	43	48	53	58	62
front crawl	25	31	37	43	49	56	62	68	74	80
back	17	21	25	30	34	38	42	47	51	55
Table tennis	14	17	20	24	28	31	34	37	41	44
Tennis	22	28	33	39	44	50	55	61	66	72
Volleyball (game)	20	25	30	35	40	45	50	55	60	65
Walking										
4 km/hr	17	19	21	23	26	28	30	32	34	36
6 km/hr	24	26	28	30	32	34	37	40	43	48

Appendix V

Glossary of Terms

Acute Exercise. A bout of physical exercise of any intensity that may last from a few seconds to a few hours.

Adapted (Adaptive) Physical Education. Subspecialty of physical education using movement, physical skills, and sports for the education and therapy of children and adolescents who are disabled emotionally, mentally, or physically.

All-out Exercise. A bout of exercise that the subject performs to exhaustion.

Anaeorobic Threshold. Exercise intensity during a progressively increasing task at which the rate of lactic acid production exceeds lactic acid removal.

Chronic Exercise. Repeated bouts of acute exercise, spreading over a period of weeks, months, or years.

Collision Sports. Sports in which hitting or ramming the opponent is integral to the sport. Examples: American football, boxing, ice hockey, rugby, karate (cf. Contact Sports).

Concentric Muscular Contraction. Shortening of the muscle during contraction.

Conditioning (Physical). The process by which repeated bouts of exercise induce morphological and functional changes in the body as a whole (cf. Training).

Contact Sports. Sports in which body contact between opponents is frequent but incidental to the sport. Examples: basketball, European handball, lacrosse, soccer (cf. Collision Sports).

Dehydration. The process of incurring fluid deficit, which leads to hypohydration.

Dynamic (Rhythmic) Exercise. An exercise pattern in which muscle contraction and relaxation alternate.

Eccentric Muscular Contraction. Lengthening of a muscle during contraction. This will occur when an external force is exerted on the muscle which is greater than the force generated by the contraction, and in opposite direction to it.

Ergometer. An apparatus used for exercise in which the mechanical power or the walking or running conditions are quantifiable and reproducible. The most commonly used ergometers are: cycle, treadmill, and a step.

Heat Strain. Physiologic and perceptual responses to heat stress. Variables commonly included are heart rate, rectal temperature, sweating rate, and rating of perceived exertion.

Heat Stress. A combination of environmental conditions that stress the heat-dissipation mechanisms. The main components are humidity, radiation, ambient temperature and wind velocity.

Hypoactivity. Physical activity level that is lower than in healthy individuals of similar age, sex, cultural, and socioeconomic background.

Hypohydration. A state of body hydration that is below optimal (cf. Dehydration).

Isometric Muscular Contraction. Contraction that does not cause a change in the muscle length, nor any skeletal movement.

Maximal Aerobic Power. The highest O_2 uptake that can be attained during exercise. (This term should be used instead of Aerobic Capacity.)

Maximal Exercise. The lowest exercise level at which O_2 uptake is maximal (cf. Supramaximal Exercise).

Maximal O_2 Uptake. The level of O_2 uptake that cannot be surpassed even when the exercise load is increased. Serves as a measure of maximal aerobic power.

Mechanical Efficiency. The ratio between external mechanical work produced by the muscle and the chemical energy utilized during the contraction.

Muscle Endurance. The ability of a muscle, or a muscle group, to sustain prolonged (static) or repeated (dynamic) contractions of high intensity.

Muscle Strength. The maximal force that a muscle, or muscle group, can exert.

Oxygen Uptake. The volume of O_2 that is consumed during a time unit by a tissue, an organ, or the whole body.

Peak Power Output. 1) The highest mechanical power that can be performed in a "progressive maximal test." 2) The highest power achieved momentarily (or up to a 5-sec period) during a supramaximal task.

Physical Working Capacity (PWC). A general term in common use to describe the working performance of an individual (often used as a synonym for maximal aerobic power). PWC is assessed by peak power output, total work, maximal O_2 uptake, or $\dot{W}_{170}$.

Prolonged Exercise. A continuous exercise bout, lasting 30 min or more.

Respiratory Quotient (RQ). The ratio between CO_2 production and O_2 uptake.

Static Exercise. An exercise pattern in which muscle contraction (usually isometric) is continuous and not alternating with relaxation.

Submaximal Exercise. An exercise level at which O_2 uptake is submaximal.

Supramaximal Exercise. An exercise level higher than maximal exercise.

Training (Physical). The process by which repeated bouts of exercise induce morphologic and functional changes in a specific tissue or body system (cf. Conditioning).

Ventilatory Equivalent. The ratio between pulmonary ventilation and O_2 uptake (i.e., the number of liters of ventilated air which facilitate 1 liter of O_2 uptake).

$\dot{W}_{170}$. The mechanical power output at which heart rate would be 170 beats/min. An index of maximal aerobic power.

Index

(Page citations with "f" refer to figures; those with "t" refer to tables.)

MIX
Papier aus verantwortungsvollen Quellen
Paper from responsible sources
FSC® C105338

If you have any concerns about our products,
you can contact us on
ProductSafety@springernature.com

In case Publisher is established outside the EU,
the EU authorized representative is:
Springer Nature Customer Service Center GmbH
Europaplatz 3, 69115 Heidelberg, Germany

Printed by Libri Plureos GmbH
in Hamburg, Germany